Topics in Applied Physics Volume 11

Topics in Applied Physics Founded by Helmut K.V. Lotsch

Digital Picture Analysis

Edited by A. Rosenfeld

With Contributions by
S.J. Dwyer R.M. Haralick C.A. Harlow
G. Lodwick R.L. McIlwain K. Preston
A. Rosenfeld J.R. Ullmann

With 114 Figures

Springer-Verlag Berlin Heidelberg New York 1976

Professor Azriel Rosenfeld

Picture Processing Laboratory, Computer Science Center
University of Maryland, College Park, MD 20742, USA

ISBN 3-540-07579-8 Springer-Verlag Berlin Heidelberg New York
ISBN 0-387-07579-8 Springer-Verlag New York Heidelberg Berlin

Library of Congress Cataloging in Publication Data. Main entry under title: Digital picture analysis. (Topics in applied physics; v. 11) Bibliography: p. Includes index. 1. Optical data processing. I. Rosenfeld, Azriel, 1931—. II. Dwyer, Samuel J. TA1630.D53 621.3819'598 75-41340

Monophoto typesetting, offset printing, and bookbinding: Brühlsche Universitätsdruckerei, Giessen

Preface

The rapid proliferation of computers during the past two decades has barely kept pace with the explosive increase in the amount of information that needs to be processed. The processing of pictorial information, obtained from both business and scientific sources, presents an especially challenging problem. Computers have found extensive use in tasks such as

a) Making weather maps from meteorological satellite television images, or land use maps from earth resources aircraft or satellite imagery;
b) Diagnosing pathological conditions by analyzing medical X-rays or microscope slides;
c) Searching for new fundamental particles by analyzing photographs of interactions in nuclear particle studies;
d) Reading handwritten or printed documents — zip codes on envelopes, credit card numbers on sales slips, etc.

This book contains review papers on the computer analysis of pictorial information in five major areas: remote sensing of the environment (via images obtained by aircraft or satellites); medical radiology; high-energy physics; cytology; and character recognition. The papers are written by recognized experts in these fields. Taken collectively, they provide a broad introduction, from a practical standpoint, to the important topic of digital picture analysis.

College Park, Maryland, USA
February 1976

AZRIEL ROSENFELD

Contents

Contributors

DWYER, SAMUEL J., III

Department of Electrical Engineering, University of Missouri-Columbia, Columbia, MO 65201, USA

HARALICK, ROBERT M.

Department of Electrical Engineering, The University of Kansas Space Technology Center, Lawrence, KS 66044, USA

HARLOW, CHARLES A.

Department of Electrical Engineering, University of Missouri-Columbia, Columbia, MO 65201, USA

LODWICK, GWILYM

Department of Radiology, University of Missouri Medical Center, University of Missouri-Columbia, Columbia, MO 65201, USA

McILWAIN, ROBERT L., JR.

Department of Physics, Purdue University, Lafayette, IN 47907, USA

PRESTON, KENDALL, JR.

Department of Electrical Engineering, Carnegie-Mellon University, Schenley Park, Pittsburgh, PA 15213, USA

ROSENFELD, AZRIEL

Picture Processing Laboratory, Computer Science Center, University of Maryland, College Park, MD 20742, USA

ULLMANN, JULIAN R.

Department of Applied Mathematics and Computing Science, University of Sheffield, Sheffield S10 2TN, Great Britain

1. Introduction

A. ROSENFELD

Modern high-speed computers can execute millions of instructions per second, but their capacity as information-processing devices is limited by the types of input data that they can accept. Much of the information that we might like our computers to process is in pictorial form; this includes alphanumeric characters (printed text, handwriting, etc.) as well as photographs and television images. Alphanumeric data can be manually converted into computer-readable form, but much faster input rates can be achieved if scanning and optical character recognition techniques are used. Photographs and images can be scanned and digitized, but it is still necessary to program the computer to analyze and interpret them automatically.

This book deals with the computer analysis of pictorial information. Each of the five chapters covers a different field of application, and is written by a recognized expert in the field. The application areas covered are

Remote sensing (Chapt. 2).
Radiology (Chapt. 3).
High-energy physics (Chapt. 4).
Cytology (Chapt. 5).
Character recognition (Chapt. 6).

Each of these areas has had specialized conferences or workshops devoted to it—in some cases, long series of them. Each has a large literature; see the individual papers for introductions to these literatures. Efforts have also been made to bring the different application areas closer together; examples of this include the Symposium on Automatic Photointerpretation, held in 1967 [1.1]; the NATO Summer Institute on Automatic Interpretation and Classification of Images, in 1968 [1.2]; the Special Issue of the IEEE Proceedings devoted to Digital Pattern Recognition, which appeared in 1972 [1.3]; and the U.S.-Japan Seminar on Picture and Scene Analysis, held in 1973 [1.4]. In addition, the many conferences on pattern recognition have included a wide variety of picture analysis applications.

Not all of the literature on computer analysis of pictures is application-oriented; over the past twenty years, a large body of general-purpose digital picture analysis techniques has been built up. A summary of this material as of the late 1960's may be found in a textbook written by the editor [1.5]; a more recent book, which also covers three-dimensional scene analysis, is [1.6]. The editor has also written a series of survey papers [1.7–11] which cover much of the English-language literature in the field. He has recently co-authored a more comprehensive textbook, treating digital picture processing and analysis in greater detail [1.12].

All of the papers in this book cover certain basic aspects of their respective application areas, including the need for image processing in the area, hardware and software requirements, and a guide to the literature. However, the emphasis differs considerably from paper to paper, due to differences in the nature and history of the areas, as well as in the writers' personal interests and styles. The treatment of image processing and recognition techniques also differs in breadth and depth from paper to paper. To provide a unifying perspective, the following paragraphs discuss some of these techniques, and indicate their roles in each of the individual papers.

Pictorial pattern recognition can have several possible goals:

1) *Matching* pictures with one another, e.g., to detect changes, or to locate objects in three dimensions.

2) *Classifying* a picture into one of a specified set of classes, based on the values of a set of "features" measured on the picture; or using these values to describe the picture.

3) *Segmenting* a picture into significant parts, and classifying or describing the parts as above.

4) *Recognizing* a picture as being composed of specified parts in specified relationships to one another.

These goals are all applicable to many different types of pictures; but they play more prominent roles in some applications than in others.

It is often necessary to "preprocess" the given picture(s), in order to improve the reliability of subsequent matching, feature measurement, or segmentation. An important type of preprocessing is *normalization*, in which the picture is brought into a standard form, in order to make feature values independent of, or at least insensitive to, various transformations which may have affected the original picture. Grayscale normalization is discussed by HARALICK (Subsect. 2.3.1); the need for such normalization is greatest in the case of remote sensor imagery, where the illumination is uncontrolled (except in the case of radar imagery). A common method of grayscale normalization is to standardize the picture's histogram (=gray level frequency distribution); on histograms see also Subsection 5.5.1. HARALICK also discusses the removal of perspective distortion from pictures (rectification; Subsect. 2.3.2), which is an important example of geometrical normalization. Size normalization (of characters that have been segmented from a document) is treated by ULLMANN (Sect. 6.5), who also discusses smoothing of the segmented characters (Sect. 6.4); these techniques are appropriate when dealing with specific, non-noisy shapes that can have a wide range of sizes. A number of basic preprocessing techniques are also mentioned by HARLOW (Subsect. 3.3.2).

Matching, or correlation, plays a major role in the recognition of characters that have known shapes (Sect. 6.6). It is also important in stereogrammetry (the location of objects in three dimensions by comparing pictures taken from two or more viewpoints); an example discussed by MCILWAIN is the spatial location of bubble tracks (Subsect. 4.3.3). Image registration and "congruencing" for matching or comparison purposes are discussed by HARALICK (Subsect. 2.3.2).

Many types of features can be used for picture classification. Simple template-like features are appropriate for classifying patterns of known shape, such as characters (Sect. 6.7). When geometrical distortions are present, however, it becomes necessary to use features that are insensitive to these distortions, e.g., topological features (Sect. 6.8). A variety of size and shape features are discussed in Subsections 5.5.2, 5.5.3, and 5.5.5, while recognition based on the occurrence of subpatterns in specific sequences ("strip classification") is treated in Section 6.9.

In cytology, radiology, and remote sensing, not only the sizes and shapes of objects (or regions) are important, but also their textures. Approaches to the measurement of texture features are discussed by HARALICK (Sect. 2.5), HARLOW (Subsect. 3.3.3), and PRESTON (Subsects. 5.5.6 and 5.5.7).

Picture segmentation is important in nearly every application of pictorial pattern recognition. In character recognition, one needs to separate the printing or writing from the background ("binarization"; Sect. 6.3), and also to segment it into individual characters (Sect. 6.5). In high-energy physics, one needs to detect tracks and vertices (Subsects. 4.3.1 and 4.3.2). Other segmentation techniques, including both thresholding and edge detection, play important roles in cytology (Subsect. 5.5.4), radiology (Subsect. 3.3.4), and remote sensing applications.

A wide variety of pattern classification techniques have been developed, including methods of classification based on given features; methods of feature selection; and methods of defining natural classes by cluster detection (or unsupervised learning). HARALICK discusses some of these ideas in Section 2.4, and they are also briefly treated by HARLOW (Subsect. 3.3.1) and PRESTON (Subsect. 5.5.9).

Many books have appeared during the past few years on pattern classification techniques. Among these, mention should be made of a recent semi-monograph, reviewing basic methods and approaches, which appeared in the Springer series *Communication and Cybernetics* [1.13]. This book can serve to provide the reader of the present volume with a general theoretical background in pattern recognition principles, most of which are broadly applicable to the field of digital picture analysis.

References

1.1 G.C. CHENG, R.S. LEDLEY, D.K. POLLOCK, A. ROSENFELD (eds.): *Pictorial Pattern Recognition* (Thompson Book Co., Washington, D. C. 1968)
1.2 A. GRASSELLI (ed.): *Automatic Interpretation and Classification of Images* (Academic Press, New York 1969)
1.3 L.D. HARMON (ed.): Proc. IEEE **60**, 1117–1233 (1972)
1.4 Computer Graphics and Image Processing **2**, 203–453 (1973)
1.5 A. ROSENFELD: *Picture Processing by Computer* (Academic Press, New York 1969)
1.6. R.O. DUDA, P.E. HART: *Pattern Classification and Scene Analysis* (Wiley, New York 1973)
1.7 A. ROSENFELD: Computing Surveys **1**, 147 (1969)
1.8 A. ROSENFELD: Computing Surveys **5**, 81 (1973)
1.9 A. ROSENFELD: Computer Graphics and Image Processing **1**, 394 (1972)
1.10 A. ROSENFELD: Computer Graphics and Image Processing **3**, 178 (1974)
1.11 A. ROSENFELD: Computer Graphics and Image Processing **4**, 133 (1975)
1.12 A. ROSENFELD, A.C. KAK: *Digital Picture Processing* (Academic Press, New York 1976)
1.13 K.S. FU (ed.): *Communication and Cybernetics*, Vol. 10: Pattern Recognition (Springer, Berlin, Heidelberg, New York 1975)

2. Automatic Remote Sensor Image Processing

R. M. HARALICK

With 10 Figures

2.1 Remote Sensing

Detecting the nature of an object by external observation, without physical contact with the object, is remote sensing. The advantages of gathering data about objects remotely are that the object is usually not disturbed; objects in inaccessible areas can be examined; and a large amount of information over any spatial area can be obtained.

The earliest and most useful form of remote sensing is photography. Here, photon energy (in the visible or near-visible portion of the spectrum) which is radiating or reflected from objects is collected by a camera (the sensor) and recorded on a light sensitive film emulsion. Aerial multiband and color photography can be used to determine the number of acres of land in different uses, such as rangeland, cropland, forest, urban, swamp, marsh, water, etc. It can help identify rock and soil type, vegetation, and surface water condition [2.1, 2].

The camera, of course, is not the only kind of remote sensor. Other types of remote sensors include the multispectral scanner, the infrared scanner, the scanning radiometer, the gamma ray spectrometer, the radar scatterometer, and the radar imager. Figure 2.1 illustrates the typical scanning sensor. A rotating mirror scans the ground scene in a line by line manner with the forward velocity of the sensing platform causing the line scans to be successively displaced. The mirror directs the received energy to a detector which converts it to a video signal which is then recorded by a film recorder to make an image. Thermal infrared scanner systems can produce imagery in daytime or nightime since their detectors are sensitive to emitted heat and not to light. The near-infrared systems are able to penetrate haze and dust making it possible to get good contrast images not achievable by aerial photography. MALILA [2.3] documented the advantages multispectral scanners have over cameras for image enhancement and discrimination. The NASA Third and Fourth Annual Earth Resources Program Reviews [2.4, 5] and the Michigan Symposia on Remote Sensing of Environment contain numerous papers on the application of multispectral or thermal scanners. See also [2.6, 7].

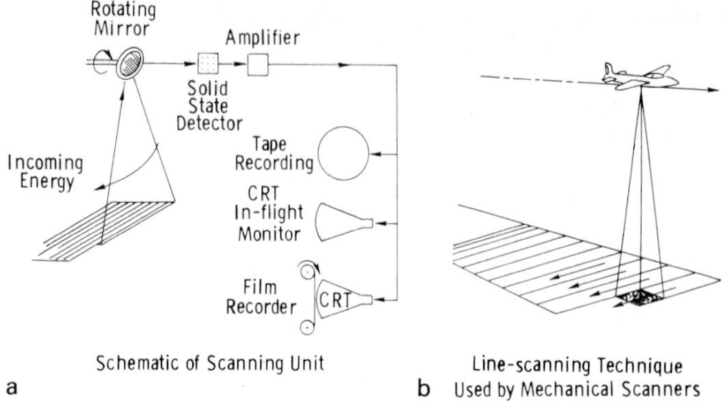

Fig. 2.1. (a) Schematic of scanning unit. (b) Line-scanning technique used by mechanical scanners.

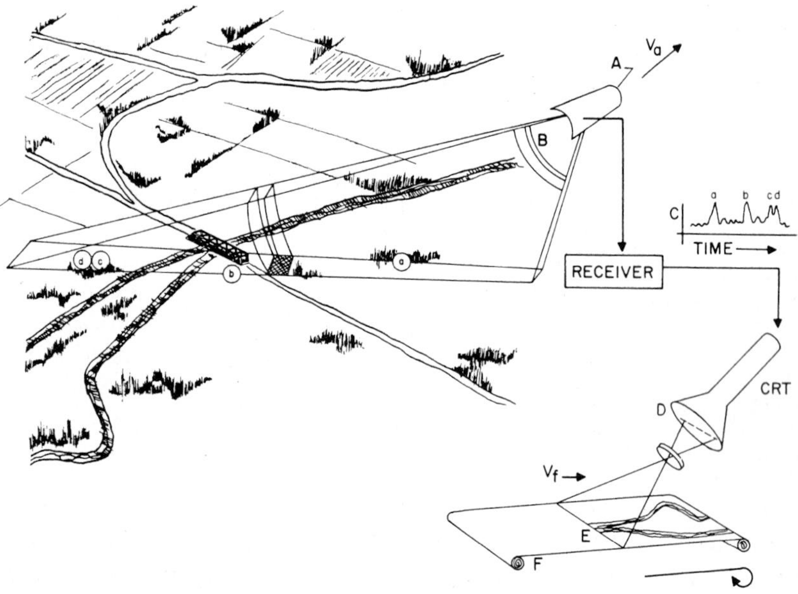

Fig. 2.2. Side-looking radar scanning

The radar imager shown in Fig. 2.2, unlike the others, is an active system. It illuminates the ground scene by transmitting its own microwave radiation and its antenna receives the reflected energy. After appropriate processing and electronic amplification the resulting video signal can be recorded by film recorder to make an image. The radar

Table 2.1. Agriculture/forestry

Application	Type of data Required	Data use
Agriculture	Inventory and distribution	Farm/forest interfaces Boundaries Topographic maps Crop type and density Crop expected yield Livestock census
	Infestation	Disease damage Insect damage Infestation patterns
	Land use	Soil texture Soil moisture and irrigation requirements Soil quality to support vegetation Farm planning
Forestry	Inventory and distribution	Forest texture Boundaries Topographic maps Tree types and count Logging yield and production Location of tree types
	Fire, disease and reclamation	Fire location and damage Pattern and discontinuity Soil moisture and texture Insect and disease damage
Conservation	Land use Grassland vigor	Maps Wildlife management

signal is influenced by ground conductivity and surface roughness. Radar images are particularly good at providing integrated landscape analysis [2.8–8a], as well as vegetation [2.9], geologic [2.10], and soil moisture [2.11] information.

Tables 2.1–5 list the variety of uses to which earth resources remote sensing has been put in the areas of agriculture, forestry, hydrology, geology, geography, and environment monitoring. MERIFIELD et al. [2.12] discussed the potential applications of spacecraft imagery.

The success of manual interpretation or automatic processing of remotely sensed imagery is constrained by the kind of physical character-

Table 2.2. Hydrology

Application	Type of data required	Data use
Water inventory	Water inflow into basins, rivers and streams	River effluents Drainage basin features Reservoir levels Ground water surveys Irrigation routes
Flood control	Excess surface water	Flood location Damage assessment Rainfall monitoring Erosion patterns
Water pollution	Natural and industrial pollution	Color Spectral signature Pollution content Salt content
Water conservation	Evaporation and transpiration	Evapotranspiration
Water resources	Seeps and springs	Temperature variation Water quality
	Glaciology	Frozen water inventory Snow surveys

istics the sensors can detect. The sensors used in the various portions of the electromagnetic spectrum are sensitive to dissimilar energy-matter interactions. They detect different characteristics and, therefore, convey different kinds of informations. In crop identification problems, for example, the molecular absorptions produce color effects in the visible region which convey information about crop type or condition; in the infrared region, the diurnal cycle of thermal response under an insolation load may give information about moisture stresses within crops; in the radar region, the backscattered return is primarily related to surface roughness and dielectric constant, which may in turn be related to crop type, percent cover, and moisture content.

For those readers interested in exploring the remote sensing applications literature further the following guide is offered. A bibliography of the earth resources remote sensing literature has been complied by NASA and is current to 1970. The third and fourth Earth Resources Program Reviews [2.4, 5] (there have been none since 1972) provide descriptions of a large number of investigations funded by NASA. The second and third Earth Resources Technology Satellite (ERTS) Symposia in 1973

Table 2.3. Geology

Application	Type of data required	Data use
Petroleum and minerals detection	Surface and sub-surface patterns	Lithology studies Outcrops Plot magnetic fields Earth folds Drainage patterns Soil compacting and stability Soil density Surface stratification and electrical conductivity
Volcano prediction	Surface feature changes	Temperature variation Lithologic identification Spatial relations
Earthquake prediction	Surface stress and discontinuities	Linear microtemperature anomalies Slope distribution Crust anomalies Soil moisture
Engineering geology	Geothermal power sources	Temperature anomalies Surface gas
	Landslide prediction	Soil moisture Slope distribution Crust anomalies

(NASA GSFC) describe much of the remote sensing done with the 4-channel multispectral scanner in the ERTS satellite.

Since 1962 the Environmental Research Institute of Michigan and the University of Michigan have co-sponsored annual remote sensing symposia with the papers published as Proceedings of the n-th International Symposium on Remote Sensing of Environment. In 1972 the University of Tennessee Space Institute also began organizing annual remote sensing symposia. The conference proceedings of these symposia are obtainable from the respective universities and contain unreviewed papers (abstracts only are reviewed). The annual proceedings of the American Society of Photogrammetry have remote sensing related application papers (also unreviewed). In 1969 the journal *Remote Sensing of Environment* was begun and it along with *Photogrammetric Engineering* and *Modern Geology* contain much of the refereed literature.

Table 2.4. Geography

Application	Type of data required	Data use
Transportation	Identify features	Locate terminals, buildings Locate roads, tracks Traffic count
	Locate new facilities	Make maps at scales of 1:25000 to 1:250000 Cultural factors Economic factors
Navigation	Topography	Make maps at scales of 1:100000 to 1:250000
Urban planning	Locate settlements	Boundary and topography
	Type of settlements	Color, texture, contrast
	Distribution of settlements	Pattern of housing density
	Occurrence of recreation areas	Color, texture, shape
	Population distribution	Population count
	Classification of facilities	Industrial planning 1:50000 scale maps Cultural/economic factors Land use intensity Spectral signature Heat budgets

Table 2.5. Environment

Application	Type of data required	Data use
Air quality monitoring	Backscatter pattern analysis Spectroscopy	SO_2 and NO_2 concentration and distribution Forecasting
Water quality monitoring	Oil slicks, Effluent Salt water intrusion	Color tones Spectral signatures
Violation detection	Spatial/temporal source data	Locate land or sea coordinates of pollution source

2.2 The Image Processing Problem

Remote sensor data systems are designed to obtain information about various aspects of an environment by remotely measuring the electromagnetic transmittance, reflective, or emissive properties of the environment. To obtain information about the "pertinent" aspects of the environment, the sensor system must preserve or mirror the "pertinent" structure of the environment onto the data structure. Automatic remote sensor data processing or pattern recognition is concerned with evaluating, inventorying, and identifying those environmental structures which are and which are not preserved through the sensor's eyes.

There are two modes of automatic processing: one mode requires special purpose hardware, uses film format or analog tape input, and operates in near real time; the other mode requires a general purpose computer, uses digital tape input, and operates relatively more slowly. The first mode—a hardware system mode—is capable of limited sorts of processing while the second mode—a software system mode—is capable of rather sophisticated processing. Neither of these modes has reached the stage of production volume pattern recognition of remotely sensed data; progress has, however, brought them out of the experimental stage, into the prototype stage, and as shown by the corn blight experiment, significant amounts of data can be processed [2.13].

Fu et al. [2.14] discussed the general pattern recognition approach to the automatic processing of remotely sensed data. Haralick et al. [2.15] described some pattern recognition techniques for crop discrimination on radar imagery. Centner and Hietanen [2.16] used automatic pattern recognition techniques to distinguish between cultivated land, urban land, wooded land, and water. Anuta and MacDonald [2.17] discussed automatic recognition of crop types from multi-band satellite photography. Hoffer et al. [2.18] described how automatic pattern recognition can supplement manual photo-interpretation methods for the recognition of crop types. Turinetti and Mintzer [2.19] discussed computer analysis of day and night thermal imagery for land-use identification. Swain [2.20] treated pattern recognition theory for remote sensing data analysis. Landgrebe [2.21] discussed the machine processing approach for remotely sensed imagery. Thus, there is no lack of data indicating that automatic processing techniques can be applied to remotely sensed images. The most recent review of digital image processing activities for remotely sensed images was prepared by Nagy [2.22].

To examine the remotely sensed imagery processing problem, we will consider the case where we have obtained a multi-image set by a combination of perhaps radar images of various frequency-polarization conditions, multispectral scanner images, or multi-band photography. Some of the images may be time-simultaneous and others time-sequential. To be realistic, we will suppose that all sensors which rely on the sun for the illumination source have obtained their pictures at perhaps different solar altitude angles, solar azimuth angles, and meridian angles (see [2.23–25] for documentation of spectral reflectivity variation with angle).

If we assume our problem is an agricultural crop identification problem, we would first like to separate out crops from non-crops (crop detection); then we would like to make an accurate identification of the detected crops and produce a thematic land use map or make estimates of the acreages of the various crops.

To be able to do this rather well-defined interpretation task automatically, we must register or congruence the individual images, calibrate or normalize the grey tones on each image, extract the relevant features required to detect and identify crops, use a decision rule which in some sense makes the best identification possible relative to the extracted features, and finally estimate the decision rule error. Section 2.3 discusses preprocessing procedures; Section 2.4 discrimination, feature extraction, and clustering methods; and Section 2.5 approaches to the quantification of image texture. Sections 2.6 and 2.7 describe some hardware and software aspects of the problem. The glossary of terms used in the remotely sensed image pattern recognition area prepared by HARALICK [2.26] is a general reference for the technical terms used in this paper.

2.3 Preprocessing Problems

2.3.1 Grey Tone Normalization

Each type of remote sensor has its own internal type of calibration or lack of calibration. For most remote sensors, this internal calibration cannot be done accurately or consistently from sensor to sensor and from day to day. Photographic sensors are particularly plagued with calibration problems since little controlled attempt is usually made to compensate for exposure, lens, film, print, developer, and off-axis illumination fall-off differences. PETTINGER [2.27], for example, documented variability in color image quality for photography flown over the Phoenix area. He ascribed the color balance variability to exposure, film (age and storage), and processing differences. The environment compounds the problem by varying atmospheric variables such as haze and cloud cover.

There are several ways to operate on such uncalibrated imagery to bring it into various normalized forms. We will discuss a few of them which are implementable in both hardware and software data processing systems.

Multi-Image Normalization for Viewing Angle, Atmospheric, and Intensity Variations

Sometimes, due to camera design, the illumination intensity across the film is not uniform. More light energy falls on the center of the film than on the edges (off-axis illumination fall-off). Sometimes, due to incorrect camera exposures or sun and cloud variations, atmospheric haze, or viewing angle variations, parts of frames or whole frames may be lighter or darker than normal. Similar sorts of problems occur with moisture variation in multi-frequency radar imagery and with cloud shadows on multispectral thermal imagery.

When the variations are only in intensity, normalization of intensity can be done. In intensity normalization, the assumption is made that the relevant information concerning crop detection and identification is in hue and not in color intensity; that is, the information is in the relationships of one emulsion or channel to another. The procedure is to take the densities of each emulsion or channel at each resolution cell and divide them by some weighted sum of the emulsion or channel densities at that location. The weights in the weighted sum are chosen so that the resulting sum is proportional to total illumination intensity. In this way the color in each resolution cell on the multi-image can be standardized to the same intensity.

When the variations are due to atmospheric aerosol and haze differences, in addition to look angle differences, the normalization is imperfect and is more complicated. The usual model relating reflectivity to energy received is

$$R_r = R_s \tau \varrho + b\,, \tag{2.1}$$

where R_r is the radiance received at the sensor, R_s is the irradiance of the source, ϱ is the reflectivity of the object, b is the backscatter and sky path radiance, and τ is the atmospheric transmissivity.

In general, all parameters may be functions of spectral waveband, viewing angle, direction, and distance. In processing remotely sensed imagery, the object reflectance ϱ is usually the feature wanted and under ideal conditions the received radiance at the sensor can indeed be proportional to ϱ. The atmospheric aerosol and haze variations make the transmissivity τ and the backscatter b variable from day to day and from spectral band to spectral band. The differences in look angle from one side of the image to the other result in different received radiances even from the same kinds of objects because reflectivity is a function of look angle.

When the spectral bandwidths are small enough, it may be assumed that the product $R_{s_1}\tau$ for one spectral band is close enough to the product $R_{s_2}\tau$ for an adjacent spectral band so that they may be considered to be equal. Then, when the backscatter in both bands is small enough to be negligible, the ratios of the viewed radiances of adjacent bands will be equal to the ratios of the object reflectivities.

$$\left.\begin{array}{l} R_{r_1}=R_{s_1}\tau_1\varrho_1+b_1, \text{ energy and reflectivity} \\ \qquad\qquad\qquad\qquad \text{relationship for band 1} \\ R_{r_2}=R_{s_2}\tau_2\varrho_2+b_2, \text{ energy and reflectivity} \\ \qquad\qquad\qquad\qquad \text{relationship for band 2} \end{array}\right\} \text{ in general,} \tag{2.2}$$

$$\left.\begin{array}{l} R_{r_1}=R_{s_1}\tau_1\varrho_1 \\ R_{r_2}=R_{s_2}\tau_2\varrho_2 \end{array}\right\} \begin{array}{l} \text{energy and reflectivity relationship} \\ \text{when backscatter is negligible,} \end{array} \tag{2.3}$$

$$\frac{R_{r_1}}{R_{r_2}}=\frac{R_{s_1}\tau_1}{R_{s_2}\tau_2}\frac{\varrho_1}{\varrho_2} \quad \begin{array}{l} \text{is ratio of received radiances when} \\ \text{backscatter is negligible,} \end{array} \tag{2.4}$$

$$\frac{R_{r_1}}{R_{r_2}}=\frac{\varrho_1}{\varrho_2} \quad \text{when} \quad R_{s_1}\tau_1=R_{s_2}\tau_2\,. \tag{2.5}$$

Thus, for a 12-channel scanner, the original 12-tuple $(x_1, x_2, \ldots, x_{12})$, x_i being the energy from the i-th channel, would be normalized to the 11-tuple $(x_1/x_2, x_2/x_3, \ldots, x_{11}/x_{12})$. Note that with this normalization procedure, one dimension is lost due to the fact that no information is obtained when the twelfth channel is normalized.

It is possible to generalize the normalization procedure to take into account the backscatter if some additional assumptions are made: the adjacent spectral bands are small enough so that the product $R_s\tau$ is equal for each three adjacent bands and the backscatter b is equal for each three adjacent bands. Under this assumption, the ratio $(R_{r_1}-R_{r_2})/(R_{r_2}-R_{r_3})$ equals $(\varrho_1-\varrho_2)/(\varrho_2-\varrho_3)$. Thus for a 12-channel scanner the 12-tuple $(x_1, \ldots, x_{12})$ would be normalized to the 12-tuple $[(x_1-x_2)/(x_2-x_3), (x_2-x_3)/(x_3-x_4), \ldots, (x_{10}-x_{11})/(x_{11}-x_{12})]$. KRIEGLER et al. [2.28] and CRANE [2.29] first reported these normalization techniques. SMEDES et al. [2.30] noted that normalization results in more accurate maps with fewer training areas. NALEPKA and MORGENSTERN [2.31] indicated that percent identification accuracy can improve especially for pixels at the edges of an image by the use of these normalizations.

Single Image Quantization

Consider now a single image in a multi-image set. Perhaps this image had been in the developer too long or it is visibly pre-fogged or it is overexposed or the film was unusually sensitive. Perhaps the scanner detector lost some of its sensitivity or the A/D converter changed its calibration. Quantization can make single images invariant to these sorts of changes.

In general, the quantization process divides the range of grey tones or grey tone levels into K intervals (λ_1, λ_2), (λ_2, λ_3),..., $(\lambda_K, \lambda_{K+1})$ determined in some fashion. A new quantized image is generated from the original image. Any resolution cell on the original image having a grey tone lying in the first interval (λ_1, λ_2), is assigned on the new image to the first quantization level; any resolution cell on the original image having a grey tone lying in the second interval (λ_2, λ_3) is assigned on the new image to the second quantization level, and so on. The quantized image, therefore, has only K possible grey tones.

There are two common ways which are used to determine the quantization intervals, depending upon the kind of assumption one is justified in making as to the nature of the uncalibrated changes. In equal interval quantizing, the assumption is that the density change can be described by a simple linear scaling and that the overall darker or lighter appearance can be described by the addition or subtraction of some grey tone. To perform equal interval quantizing, the lightest grey tone λ_1 and the darkest grey tone λ_{K+1} on the original uncalibrated image are determined. The range $\lambda_{K+1} - \lambda_1$ of grey tones is then divided into K equal length intervals, each of length $(\lambda_{K+1} - \lambda_1)/K$. The interval end points are determined by

$$\lambda_i = \lambda_1 + (i-1)(\lambda_{K+1} - \lambda_1)/K\,, \qquad i = 1, 2, \ldots, K-1\,.$$

In this way, two images which are the same except for a linear scaling and grey tone translation will produce the same equal interval quantized image. Thus, equal interval quantizing is invariant with respect to linear transformations.

Another normalization procedure that achieves the same kind of invariance as discrete equal interval quantizing is to determine the mean and standard deviation of the grey tones on the original image and generate a normalized image by subtracting the mean and dividing by the standard deviation.

In equal probability quantizing, it is assumed that the original image differs from the ideal calibrated image by some monotonic function which can be linear or nonlinear; this assumption implies that any object which is darker than another object on the original uncalibrated image will also be so for the ideal calibrated image. To perform equal probability quantizing, the range of grey tones is divided into K intervals, each of equal probability. In other words, for any interval $(\lambda_i, \lambda_{i+1})$ of grey tones, the number of resolution cells on the original image having a grey tone in that interval is the same as for any other interval and is in fact equal to $(1/K)$-th of the total number of resolution cells on the image.

If equal probability quantizing were applied to two images, one a calibrated image and the other an uncalibrated image differing from the calibrated image only by a monotonic grey tone transformation, then the two equal probability quantized images produced from them would be identical. Hence, equal probability quantizing is invariant with respect to any monotonic grey tone transformation.

2.3.2 Image Registration, Congruencing, Rectification

In order to create a multi-image set from separate images which might have been taken of the same area but perhaps at different times or with different sensors, each individual image must be aligned or spatially fit with every other image in the set so that there is no geometric distortion and all corresponding points match. This alignment problem occurs for multi-image sets which have time simultaneous image combinations from different sensors (such as radar imagery with multi-band, or multi-band with scanner imagery), or for time sequential imagery from any sensor(s). When the individual images are of the same scale and geometry, the alignment can be done by rotating and translating one image with respect to another until the cross-correlation between the images (or between the derivatives of the images) is maximal. The process of aligning the image by translation and rotation is called registration and devices which do such translation and rotation are often called image correlators.

When the individual images in the multi-image set have different geometry, the alignment process is more difficult. Geometrical irregularity can be caused by 1) a multi-image set containing a combination of images taken by different types of sensors, each introducing its own kind of geometric distortion due to the way the sensor operates, or 2) the same type of sensors having slightly different "look angles" or altitudes, or 3) various sorts of uncompensated platform motion. In this case the alignment process is called congruencing, since it must literally bring points into a point by point correspondence. When the congruencing process has the additional constraint that the image geometry after congruencing must be planimetric, the process is called image rectification.

Often congruencing or recitification for images in photographic form is done by one or a sequence of projective transformations with optical, electro-optical, or electronic equipment. The basic idea of a projective transformation can be easily visualized by thinking of the image being on a flat rubber surface. The projective transformation places a quadrilateral frame on the surface and then pulls or pushes each side of the frame at a different orientation angle to obtain the desired geometry. The optical devices perform the projective transformation on the entire image, while the electro-optical devices are capable of dividing the image into regions and operating on each region with a different transformation. The parameters for the transformation can be determined from supplied ground control points, or by an automatic cross-correlation scheme such as that employed by the BAI Image correlator [2.32]. McEwen [2.32a] evaluated various analog techniques for image registration.

A greater degree of flexibility is afforded by the use of a digital computer to do digital image registration or image congruencing. Projective transformations, affine transformations, and polynomial transformations are all possible. Let R and C be the index sets for the rows and columns of the image and G be the set of grey tones for the image. Correcting for geometric distortion, or spatially fitting one image to another by registering or congruencing, corresponds to constructing from the input image I, $I: R\times C\to G$, an output image J, $J: R\times C\to G$ by some transformation f from the spatial coordinates of the output image to the spatial coordinates of the input image, i.e. $f: R\times C\to R\times C$, so that the congruenced image can be determined by

$$J(r,c)=I(f(r,c)).$$

This equation says that for each (row, column) coordinates (r,c) in the output image, the grey tone which we put there is $J(r,c)$, and this grey tone is obtained as the grey tone appearing on the input image I at coordinates (r',c'), where $(r',c')=f(r,c)$. For affine transformations [2.33, 34]

$$\begin{aligned} r' &= [a_{11}r+a_{12}c+a_{13}] \\ c' &= [a_{21}r+a_{21}c+a_{23}]. \end{aligned} \tag{2.6}$$

For projective transformations [2.35, 36]

$$\begin{aligned} r' &= \left[\frac{a_{11}r+a_{12}c+a_{13}}{a_{31}r+a_{32}c+a_{33}}\right] \\ c' &= \left[\frac{a_{21}r+a_{22}c+a_{23}}{a_{31}r+a_{32}c+a_{33}}\right]. \end{aligned} \tag{2.7}$$

For K-th order polynomial transformations [2.37]

$$r' = \left[\sum_{i=0}^{K} \sum_{j=0}^{K-i} a_{ij} r^i c^j\right]$$
$$c' = \left[\sum_{i=0}^{K} \sum_{j=0}^{K-i} b_{ij} r^i c^j\right]. \tag{2.8}$$

(Notation: $[x]$ means the nearest integer to x.)

These kinds of spatial transformations are intended only to account for the low (spatial) frequency, sensor-associated spatial distortions (centering, size, skew, pincushion) and for distortions due to earth's curvature, sensor attitude and altitude deviations. Note that since the pure projective affine or polynomial transformations do not, in general, produce r' and c' as integers, we take whatever values these transformations give for r' and c' and either interpolate on the grey tones surrounding these coordinates, or, as shown in our equations, convert them to the corresponding nearest integers and take the nearest neighbor grey tone. This means that in actual implementation, all the resolution cells of the output image are examined and for each one the corresponding coordinates on the input image are determined. Then in the nearest neighbor interpolation approach, for example, the grey tone on the output image is defined to be the grey tone appearing in the pixel of the input image closest to the determined corresponding coordinates.

Having seen how the transformation is implemented, we must now discuss how the parameters of the transformation are determined. When the transformation is a simple translation

$$r' = r + a$$
$$c' = c + b,$$

the translation parameters a and b can be determined automatically using cross correlation or distance measures. If I is the image which needs to be translationally registered so that it fits image J, the parameters a and b can be chosen so that

$$\frac{\{\sum_{(r,c)\in R\times C} I(r+a, c+b) J(r,c)\}^2}{\sum_{(r,c)\in R\times C} I^2(r+a, c+b) \sum_{(r,c)\in R\times C} J^2(r,c)}$$
$$\geqq \frac{\{\sum_{(r,c)\in R\times C} I(r+\alpha, c+\beta) J(r,c)\}^2}{\sum_{(r,c)\in R\times C} I^2(r+\alpha, c+\beta) \sum_{(r,c)\in R\times C} J^2(r,c)}$$

for all α, β (cross correlation measure),

or (2.9)

$$\sum_{(r,c)\in R\times C} |I(r+a, c+b) - J(r,c)|^P \leqq \sum_{(r,c)\in R\times C} |I(r+\alpha, c+\beta) - J(r,c)|^P$$

for all α, β (distance measure).

The above summations over rows and columns are actually taken only over those points $(r,c)\in R\times C$ such that $(r+a,c+b)\in R\times C$. ANUTA [2.38] discussed the spatial registration problem from the cross-correlation point of view and implemented the algorithm using the fast Fourier transform technique. BARNEA and SILVERMAN [2.39] described a sequential similarity detection algorithm for translational registration. Essentially it is a distance approach implemented in a fast way by not requiring all terms in the summation to be computed for each translation (α,β). WEBBER [2.40] combined an affine transformation with a sequential similarity detection algorithm to determine all the parameters of the affine transformation.

When the image registration or congruencing problem is more involved than simple translation it is more difficult to determine the spatial transformation parameters automatically. Usually, corresponding ground control points are identified by normal photo interpretation methods. Care is taken so that the ground control points are not concentrated in any one area of the image but to the best possible degree are spread uniformly across the image. Then a least squares fit is performed to determine the parameters. In the K-th order polynomial registration method for $K\geqq 2$, the coefficients of the high-order terms can become strong functions of the location of the matching ground control points. This is undesirable. YAO [2.41] discussed a piecewise rubber sheeting process which is less sensitive to ground control point placement.

MARKARIAN et al. [2.37] developed the polynomial approach to spatial registration and discussed how a point shift algorithm can be used to speed up the calculations when the actual amount of geometric correlation needed is small. RIFFMAN [2.42] compared nearest neighbor, bilinear, and cubic interpolation methods as regards the problem of generating the grey tone on the output image from non-integer spatial coordinates on the input image. As expected, the nearest neighbor interpolation method yields pixel jitter especially for high contrast areas. The bilinear interpolation is free from pixel jitter but reduces resolution. The cubic convolution chosen to approximate a $(\sin x)/x$ kernel has no pixel jitter and does not reduce resolution but requires more computation time.

SZETO [2.35] described a linear interpolation scheme based on projective transformations for the rectification of digitized images. He decomposed the image into quadrilateral subimages and derived a bound on the maximum error between the interpolation scheme and the true projection within each quadrant. When the precomputed error in any quadrilateral subimage exceeds a prespecified limit, the subimage is further divided and the calculation is repeated.

2.4 Image Pattern Recognition

2.4.1 Category and Training Data Selection

Once the image data set has been registered or congruenced and the images appropriately normalized and/or quantized, the most time consuming task of processing may begin. This task involves familiarization with the data and the selection of the categories and appropriate "training data" for them.

The investigator usually has some kinds of land-use categories in mind between which he would like to distinguish. Hopefully, the sensor(s) are sensitive enough to spectral, tone, or texture differences so that the categories can, in fact, be distinguished to a large degree purely on the basis of the images in the multi-image data set. If this is not the case, there is no point to going on and either a redefinition of more distinguishable and reasonable categories must be made or different, more sensitive sensors must be used.

Determining the categories to be used is more than a simple matter of naming them. For example, the naming of the category "wheat" in an agricultural remote sensing problem is not enough. Not only are there many kinds of wheat, but the percent of covering of wheat over the ground for wheat of the same maturity may vary from field to field, and the percent of weeds in the wheat field may vary. Does a small-area ground patch the size of a resolution cell consisting of 25% weeds and 75% wheat fall into the wheat category, the weed category, or a mixed category? What percent cover must the wheat have for the category identification to change from bare ground to wheat? How are these variations reflected in the spectral or temporal signatures of these categories?

How can the investigator tell if his categories form a reasonable set? The first step is to examine the data structure by looking at histograms or scattergrams of sampled regions from the different categories. If these histograms or scattergrams show little overlap between data points of different categories, then the categories are resonable. If a pair of categories shows significant overlap for histograms of each channel and/or scattergrams for each pair of channels, then there is a definite problem with distinguishing that pair of categories.

Another way of determining possible categories is by cluster analysis (a topic which we will discuss at greater length in Subsect. 2.4.5) [2.43]. Clustering the training data can determine subclasses having characteristic centroids and typically unimodal distributions over the measurement space. A comparison of the data points in each cluster with their corresponding ground truth characteristics will allow the investigator to choose category classes whose data points fall in distinct clusters. He thereby assures himself that data points from each category are likely to be discriminated correctly by the decision rule in the pattern identification or classification process. DAVIS and SWAIN [2.44] advocated the clustering approach to category selection.

When the investigator is satisfied with his category choices, he is ready to choose his training data from the multi-image. The training data is given by a specially prepared subset of pixels from the multi-image. The preparation consists of labeling each point in the training data with its true category identification. This identification can come from either ground truth observation or photo interpretation.

Since the training data are to be used to determine a decision rule which will make a category assignment for each pixel in the whole image data set, there are two points to which attention must be paid: 1) the training data for each category must be representative of all data for that category; 2) the training data for each category must come close to fitting the distributional assumptions on which the decision rule is based. For example, point 1) says that if one category is wheat, and there are some dry and wet wheat fields which are distinguishably different on at least one image of the multi-image set, then data points from both kinds of wheat fields must be included in the training data for wheat. Point 2) says that if the histograms or scattergrams of wheat indicate that wheat has a bimodal distribution, one mode corresponding to the wet fields and one mode corresponding to the dry fields, and if the decision rule is to be based on some unimodal distributional assumption (such as normal or Gaussian), then for the purposes of the decision rule determination, the wheat category ought to be split into two subcategories: dry wheat and wet wheat. Then after the decision rule has made its identification assignments for the whole data set, the wet wheat and dry wheat subcategories can be collapsed to the one category, wheat.

The process of defining training set subcategories when using a unimodal distributional assumption is important not only for categories which have obvious subclasses but also for those categories which may not seem to have subclasses. This observation is prompted by the fact that in an agricultural context, multispectral data from a single crop field usually exhibit a unimodal structure, yet the pooled data from several fields of the same species often exhibit a multimodal character, the number of modes being nearly equal to the number of fields. Thus while the unimodal assumption might be appropriate for a single field it may not be appropriate for a group of fields.

Monte Carlo experimental results by WACKER [2.45] using the Gaussian classifier verify that increasing the number of subcategories defined for each category increases the average classifier accuracy over many fields and in a pronounced way decreases the classifier variability on a field by field basis. In other words, having subcategory training sets when using a Gaussian classifier simultaneously increases identification accuracy over the entire image and more evenly or randomly distributes the identification errors.

A final item to which attention must be paid is the number of samples for each (sub)category in the training set. Parametric classifiers which make the unimodal distributional assumption should have a number of samples per category which at a minimum is between 3 and 10 times the dimensionality of the pattern vector [2.46]. Non-parametric classifiers such as a discrete table look-up rule can require orders of magnitude more samples per category than the parametric classifiers. KANAL and CHANDRASEKARAN [2.47], and FOLEY [2.46] discussed the relationship between sample size and dimensionality.

2.4.2 Decision Rule Determination

We first discuss how the training data determine a distribution-free decision rule, for it is here that we can focus on the essence of decision rules. Then we will describe some decision rules determined from certain kinds of distributional assumptions or approximations. In what follows we denote by "d" the typically 3- or 4-tuple multi-spectral feature vector and by "c" a land use category. The set of all features is D and the set of all categories is C.

To determine a distribution free decision rule, the first step is to estimate the set of conditional probabilities $\{P_d(c)|d \in D, c \in C\}$. $P_d(c)$ denotes the probability of category "c" being the true category identification of a small-area ground patch given that measurements of it generated feature vector "d". The conditional probability of category "c" given feature "d", $P_d(c)$, can be estimated by the proportion of those data points having true category identification "c" in that subset of the training data whose measurements generate feature "d".

Once these conditional probabilities have been computed, common sense should reveal the optimal decision rule. Consider the decision rule's problem when it tries to assign a category identification to a data resolution cell with feature vector or pattern "d". The conditional probabilities $P_d(c_1)$, $P_d(c_2)$,..., $P_d(c_K)$ have been estimated for categories $c_1, c_2, \ldots c_K$. To decide the category identification for a data point with feature "d" is an easy matter. Assign it to any category "c_k" having highest conditional probability; that is, assign it to a category "c_k" if and only if

$$P_d(c_k) \geqq P_d(c_i), \quad i=1, 2, \ldots, K.$$

Such an optimal decision rule is called a simple Bayes rule.

Sometimes it is more convenient to estimate the set of conditional probabilities $\{P_c(d)|d \in D, c \in C\}$. $P_c(d)$ is the probability of obtaining a feature "d" from a small-area ground patch given that the small-area ground patch has true category identification "c". The conditional probability of feature "d" given category "c", $P_c(d)$, can be estimated by the proportion of training data points having feature "d" in the subset of training data points having true category identification "c". Using these conditional probabilities, the only logical assignment a decision rule can make of a data point with feature "d" is to assign it to any category "c_k" such that

$$P_{c_k}(d) \geqq P_{c_i}(d), \quad i=1, 2, \ldots, K.$$

Such a decision rule is called a maximum likelihood rule.

The relationship between the maximum likelihood rule and the simple Bayes rule is easy to develop. Let us denote by $P(d)$ the proportion of measurements in the training data having feature "d", by $P(c)$ the proportion of measurements in the training data having true category identification "c", and by $P(d, c)$ the proportion of measurements in the training data having feature "d" and true category identification "c". Then by definition of conditional probability

$$P_d(c) = \frac{P(d, c)}{P(d)}, \quad P_c(d) = \frac{P(d, c)}{P(c)}. \tag{2.10}$$

By multiplying the inequality $P_d(c_k) \geqq P_d(c_i)$ on both sides by $P(d)$ we obtain

$$P_d(c_k)P(d) \geqq P_d(c_i)P(d). \tag{2.11}$$

But, $P_d(c_k)P(d) = P(d, c_k) = P_{c_k}(d)P(c_k)$. Hence,

$$P_d(c_k) \geqq P_d(c_i) \quad \text{if and only if} \quad P_{c_k}(d)P(c_k) \geqq P_{c_i}(d)P(c_i). \tag{2.12}$$

When the category prior probabilities are all equal, $P(c_j) = 1/K$ $(j=1, 2, \ldots, K)$, we obtain $P_d(c_k) \geqq P_d(c_i)$ if and only if $P_{c_k}(d) \geqq P_{c_i}(d)$. Therefore, the simple Bayes rule and maximum likelihood rule are identical if the prior probabilities $P(c)$, $c \in C$, are all equal.

Use of a distribution-free rule in the digital computer is only possible under those circumstances in which it is possible to store the set of conditional probabilities. This storage possibility is strongly conditioned by the number of categories and the number of quantized levels to which each component of the feature vector is expressed. For example, if there are 10 categories, and if the feature vector has 4 components, and each component is quantized to 5 levels, then a total of $10 \times 5^4 = 6250$ words is needed to store the conditional probabilities. This is a reasonable storage requirement. However, if each component were to be quantized to 10 levels, then a total of $10 \times 10^4 = 10^5$ words of storage are needed to store the conditional probabilities. This is usually an unreasonable storage requirement.

There are two approaches to handling the unreasonable storage problem: The table look-up approach reduces the storage requirement by careful storage and use of only essential information; the parametric approach assumes that the conditional probability $P_c(d)$ can be expressed by a formula having only a few parameters which need to be estimated. The table look-up approach uses more storage than the parametric approach but is much quicker in performing category assignments than the more computationally complex parametric approach. The two approaches are consistent with the observation that memory storage often can be traded for computational complexity.

Table Look-Up Approach

BROONER et al. [2.47a] used a table look-up approach on high altitude multiband photography flown over Imperial Valley, California, to determine crop types. Their approach to the storage problem was to perform an equal probability quantizing from the original 64 digitized grey levels to ten quantized levels for each of the three bands: green, red, and near infrared. Then after the conditional probabilities were empirically estimated, they used a Bayes rule to assign a category to each of the 10^3 possible quantized vectors in the 3-dimensional measurement space. Those vectors which occurred too few times in the training set for any category were deferred assignment. Figure 2.3 illustrates the decision regions associated with such a table look-up discrete Bayes decision rule. Notice how the quantized multispectral measurement vector can be used as an address in the 3-dimensional table to look up the corresponding category assignment.

The rather direct approach employed by BROONER et al. has the disadvantage of requiring a rather small number of quantization levels. Furthermore, it cannot be used with measurement vectors of dimension greater than four; for if the number of quantization levels is about 10, then the curse of dimensionality forces the number of possible quantized vectors to an unreasonably large size. Recognizing the grey level precision restriction forced by the quantizing coarsening effect, EPPLER et al. [2.48] suggested a way to maintain greater quantizing precision by defining a

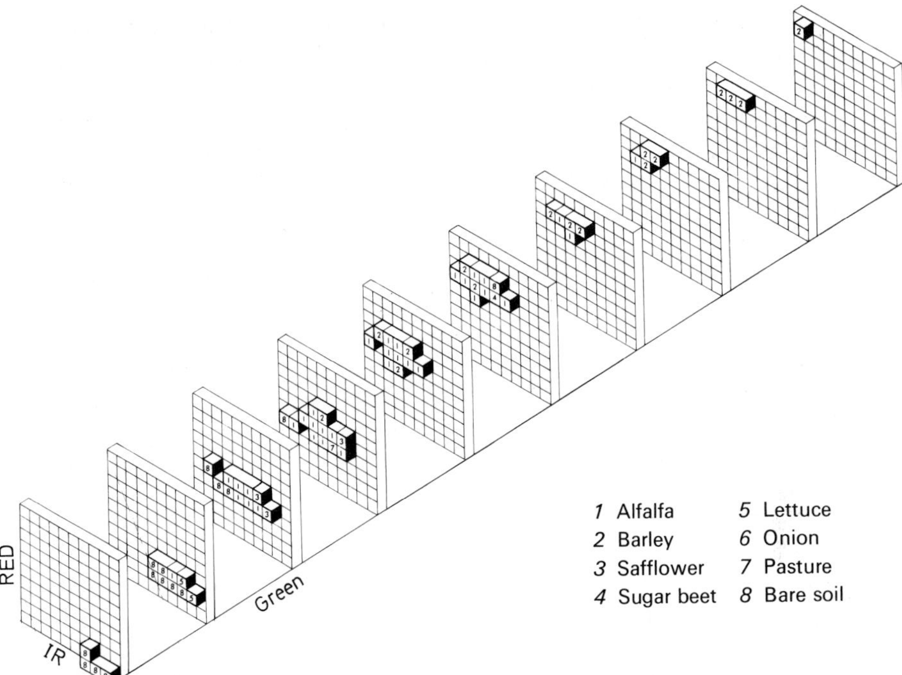

Fig. 2.3. Viewed as an expanded cube, each dimension representing the spectral region of the three multiband images whose density values have been quantized to ten equally spaced levels, this sketch depicts the decision rule boundaries for each land-use category used by the discrete Bayes rule

quantization rule for each category and measurement dimension as follows:

1) fix a category and a measurement dimension component;
2) determine the set of all measurement patterns which would be assigned by the decision rule to the fixed category;
3) examine all the measurement patterns in this set and determine the minimum and maximum grey levels for the fixed measurement component;
4) construct the quantizing rule for the fixed category and measurement dimension pair by dividing the range between the minimum and maximum grey levels into equal spaced quantizing intervals.

This multiple quantizing rule in effect determines for each category a rectangular parallelepiped in measurement space which contains all the measurement patterns assigned to it. Then, as shown in Fig. 2.4, the equal interval quantizing puts a grid over the rectangular parallelepiped.

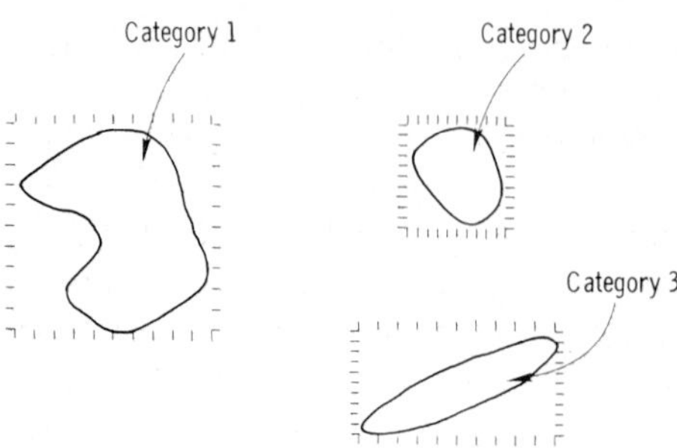

Fig. 2.4. Illustrates how quantizing can be done differently for each category, thereby enabling more accurate classification by the following table look-up rule: 1) quantize the measurement by the quantizing rule for category *1*, 2) use the quantized measurement as an address in a table and test if the entry is 1 or 0, 3) if it is 1 assign the measurement to category *1*; if it is 0, repeat the procedure for category *2*, etc.

Notice how for a fixed number of quantizing levels, the use of multiple quantizing rules in each band allows greater grey level quantizing precision compared to the single quantization rule for each band.

A binary table for each category can be constructed by associating each entry of the table with one corresponding cell in the gridded rectangular parallelepiped. Then we define the entry to be 1 if the decision rule has assigned a majority of the measurement patterns in the corresponding cell to the specified category; otherwise, we define the entry to be 0.

The binary tables are used in the implementation of the multiple quantization rule table look-up in the following way. Order the categories in some meaningful manner such as by prior probability. Quantize the multispectral measurement pattern using the quantization rule for category c_1. Use the quantized pattern as an address to look up the entry in the binary table for category c_1 to determine whether or not the prestored decision rule would assign the pattern to category c_1. If the decision rule makes the assignment to category c_1 the entry is 1 and all is finished. If the decision rule does not make the assignment to category c_1, the entry is 0 and the process repeats in a similar manner with the quantization rule and table for the next category.

One advantage of this form of the table look-up decision rule is the flexibility of being able to use different subsets of bands for each category look-up table and thereby take full advantage of the feature selecting capability to define an optimal subset of bands to discriminate one category from all the others. A disadvantage of this form of rule is the large amount of computational work required to determine the rectangular parallelepipeds for each category and the still large amount of memory storage required (about 5000 8-bit bytes per category).

EPPLER [2.49] discussed a modification of the table look-up rule which enables memory storage to be reduced by five times and decision rule assignment time to be decreased by two times. Instead of pre-storing in tables a quantized measurement space image of the decision rule, he suggested a systematic way of storing in tables the boundaries or end-points for each region in measurement space satisfying a regularity condition and having all its measurement pattern assigned to the same category.

Let $D=D_1 \times D_2 \times \ldots \times D_N$ be the measurement space. A subset $R \subseteq D_1 \times D_2 \times \ldots \times D_N$ is a regular region if and only if there exist constants L_1 and H_1 and functions $L_2, L_3, \ldots, L_N, H_2, H_3, \ldots, H_N$

$$\begin{aligned}&(L_n: D_1 \times D_2 \times \ldots \times D_{n-1} \rightarrow (-\infty, \infty);\\&H_n: D_1 \times D_2 \times \ldots \times D_{n-1} \rightarrow (-\infty, \infty))\end{aligned} \tag{2.13}$$

such that

$$\begin{aligned}R=\{(x_1, \ldots, x_N) \in D \mid &L_1 \leqq x_1 \leqq H_1\\&L_2(x_1) \leqq x_2 \leqq H_2(x_1)\\&\vdots\\&L_N(x_1, x_2, \ldots, x_{N-1}) \leqq x_N \leqq H_N(x_1, x_2, \ldots, x_{N-1})\}\end{aligned} \tag{2.14}$$

From the definition of a regular region, it is easy to see how this table look-up by boundaries decision rule can be implemented. Let $d=(d_1, \ldots, d_N)$ be the measurement pattern to be assigned a category. To determine if d lies within a regular region R associated with category c we look up the numbers L_1 and H_1 and test to see if d_1 lies between L_1 and H_1. If so, we look up the numbers $L_2(d_1)$ and $H_2(d_1)$ and so on. If all the tests are satisfied, the decision rule can assign measurement pattern d to category c. If one of the tests fails, tests for the regular region corresponding to the next category can be made.

The memory reduction in this kind of table look-up rule is achieved by only storing boundary or end-points of decision regions, and the speed-up is achieved by having one-dimensional tables whose addresses are easier to compute than the three- or four-dimensional tables required by the initial table look-up decision rule. However, the price paid for these advantages is the regularity condition imposed on the decision regions for each category. This regularity condition is stronger than set connectedness but weaker than set convexity.

Another approach to the table look-up rule can be based on ASHBY's [2.50] technique of constraint analysis. ASHBY suggested representing in an approximate way subsets of Cartesian product sets by their projections on various smaller dimensional spaces. Using this idea for two-dimensional spaces we can formulate the following kind of table look-up rule.

Let $D = D_1 \times D_2 \times \ldots \times D_N$ be the measurement space, C be the set of categories, and $J \subseteqq \{1, 2, \ldots, N\} \times \{1, 2, \ldots, N\}$ be an index set for the selected two-dimensional spaces. Let the probability threshold α be given. Let $(i, j) \in J$; for each $(x_1, x_2) \in D_i \times D_j$ define the set $S_{ij}(x_1, x_2)$ of categories having the highest conditional probabilities given (x_1, x_2) by $S_{ij}(x_1, x_2) = \{c \in C | P_{x_1, x_2}(c) \geqq \alpha_{ij})\}$, where α_{ij} is the largest number which satisfies

$$\sum_{c \in S_{ij}(x_1, x_2)} P_{x_1, x_2}(c) \geqq \alpha . \tag{2.15}$$

$S_{ij}(x_1, x_2)$ is the set of likely categories given that components i and j of the measurement pattern take the values (x_1, x_2).

The sets S_{ij}, $(i, j) \in J$, can be represented in the computer by tables. In the (i, j)-th table S_{ij} the (x_1, x_2)-th entry contains the set of all categories of sufficiently high conditional probability given the marginal measurements (x_1, x_2) from measurement components i and j, respectively. This set of categories is easily represented by a one-word table entry; a set containing categories c_1, c_7, c_9, and c_{12}, for example, would be represented by a word having bits 1, 7, 9, and 12 on and all other bits off.

The decision region $R(c)$ containing the set of all measurement patterns to be assigned to category c can be defined from the S_{ij} sets by

$$\begin{aligned} R(c) = \{(d_1, d_2, \ldots, d_N) \in D_1 \times D_2 \times \ldots \times D_N | \\ \{c\} = \textstyle\bigcap_{(i,j) \in J} S_{ij}(d_i, d_j)\} . \end{aligned} \tag{2.16}$$

This kind of table look-up rule can be implemented by using successive pairs of components (defined by the index set J) of the (quantized) measurement patterns as addresses in the just mentioned two-dimensional tables. The set intersection required by the definition of the decision region $R(c)$ is implemented by taking the Boolean AND of the words obtained from the table look-up for the measurement to be assigned a category. Note that this Boolean operation makes full use of the natural parallel computation capability that the computer has on bits of a word. If the k-th bit is the only bit which remains on in the resulting word, then the measurement pattern is assigned to category c_k. If there is more than one bit on, or no bits are on, then the measurement pattern is deferred assignment (reserved decision). Thus we see that this form of a table look-up rule utilizes a set of "loose" Bayes rules in the lower dimensional projection spaces and intersects the resulting multiple category assignment sets to obtain a category assignment for the measurement pattern in the full measurement space.

Because of the natural effect which the category prior probabilities have on the category assignments produced by a Bayes rule it is possible for a measurement pattern to be the most probable pattern for one category yet be assigned by the Bayes rule to another category having much higher prior probability. This effect will be pronounced in the table look-up rule just described because the elimination of such a category assignment from the set of possible categories by one table look-up will completely eliminate it from consideration because of the Boolean AND or set intersection operation. However, by using an appropriate combination of maximum likelihood and Bayes rules, something can be done about this.

For any pair (i, j) of measurement components, fixed category c, and probability threshold β, we can construct the set $T_{ij}(c)$ of the most probable pairs of measurement values from components i and j arising from category c. The set $T_{ij}(c)$ is defined by

$$T_{ij}(c) = \{(x_1, x_2) \in D_i \times D_j | P_c(x_1, x_2) \geqq \beta_{ij}(c)\}, \tag{2.17}$$

where $\beta_{ij}(c)$ is the largest number which satisfies

$$\sum_{(x_1, x_2) \in T_{ij}(c)} P_c(x_1, x_2) \geqq \beta .$$

Tables which can be addressed by (quantized) measurement components can be constructed by combining the S_{ij} and T_{ij} sets. Define $Q_{ij}(x_1, x_2)$ by

$$Q_{ij}(x_1, x_2) = \{c \in C | (x_1, x_2) \in T_{ij}(c)\} \cup S_{ij}(x_1, x_2). \tag{2.18}$$

The set $Q_{ij}(x_1, x_2)$ contains all the categories whose respective conditional probabilities given measurement values (x_1, x_2) of components i and j are sufficiently high (a Bayes rule criterion) as well as all those categories whose most probable measurement values for components i and j, respectively, are (x_1, x_2) (a maximum likelihood criterion). A decision region $R(c)$ containing all the (quantized) measurement patterns can then be defined as before using the Q_{ij} sets

$$\begin{aligned} R(c) = \{(d_1, d_2, \ldots, d_N) \in D_1 \times D_2 \ldots \times D_N | \\ \{c\} = \bigcap_{(i,j) \in J} Q_{ij}(d_i, d_j)\}. \end{aligned} \tag{2.19}$$

A majority vote version of this kind of table look-up rule can be defined by assigning a measurement to the category most frequently selected in the lower dimensional spaces

$$R(c)=\{(d_1, d_2, \ldots, d_N)\in D_1 \times D_2 \times \ldots \times D_N | \\ \#\{(i,j)\in J | c\in Q_{ij}(d_i, d_j)\} \geqq \#\{(i,j)\in J | c'\in Q_{ij}(d_i, d_j) \\ \text{for every} \quad c'\in C-\{c\}\}\}. \tag{2.20}$$

Multivariate Normal Assumption

The other approach to easing the storage requirement of the decision rule entails making assumptions about the form the conditional probability $P_c(d)$ takes. By expressing $P_c(d)$ as a formula with only a few parameters which have to be estimated, storage is saved and computational complexity is increased. The most frequent distributional assumption made is the N-dimensional multivariate normal or Gaussian one. Here, the conditional probability is a unimodal function and represented as a formula of the form

$$P_c(d)=\frac{1}{(2\pi)^{N/2}|\Sigma_c|^{1/2}}\,\mathrm{e}^{-(x-\mu_c)'\Sigma_c^{-1}a(x-\mu_c)/2}, \tag{2.21}$$

where μ_c is the mean vector of category c, and Σ_c is the covariance matrix for category c. The maximum likelihood rule then takes the form: a small-area ground patch having feature "d" is assigned to any category "c_k" where

$$\frac{1}{(2\pi)^{N/2}|\Sigma_{c_k}|^{1/2}}\,\mathrm{e}^{-(x-\mu_{c_k})'\Sigma_{c_k}^{-1}(x-\mu_{c_k})/2} \\ \geqq \frac{1}{(2\pi)^{N/2}|\Sigma_{c_i}|^{1/2}}\,\mathrm{e}^{-(x-\mu_{c_i})'\Sigma_{c_i}^{-1}(x-\mu_{ci})/2}. \qquad i=1,2,\ldots,K. \tag{2.22}$$

The simple Bayes rule takes the form: a small-area ground patch having feature "d" is assigned to any category "c_k" where

$$\frac{P(c_k)}{(2\pi)^{N/2}|\Sigma_{c_k}|^{1/2}}\,\mathrm{e}^{-(x-\mu_{c_k})'\Sigma_{c_k}^{-1}(x-\mu_{c_k})/2} \\ \geqq \frac{P(c_i)}{(2\pi)^{N/2}|\Sigma_{c_i}|^{1/2}}\,\mathrm{e}^{-(x-\mu_{c_i})'\Sigma_{c_i}^{-1}(x-\mu_{c_i})/2}. \qquad i=1,2,\ldots,K. \tag{2.23}$$

Taking logarithms on both sides of the inequality we can see that the Bayes rule under a multivariate normal assumption takes the form of a quadratic computation. When the covariance matrices are all equal, there is considerable simplification in the maximum likelihood rule which then takes a linear form: a small-area ground patch having feature "d" is assigned to any category "c_k" where

$$a'_{ki}d - t_{ki} \geqq 0, \quad \text{for each} \quad i = 1, 2, \ldots, K, \tag{2.24}$$

where

$$a_{ki} = \Sigma^{-1}(\mu_{c_k} - \mu_{c_i})$$

and

$$t_{ki} = (1/2)(\mu_{c_k} - \mu_{c_i})' \Sigma^{-1}(\mu_{c_k} + \mu_{c_i}).$$

The functions $f_{ki}(d) = a_{ki}d - t_{ki}$, $i = 1, 2, \ldots, K$, $k = 1, 2, \ldots, K$ are called linear discriminant functions. More detailed discussions of linear and quadratic decision rules may be found in [2.51] and in the review article by NAGY [2.52]. The book by SEBESTYEN [2.53] is also a good general reference. See ANDERSON and BAHADUR [2.54] for a discussion of the optimal linear decision rule under the multivariate normal assumption for unequal covariance matrices.

2.4.3 Feature Selection

Multi-image data sets, whether stored in analog form on video tape or on film, or stored in digital form on digital magnetic tape, typically contain on the order of 10^6 distinctly resolvable data points with each data point being an N-tuple, $N = 3$ for color film and up to $N = 12$ or more for a multispectral scanner. Most processing algorithms treat each data point independently and this implies that there are on the order of 10^6 separate decisions which the decision rule must make. In order to enable the decision rule to do its job quickly and simply, it is prudent to give the decision rule the least amount of simple non-redundant information which enables it to do its job well. The process of selecting what information to present to the decision rule is called feature selection.

Feature selection algorithms for multi-image data sets are usually based on the fact that the images in the set contain highly redundant information. This is easily understood if we consider the fact that images in neighboring spectral bands taken at the same time or images taken within short time intervals (a few days or weeks) in the same spectral band are highly correlated. Feature selection can then be thought of as choosing that set of, say, three or four images of the multi-image set which minimize the errors which the decision rule makes.

Let I be an index set for the different bands of a multi-image. Let n, the number of bands we wish the decision rule to work with, be given.

Then for each subset S of I indexing exactly n bands of the multi-image, application of a decision rule will produce a probability of error $P_e(S)$. $P_e(S)$ is the probability that the decision rule will make an incorrect assignment on some data point when allowed to use only those bands indexed by S. Clearly, the best set of bands to use is that set S which minimizes the probability of error; that is,

$$P_e(S) \leqq P_e(S') \quad \text{for all} \quad S' \subseteq I \quad \text{satisfying} \quad \# S' = n .$$

It often is the case that the error probabilities are difficult to compute. This is especially so when the domains of the parametric probability density functions are large dimensional spaces. Thus measures have been sought which are easier to compute for parametric functions and which are closely related to probability of error.

If we let $P(i, j; S)$ be the joint probability that a data point from category j will be assigned by the decision rule to category i when the decision rule is only allowed to use the n bands indexed by the set S, then the probability of error $P_e(S)$ can be written as the sum of all the above joint probabilities for which $i \neq j$:

$$P_e(S) = \sum_{i=1}^{K} \sum_{\substack{j=1 \\ j \neq i}}^{K} P_e(i, j; S) . \tag{2.25}$$

If we consider the probability of error to be the probability of error incurred by a Bayes decision rule, then dropping the notational dependence on the index set S we may express $P_e(i, j)$ by

$$P_e(i, j) = \sum_{g \in A_i} P(g, c_j) , \tag{2.26}$$

where $P(g, c_j)$ is the joint probability of a pattern being in class c_j and having value g and

$$A_i = \{g \in G | P(g, c_i) \geqq P(g, c_n), n = 1, 2, \ldots, K\} . \tag{2.27}$$

Noticing that the set B_{ij} defined by

$$B_{ij} = \{g \in G | P(g, c_i) \geqq P(g, c_j)\} \tag{2.28}$$

contains the set of points A_i we must have

$$P_e(i, j) = \sum_{g \in A_i} P(g, c_j) \leqq \sum_{g \in B_{ij}} P(g, c_j) . \tag{2.29}$$

Hence, $P_e(i,j)+P_e(j,i) \leqq \sum_{g \in B_{ij}} P(g,c_j) + \sum_{g \in B_{ji}} P(g,c_i) = \sum_{g \in G} \min\{P(g,c_j), P(g,c_i)\}$. Since $\min\{a,b\} \leqq \sqrt{ab}$, we have

$$\begin{aligned} P_e(i,j)+P_e(j,i) &\leqq \sum_{g \in G} \sqrt{P(g,c_i)P(g,c_j)} \\ &\leqq \sqrt{P(c_i)P(c_j)} \sum_{g \in G} \sqrt{P_{c_i}(g)P_{c_j}(g)}\,. \end{aligned} \tag{2.30}$$

The number

$$\varrho_{ij} = \sum_{g \in G} \sqrt{P_{c_i}(g)P_{c_j}(g)} \tag{2.31}$$

is called the pairwise Bhattacharyya coefficient for categories c_i and c_j and provides an upper bound on the error probability [2.55]

$$P_e(S) \leqq \sum_{i=1}^{K-1} \sum_{j=i+1}^{K} \sqrt{P(c_i)P(c_j)}\varrho_{ij}(S) \leqq \tfrac{1}{2} \sum_{i=1}^{K-1} \sum_{j=i+1}^{K} \varrho_{ij}(S)\,. \tag{2.32}$$

Using this inequality, one way of choosing the feature index set S having n elements is to compute the upper bound over all possible index sets having n elements and choose the one having the smallest upper bound. When the number of bands is large, this exhaustive search is rather expensive in terms of the number of possibilities it must account for. A suboptimal procedure is to construct S iteratively. For the first element of S, choose that feature which alone yields the lowest upper bound on error probability. For the next element of S, take that feature which when used with the previously chosen features yields the lowest upper bound on error probability, and so on [2.56].

For categories having normal distributions, ϱ_{ij} can be computed in terms of the mean and covariance matrices

$$\begin{aligned} -\ln \varrho_{ij} &= \tfrac{1}{8}(\mu_i-\mu_j)'\Sigma^{-1}(\mu_i-\mu_j) \\ &\quad + \frac{1}{2}\ln\frac{|\Sigma|}{|\Sigma_i|\,|\Sigma_j|}\,, \end{aligned} \tag{2.33}$$

where $\Sigma = \frac{1}{2}(\Sigma_i+\Sigma_j)$.

The Battacharyya coefficient is not the only measure of category probability distribution separability employed to find bounds on error probabilities. The use of divergence to measure separability between distributions was first suggested by KULLBACK [2.57]. MARILL and GREEN [2.58] proposed it for use in recognition systems, and KAILATH [2.59] suggested it could be used in signal processing applications and showed its close relationship to probability of error in pattern discrimination problems. The divergence between two conditional probability functions for categories c_i and c_j is the number d_{ij} defined by

$$d_{ij} = \sum_{g \in G} [P_{c_i}(g) - P_{c_j}(g)] \ln \frac{P_i(g)}{P_j(g)} . \tag{2.34}$$

For categories having normal distributions, the divergence d_{ij} can be computed in terms of the mean and covariance matrices:

$$\begin{aligned} d_{ij} = {} & \tfrac{1}{2} \operatorname{trace}(\Sigma_i - \Sigma_j)(\Sigma_j^{-1} - \Sigma_i^{-1}) \\ & + \tfrac{1}{2}(\mu_i - \mu_j)'(\Sigma_i^{-1} + \Sigma_j^{-1})(\mu_i - \mu_j) . \end{aligned} \tag{2.35}$$

Under the normal assumption, the Bhattacharyya coefficient and divergence are related by an inequality

$$\varrho_{ij} \geqq \mathrm{e}^{-d_{ij}/8}, \quad \text{with equality when} \quad \Sigma_i = \Sigma_j ,$$

so that

$$\sum_{i=1}^{K-1} \sum_{j=i+1}^{K} \varrho_{ij} \geqq \sum_{i=1}^{K-1} \sum_{j=i+1}^{K} \mathrm{e}^{-d_{ij}/8} . \tag{2.36}$$

Since $P_e(S) \leqq \frac{1}{2} \sum_{i=1}^{K-1} \sum_{j=i+1}^{K} \varrho_{ij}(S)$,

the inequalities go the wrong way to let the divergence establish a bound on the error probability. Nevertheless, the criteria of maximizing the minimum of the pairwise divergence or maximizing the average divergence d_{ave} have been proposed [2.60–62]

$$d_{\mathrm{ave}} = \frac{2}{K(K-1)} \sum_{i=1}^{K-1} \sum_{j=i+1}^{K} d_{ij} . \tag{2.37}$$

Comparison between the use of the Bhattacharyya coefficient and the average divergence for band or feature selection reveals the superiority of the Bhattacharyya coefficient [2.63]. SWAIN and KING [2.63a] have observed that minimizing the average of the transformed divergence d_{T} defined by

$$d_{\mathrm{T}} = \frac{2}{K(K-1)} \sum_{i=1}^{K-1} \sum_{j=i+1}^{K} \mathrm{e}^{-d_{ij}/8} \tag{2.38}$$

is a better feature selection criterion than minimizing the average divergence. The exponential form is, of course, motivated by the form of the inequality relating divergence and the Bhattacharyya coefficient.

FU et al. [2.14] discussed a general separability measure which relates more closely than divergence does to the probability of error achievable by a minimax linear decision rule under a multivariate normal assumption. They define the separability measure S_{ij} between the i-th and j-th classes by

$$S_{ij}=\frac{b'_{ij}(\mu_i-\mu_j)}{(b'_{ij}\Sigma_i b_{ij})^{1/2}+(b'_{ij}\Sigma_j b_{ij})^{1/2}}, \tag{2.39}$$

where the vector b_{ij}, the difference between the i-th and j-th linear discriminant functions, takes the form

$$b_{ij}=(\lambda_{ij}\Sigma_i+(1-\lambda_{ij})\Sigma_j)^{-1}(\mu_i-\mu_j) \tag{2.40}$$

and λ_{ij} is the Lagrange multiplier satisfying

$$b'_{ij}(\lambda_{ij}^2\Sigma_i-(1-\lambda_{ij})^2\Sigma_j)b_{ij}=0, \quad 0\leqq\lambda_{ij}\leqq 1\,. \tag{2.41}$$

Then

$$P_e(S)\leqq \sum_{i=1}^K \sum_{\substack{j=1\\ j\neq i}}^K P_e(i,j)\leqq \sum_{i=1}^K \sum_{\substack{j=1\\ j\neq i}}^K P(c_i)P(c_j)\int_{S_{ij}}^{\infty} e^{-x^2/2}dx \tag{2.42}$$

so that the separability measure can be used as a bound on the error probability achievable by a minimax linear decision rule and feature selection can be done by choosing features which minimize

$$\sum_{i=1}^K \sum_{\substack{j=1\\ j\neq i}}^K P(c_i)P(c_j)\int_{S_{ij}}^{\infty} e^{-x^2/2}dx\,. \tag{2.43}$$

A different form, slightly easier to compute than the error bound indicated above, is the average separability which FU et al. [2.14] suggested maximizing

$$\sum_{i=1}^{K-1} \sum_{j=i+1}^K P(c_i)P(c_j)S_{ij}\,. \tag{2.44}$$

Other distance or distance-like separability measures suggested for feature selection have been:

1) The Jeffreys-Matusita distance defined by [2.64, 65]

$$\{\textstyle\sum_{g\in G}[\sqrt{P_{c_i}(g)}-\sqrt{P_{c_j}(g)}]^2\}^{1/2}\,. \tag{2.45}$$

2) The Kolmogorov variational distance defined by [2.59]

$$\tfrac{1}{2}\textstyle\sum_{g\in G}|P(g,c_i)-P(g,c_j)|\,. \tag{2.46}$$

3) The Kullback-Liebler number defined by [2.66]

$$\sum_{g\in G}\left[\ln\frac{P_{c_i}(g)}{P_{c_j}(g)}\right]P_{c_i}(g)\,. \tag{2.47}$$

4) The Mahalanobis distance defined by [2.67]

$$(\mu_i-\mu_j)'\sum{}^{-1}(\mu_i-\mu_j)\,. \tag{2.48}$$

5) The Euclidean distance defined by

$$\{\sum_{g\in G}[P_{c_i}(g)-P_{c_j}(g)]^2\}^{1/2}\,. \tag{2.49}$$

The Euclidean distance has been related to an upper bound on probability of error [2.68] which is looser than the upper bound obtained by the Bhattacharyya coefficient. The Kolmogorov variational distance is intimately related to probability of error [2.59]

$$P_e(i,j)=\tfrac{1}{2}(P(c_j)+P(c_i))-\tfrac{1}{2}\sum_{g\in G}|P(g,c_i)-P(g,c_j)|\,. \tag{2.50}$$

In this subsection we have used the criterion that the best features are those which minimize the error probability. Clearly this is an optimal criterion for selecting which bands to use and which ones not to use. A different kind of feature selection problem can be posed by asking what are the combinations of the multispectral bands that can best represent, in a mean square error sense, the information contained in all the bands. In this kind of problem a function is sought which represents each measurement pattern in a compressed form in a smaller dimensional space in such a way that if the original measurement pattern were to be reconstructed from the compressed measurement, the average squared error between the reconstruction and the original would be minimized. The transformation which achieves this minimum turns out to be linear and is obtained as the $K\times N$ matrix whose rows are the K eigenvectors with corresponding largest eigenvalues of the second moment matrix of the entire data set. The representation is called the principal components representation ([2.69]). VIGLIONE [2.70] discussed a feature selection process which uses normality assumptions to determine optimum quadratic discriminant functions in the subspace, and then an interactive process based on minimizing an exponential loss function using sample pattern separation to select, evaluate, and weight the quadratic discriminant features and form a piecewise quadratic discriminant boundary in the original signal space.

Principal component feature selection has the advantage that it does not depend on the choice of categories, and although examples can be constructed illustrating how minimum mean square error may not preserve distinctions between categories, READY et al. [2.71] provided evidence that this tends not to happen with remote sensing spectral data. They indicated that 3 or 4 principal component features may be used from 12-channel multispectral data with essentially no decrease in the probability of correct category identification. HARALICK and DINSTEIN [2.72] used principal components as a feature selection method for their iterative clustering procedure applied to remote sensing multispectral scanner imagery. READY and WINTZ [2.73] indicated that spectral feature selection by principal components results in feature patterns with a better signal to noise ratio than in the original image bands. TAYLOR [2.74] described how the principal components representation can be used to enhance false color combinations of multiband imagery.

2.4.4 Estimating Decision Rule Error

Once a decision rule has been determined, it is important to estimate the percentage of errors which the decision rule makes. To do this another subset of data from the multi-image must be chosen and each point in this data set must be labeled with its true category identification. This data set is called the "prediction data", and for the error estimates to be accurate, the prediction data set must be 1) as representative as the training data set was required to be, and 2) independent of the training data.

For any given category "c" there are two types of errors which the decision rule can make: it can assign a small-area ground patch whose true category identification is "c" to some category $c' \neq c$, an error of misidentification or misdetection for category "c"; or, it can assign a small-area ground patch whose true category identification is "c'", $c' \neq c$, to category "c", an error of false detection for category "c". Both these kinds of errors are easily estimated by applying the decision rule to the prediction data and comparing the true category identification of each point with the category identification assigned by the decision rule.

Let us denote by P_{ij} the proportion of points in the prediction data having true category identification "c_i" and decision rule assigned identification "c_j". The total probability of correct identification can be estimated by

$$\sum_{i=1}^{K} P_{ii} \Big/ \sum_{j=1}^{K} \sum_{i=1}^{K} P_{ij} \, . \tag{2.51}$$

Given that a point has true category identification "c_i", the probability that it will be misidentified is

$$\sum_{\substack{j=1\\j\neq i}}^{K} P_{ij} \Big/ \sum_{j=1}^{K} P_{ij}\,. \tag{2.52}$$

Given that a point has true category identification "c_k", $c_k \neq c_i$, the probability that it will be falsely identified as category "c_i" can be estimated by

$$\sum_{\substack{j=1\\j\neq i}}^{K} P_{ji} \Big/ \Big[1 - \sum_{j=1}^{K} P_{ij}\Big]\,. \tag{2.53}$$

Sometimes a more easily interpreted false identification error probability can be computed as $\sum_{\substack{j=1\\j\neq i}}^{K} P_{ji} \Big/ \sum_{j=1}^{K} P_{ji}$ which is the proportion of those small-area ground patches falsely identified as category "c_i" out of the set of all small-area ground patches identified correctly or incorrectly as category "c_i".

2.4.5 Clustering[1]

We introduce our discussion of clustering by contrasting it to the decision rule algorithms used in pattern discrimination. With pattern discrimination techniques, a training set of data is gathered for which the correct category identification of each distinct entity in the data is known. Then estimates are made of the required category conditional probability distributions and a decision rule is determined from them. The decision rule can then be employed to identify any other data set gathered under similar conditions. With clustering techniques there is no training data set or decision rule. Rather, natural data structures are determined. Distinct structures are then interpreted as corresponding to distinct objects or environmental processes.

The advantage of the discrimination techniques is that the scientist is able to decide the types of environmental categories among which he wishes to distinguish. The decision rule then determines, as well as possible, to which environmental category an arbitrary datum belongs. The disadvantage of the discrimination techniques is that they are sensitive to mis-calibrations which cannot be normalized out. Any difference between the sensor calibrations or state of environment for the training data and the new data will cause error.

[1] See also [2.75].

The advantage of the clustering techniques is that they are not sensitive to calibration errors which shift or stretch measurement space in minor ways. Two small-area patches of corn growing in the same field will be detected as being similar because they have similar grey tones associated with them. The disadvantage of the clustering techniques is that they are not able to name the distinct environment structures they determine.

A wide variety of iterative procedures can be used to cluster remotely sensed image data and there is no general theory of clustering [2.76–79]. Each scheme involves grouping together of similar data. Perhaps the most popular clustering technique for remotely sensed data is the K-means or ISODATA technique [2.80–83]. Here each data point is put into that cluster for which the squared distance between it and the cluster mean is least. Then new cluster means are computed and the whole procedure repeated. On each repetition the total squared distance between the data points and the cluster means is guaranteed not to increase. There are numerous variants concerned with what to do when a cluster is too big, too small, etc.

Many clustering procedures deal with measurements in ways that do not consider the natural order or arrangement of the measurements. Indeed, for many problems, the order in which the measurements are taken or the spatial distribution of the units under consideration are irrelevant to the clustering process. This is not the case for remotely sensed image data. When the resolution of the image data is properly selected, each object of interest yields many measurements. Therefore, resolution cells in the same neighborhood are likely to belong to the same object. We call procedures that take into account these kinds of spatial relations between neighboring resolution cells spatial clustering procedures.

Some of the spatial clustering ideas being explored by investigators in the remotely sensed data area closely parallel those of investigators in the artificial intelligence community, which has been an active user of spatial information from scene data. Much artificial intelligence work has gone into the definition of homogeneous region and into edge detection. BRICE and FENNEMA [2.84] described a procedure for partitioning an image into a large set of primitive regions each of which is typically a connected component having the same grey tone. Then a merging algorithm can be applied to group together those regions having similar tone. Gradient and derivative algorithms can be used to find edges [2.85–88]. So we see that scene analysis and spatial clustering can have much in common.

HARALICK and KELLY [2.89] discussed a spatial clustering procedure designed to take into account the spatial dependencies which nearby pixels have in remotely sensed imagery. Spatial clusters are grown around those resolution cells having grey tone n-tuples of high relative frequency in their own local neighborhoods. The spatial growing proceeds by taking in resolution cells spatially neighboring those resolution cells already in the cluster and having grey tone n-tuples which neighbor in measurement space some of the grey tone n-tuples of the resolution cells already in the cluster.

NAGY et al. [2.90] based a spatially oriented clustering procedure on a chaining idea proposed by BONNER [2.47]. In the first step, points are assembled into row strips. This is based upon the assumption that "spatially adjacent grey tone n-tuples tend to belong to the same type of ground cover except at field boundaries". Examining the resolution cells along the scan lines, similar resolution cells are assigned to strips. Each strip is terminated only when the addition of the grey tone n-tuple of the next resolution cell would increase the internal scatter of the strip above a given threshold. At that point, the formed strip is assigned to a cluster (or designated to start a new cluster), and the formation of a new strip begun. The assignment of a strip to a cluster is done by a measurement space comparison of the strip to the cluster centers. The search for a cluster is done in decreasing order of cluster size in order to save computation time and to eliminate small groups of abnormal grey tone n-tuples.

JAYROE [2.91] introduced a three step spatial clustering procedure for multi-images. Two gradient images are obtained by computing the measurement space Euclidean distance between nearest neighbors in the horizontal and vertical directions. A boundary map is then prepared by thresholding of gradient images. In the second step, clusters are formed by scanning the boundary map with a fixed size square of resolution cells. When the square hits a region in which there are no boundary cells, that region is assigned to cluster 1. The square is then moved farther, and if no boundary cells are encountered, all the resolution cells within the square are assigned to cluster 1. The scanning continues until all possible cells are assigned to that cluster. Next, the square is moved until it hits a new region with no boundary cells, and the process is repeated. In the third step, clusters are merged according to their spectral measurements.

ROBERTSON et al. [2.92] describe a procedure for successively partitioning a multi-image into rectangular blocks. However, the final clustered images produced by the algorithm yield images which suffer from excessive blockiness.

HARALICK and DINSTEIN [2.72], like JAYROE [2.91], discussed a spatial clustering procedure based on gradient images. In the first step the gradient image is computed, thresholded, and cleaned, thereby creating a binary image showing homogeneous areas. In the second step, the maximally connected spatial regions of the cleaned image are computed. In the third step a measurement space clustering procedure is applied to the distinctly labelled connected regions to determine spatial regions of similar spectral character.

2.5 Image Texture

Spatial environments can be understood as being spatial distributions of various area-extensive objects having characteristic size and reflective or emissive qualities. The spatial organization and relationships of the objects appear as spatial distributions of grey tone on imagery taken of the environment. We call the pattern of spatial distributions of grey tone, texture.

Figure 2.5, taken from [2.93], illustrates how texture relates to geomorphology. It shows plains, low hills, high hills, and mountains in the Panama and Columbia area, as seen by the Westinghouse AN/APQ97 K-band radar imaging system. The plains have an apparent relief of 0–50 m, the hills 50–350 m, and the mountains more than 350 m. The low hills have little dissection and are generally smooth convex surfaces, whereas the high hills are highly dissected and have prominent ridge crests.

The mountain texture is distinguishable from the hill texture on the basis of the extent of radar shadowing (black tonal areas). The mountains have shadowing over more than half the area and the hills have shadowing over less than half the area. The hills can be ranked from low to high on the basis of the abruptness of tonal change from terrain front slope to terrain back slope.

Figure 2.6, taken from [2.94], illustrates how texture relates to geology. It shows igneous and sedimentary rocks in Panama, as seen by the same radar system. Figure 2.6i, k, l show a fine-textured drainage pattern which is indicative of non-resistant fine-grained sedimentary rocks. The coarser texture of Figure 2.6h, left half, is indicative of coarser-grained sediments. A massive texture with rugged and peaked divides (Fig. 2.6a–e) is indicative of igneous rocks. When erosion has nearly base-leveled an area, the texture takes on the hummocky appearance of Fig. 2.6g.

Figure 2.7, taken from [2.95], illustrates how texture relates to land use categories. Here, there are six land use categories as they appear on panchromatic aerial photography. Notice how the texture of the wooded area is coarser and more definite than that of the scrub area. The swamp

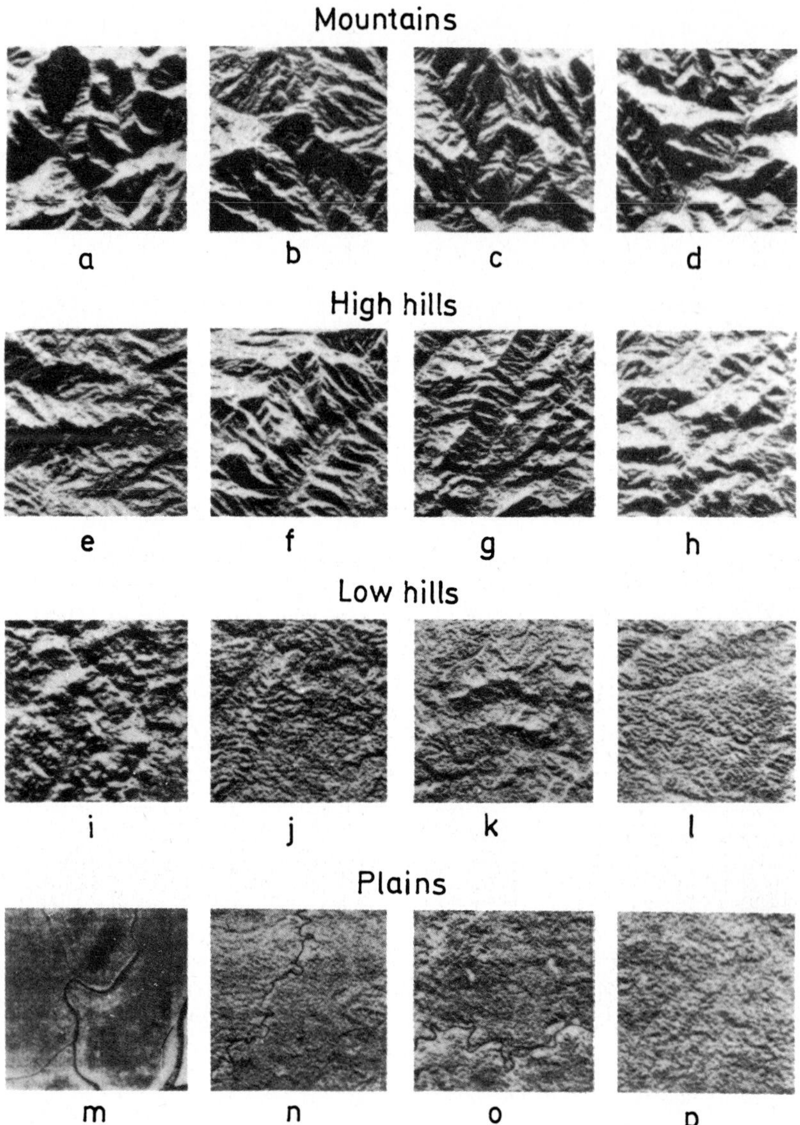

Fig. 2.5a–p. Illustrates how texture relates to geomorphology. (Taken from [2.93])

and marsh generate finer textures than the wood or scrub areas. The swamp texture is finer than and shows more gradual grey tone change than the marsh texture.

Figure 2.8 is taken in the Pisgah Crater area and shows some examples where the same type of terrain generates a variety of textures within the

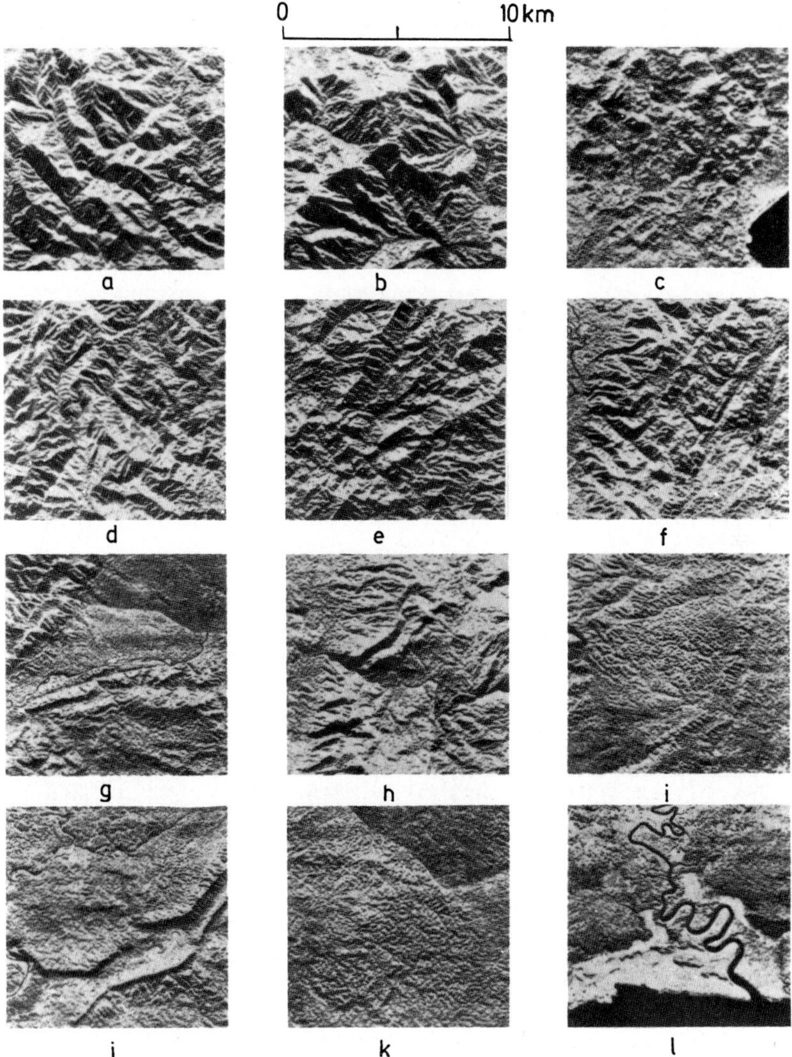

Fig. 2.6a–l. Illustrates how texture relates to geology. (Taken from [2.94])

same texture family. Here, the texture changes are due to the way the vegetation increases in size and disperses.

There have been six basic approaches to the measurement and characterization of image texture: autocorrelation functions [2.96], optical transforms [2.97], digital transforms [2.98–100], edgeness [2.88],

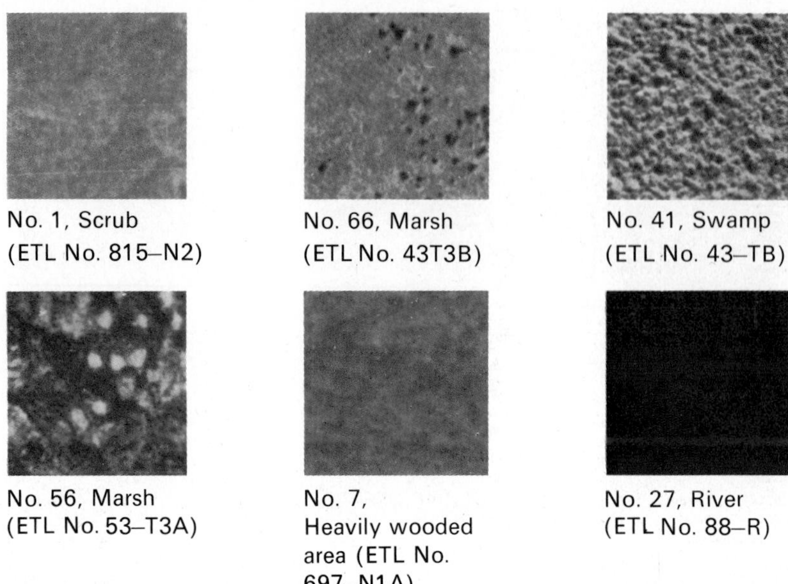

Fig. 2.7. Illustrates how texture relates to land use categories. (Taken from [2.95])

structural elements [2.101, 102], and spatial grey tone co-occurrence probabilities [2.103, 104]. The first three of these approaches are related in that they all measure spatial frequency directly or indirectly. Spatial frequency is related to texture because fine textures are rich in high spatial frequencies while coarse textures are rich in low spatial frequencies.

An alternative to viewing texture as spatial frequency distribution is to view texture as amount of edge per unit area. Coarse textures have a small number of edges per unit area, while fine textures have a high number of edges per unit area.

The structural element approach uses a matching procedure to detect the spatial regularity of shapes called structural elements in a binary image. When the structural elements themselves are single resolution cells, the information provided by this approach is the autocorrelation function of the binary image. By using larger and more complex shapes, a more generalized autocorrelation can be computed.

The grey tone co-occurrence approach characterizes texture by the spatial distribution of its grey tone. Coarse textures are those for which the distribution changes only slightly with distance and fine textures are those for which the distribution changes rapidly with distance.

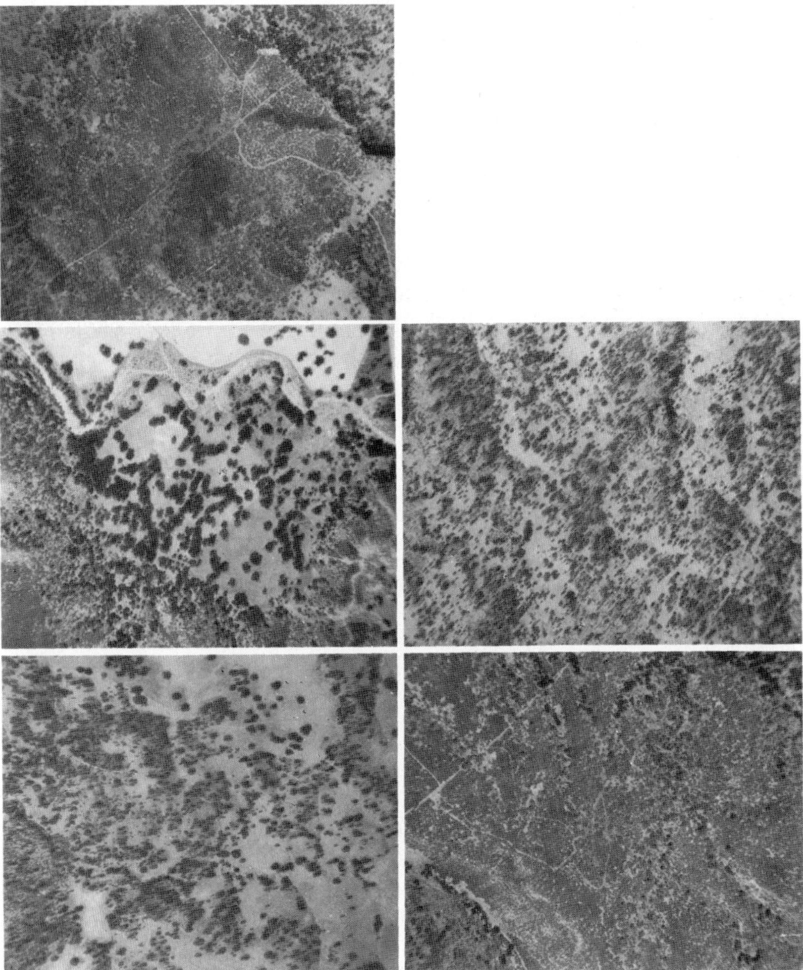

Fig. 2.8. Shows some examples where the same type of terrain generates a variety of textures within the same texture family

2.5.1 Optical Processing Methods and Texture

O'NEILL's [2.105] article on spatial filtering introduced the engineering community to the fact that optical systems can perform filtering of the kind used in communication systems. In the case of the optical systems, however, the filtering is two-dimensional. The basis for the filtering capability of optical systems lies in the fact that the light amplitude distributions at the front and back focal planes of a lens are Fourier transforms of one another. The light distribution produced by the lens is more commonly known as the Fraunhofer diffraction pattern. Thus, optical methods facilitate two-dimensional frequency analysis of images.

The paper by CUTRONA et al. [2.106] provided a good review of optical processing methods for the interested reader. More recent books by GOODMAN [2.107], SHULMAN [2.108], PRESTON [2.109] and HUANG [2.110] comprehensively survey the area.

In this section, we describe the experiments done by LENDARIS and STANLEY [2.97], EGBERT et al. [2.111], and SWANLUND [2.112] using optical processing methods for aerial or satellite imagery. LENDARIS and STANLEY illuminated small circular sections of low altitude aerial photography and used the Fraunhofer diffraction pattern to derive features for identifying the sections. The circular sections represented circular areas on the ground 750 feet in diameter. The major category distinction they were interested in making was man-made versus non-man-made. They further subdivided the man-made category into roads, road intersections, buildings, and orchards.

The pattern vectors they used from the diffraction pattern consisted of 40 components. Twenty components were averages of the energy in rings on the diffraction pattern, and 20 were averages of the energy in 9° wedges on the diffraction pattern. They obtained over 90% identification accuracy.

EGBERT et al. [2.111] used an optical processing system to examine the texture on ERTS imagery of Kansas. They used circular areas corresponding to a ground diameter of about 35 km and looked at the diffraction patterns for the areas when they were snow covered and when they were not snow covered. They used a diffraction pattern sampling unit having 32 sector wedges and 32 annular rings to sample and measure the diffraction patterns. (See [2.113] for a description of the sampling unit and its use in coarse diffraction pattern analysis.) They were able to interpret the resulting angular orientation graphs in terms of dominant drainage patterns and roads, but were not able to interpret the spatial frequency graphs, which all seem to have had the same character: the higher the spatial frequency, the less the energy in that frequency band.

SWANLUND [2.112] has done work using optical processing of aerial images to identify species of trees. Using imagery obtained from Itasca State Park in northern Minnesota, photo-interpreters identified five (mixture) species of trees on the basis of texture: Upland hardwoods, Jack pine overstory/Aspen understory, Aspen overstory/Upland hardwoods understory, Red pine overstory/Aspen understory, and Aspen. They achieved classification accuracy of over 90%.

2.5.2 Texture and Edges

The autocorrelation function, the optical transforms, and the fast digital transforms (FFT and FHT) basically all relate texture to spatial frequency. ROSENFELD and THURSTON [2.88] conceived of texture not in terms of spatial frequency but in terms of edgeness per unit area. An edge passing through a resolution cell is detected by comparing the values for local properties obtained in pairs of nonoverlapping neighborhoods bordering the resolution cell. To detect microedges, small neighborhoods must be used. To detect macroedges, large neighborhoods must be used.

The local property which ROSENFELD and THURSTON suggested was the quick ROBERTS gradient (the sum of the absolute values of the differences between diagonally opposite neighboring pixels). Thus, a measure of texture for any subimage is obtained by computing the ROBERTS gradient image for the subimage and from it determining the average value of the gradient in the subimage. TRIENDL [2.114] used the Laplacian instead of the ROBERTS gradient.

SUTTON and HALL [2.115] extended the ROSENFELD and THURSTON idea by making the gradient a function of the distance between the pixels. Thus, for every distance d and subimage I defined over a neighborhood N of resolution cells, they compute

$$\begin{aligned} g(d) = \textstyle\sum_{(i,j)\in N} \{ & |I(i,j)-I(i+d,j)| + |I(i,j)-I(i-d,j)| \\ & + |I(i,j)-I(i,j+d)| + |I(i,j)-I(i,j-d)| \} . \end{aligned} \tag{2.54}$$

The graph of $g(d)$ is like the graph of minus the autocorrelation function translated vertically. SUTTON and HALL [2.115] applied this textural measure in a pulmonary disease identification experiment using radiographic imagery and obtained identification accuracy in the 80 percentile range for discriminating between normal and abnormal lungs when using a 128×128 subimage.

2.5.3 Digital Transform Methods and Texture

In the digital transform method of texture analysis, the digital image is typically divided into a set of non-overlapping small square subimages. Suppose the size of the subimage is $n \times n$ resolution cells; then the n^2 grey tones in the subimage can be thought of as the components of an n^2-dimensional vector. In the transform technique, each of these vectors is re-expressed in a new coordinate system. The Fourier transform uses the sine-cosine basis set, the Hadamard transform uses the Walsh function basis set, etc. The point of the transformation is that the basis vectors of the new coordinate system have an interpretation that relates to spatial frequency (sequency), and since frequency (sequency) is a close relative of texture, we see that such transformations can be useful.

GRAMENOPOULOS [2.98] used a transform technique using the sine-cosine basis vectors (and implemented with the FFT algorithm) on ERTS imagery to investigate the power of texture and spatial pattern for terrain type recognition. He used subimages of 32 by 32 resolution cells and found that on Phoenix, Arizona, ERTS image 1049-17324-5, spatial frequencies larger than 3.5 cycles/km and smaller than 5.9 cycles/km contain most of the information needed to discriminate between terrain types. The terrain classes were: clouds, water, desert, farms, mountains, urban, riverbed, and cloud shadows. He achieved an overall identification accuracy of 87%.

HORNUNG and SMITH [2.99] have done work similar to that of GRAMENOPOULOS, but with aerial multispectral scanner imagery instead of ERTS imagery. MAURER [2.115a] used Fourier series analysis on color aerial film to obtain textural features to help determine crop types.

KIRVIDA and JOHNSON [2.100] compared the fast Fourier, Hadamard, and slant transforms for textural features on ERTS imagery over Minnesota. They used 8×8 subimages and five categories: Hardwoods, Conifers, Open, Water, City. Using only spectral information, they obtained 74% correct identification accuracy. When they added textural information obtained from the slant transform, they increased their identification accuracy on the training set to 99%. They found little difference between the different transform methods, although performance using 4×4 subimages was poorer than that using 8×8 subimages.

2.5.4 Spatial Grey Tone Dependence: Co-Occurrence

One aspect of texture is concerned with the spatial distribution and spatial dependence among the grey tones in a local area. BIXBY et al. [2.116] used restricted first and second order Markov meshes. DARLING and JOSEPH [2.117] used statistics obtained from the nearest neighbor grey tone transition matrix to measure this dependence for satellite images of clouds and were able to identify cloud types on the basis of their texture. READ and JAYARAMAMURTHY [2.118] divided an image into all possible (overlapping) subimages of reasonably small and fixed size and counted the frequency for all the distinct grey tone patterns. This is one step more general than DARLING, but requires too much memory if the grey tones can take on very many values. HARALICK and co-workers [2.104, 119, 120] suggested an approach which is a compromise between the two: to measure the spatial dependence of grey tones in a co-occurrence matrix for each fixed distance and/or angular spatial relationship, and use statistics of this matrix as measures of image texture. JULESZ [2.121] provided evidence that the human visual perception system discriminates texture on the basis of co-occurrence statistics.

The co-occurrence matrix $P=(p_{ij})$ has its (i, j)-th entry p_{ij} defined as the number of times grey tone i and grey tone j occur in resolution cells of a subimage having a specified spatial relation, such as being distance 1 neighbors. The textural features for the subimage are obtainable from the co-occurrence matrix by measures such as

$$\sum_i \sum_j p_{ij}^2$$

and

$$\sum_i \sum_j \frac{p_{ij}}{1+|i-j|}\,.$$

HARALICK et al. [2.103, 104] listed 14 different such measures.

Using statistics derived from the co-occurrence matrix, HARALICK performed a number of identification experiments. On a set of aerial imagery of eight terrain classes (old residential, new residential, lake, swamp, marsh, urban, railroad yard, scrub and wooded), he obtained 82% correct identification with 64×64 subimages. On an ERTS Monterey Bay, California, image, he obtained 84% correct identification using 64×64 subimages and both spectral and textural features on seven terrain classes: coastal forest, woodlands, annual grasslands, urban areas, large irrigated fields, small irrigated fields, and water. On a set of sandstone photomicrographs, he obtained 89% correct identification on five sandstone classes: Dexter-L, Dexter-H, St. Peter, Upper Muddy, Gaskel. The wide variety of images on which it has been found that grey tone co-occurrence carries much of the texture information is probably indicative of the power and generality of this approach.

2.5.5 A Textural Transform

Each of the approaches described above for the quantification of textural features has the common property that the textural features were computed for subimages of sizes such as 8×8, 16×16, 32×32, and 64×64 resolution cells. To determine textural features for one pixel we would center a subimage on the specified resolution cell and compute the textural features for the subimage. If we had to determine textural features for each pixel in an image, this would require a lot of computation work and would significantly increase the size of our data set. Thus the usual approach has been to divide the image into mutually exclusive subimages and compute textural features on the selected subimages. Unfortunately, this procedure produces textural features at a coarser resolution than the original image.

In this section we use grey tone co-occurrence textural matrices to define a textural transform, and show how by only doubling or tripling the computation time required to determine the matrix it is possible to produce a resolution-preserving textural transform in which each pixel in the transformed image has textural information about its own neighborhood derived from local and global grey tone co-occurrences in the image. This kind of textural transform belongs to the class of image dependent non-linear spatial filters.

Let $Z_r \times Z_c$ be the set of resolution cells of an image I (in row-column coordinates). Let G be the set of grey tones that can appear on image I. Then $I: Z_r \times Z_c \to G$. Let R be a binary relation on $Z_r \times Z_c$ pairing together all those resolution cells that are in the desired spatial relationship. The co-occurrence matrix P, $P: G \times G \to [0, 1]$, for image I and binary relation R is defined by

$$P(i, j) = \frac{\#\{((a, b), (c, d)) \in R \mid I(a, b) = i \text{ and } I(c, d) = j\}}{\# R} \tag{2.55}$$

The textural transform J, $J: Z_r \times Z_c \to (-\infty, \infty)$, of image I relative to function f, is defined by

$$J(y, x) = \frac{1}{\# R(y, x)} \sum_{(a,b) \in R(y,x)} f[P(I(y, x), I(a, b))] \,. \tag{2.56}$$

Assuming f to be the identity function, the meaning of $J(y, x)$ is as follows. The set $R(y, x)$ is the set of all those resolution cells in $Z_r \times Z_c$, that are in the desired spatial relation to resolution cell (y, x). For any resolution cell $(a, b) \in R(y, x)$, $P(I(y, x), I(a, b))$ is the relative frequency by which the grey tone $I(y, x)$, appearing at resolution cell (y, x), and the grey tone $I(a, b)$, appearing at resolution cell (a, b), co-occur together in the desired spatial relation on the entire image. The sum

$$\sum_{(a,b) \in R(y,x)} P(I(y, x), I(a, b))$$

is just the sum of the relative frequencies of grey tone co-occurrence over all resolution cells that are in the specified relation to resolution cell (y, x). The factor $[\# R(y, x)]^{-1}$, the reciprocal of the number of resolution cells in the desired spatial relation to (y, x) is just a normalizing factor. Figure 2.9 illustrates the 0.82 to 0.88 micrometer band of some multi-

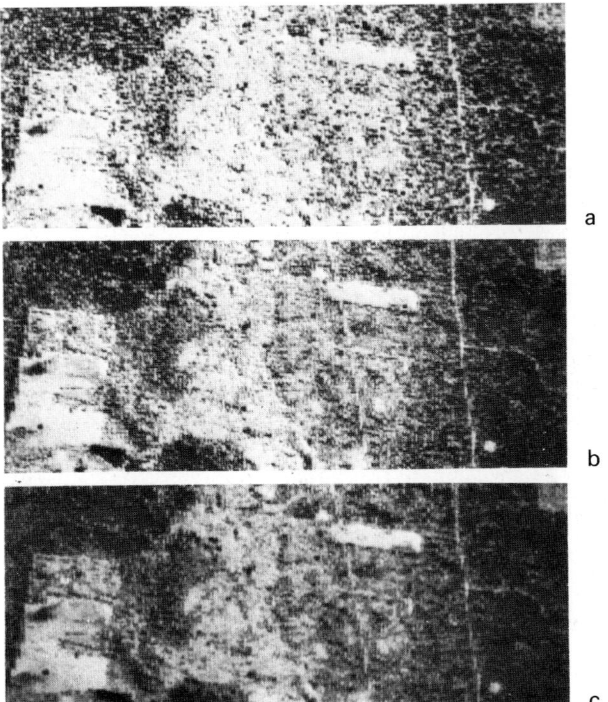

Fig. 2.9a–c. Illustrates the 0.82 to 0.88 μm band of some multispectral scanner imagery taken at 10000 feet over the Sam Houston National Forest, March 21, 1973. The image is shown in its original form, and after a 2×2 and 3×3 rectangular convolution. (a) 1×1 original of band 9 in edit No. 9, SAMH3 S19; (b) 2×2 convolution of band 9 in edit No. 9, SAMH3 S29; (c) 3×3 convolution of band 9 in edit No. 9, SAMH3 S39

spectral scanner imagery taken at 10000 feet over the Sam Houston National Forest, March 21, 1973. The image is shown in its original form, and after a 2×2 and 3×3 rectangular convolution. Figure 2.10 shows the respective textural transforms of the three images of Fig. 2.9 where the spatial relation R consists of all pairs of 8-neighboring resolution cells in $Z_r \times Z_c$. HARALICK [2.122] indicates that classification accuracy improves when both spectral and textural transform features are used on a Skylab S-192 image.

ZUCKER et al. [2.123] discussed a different kind of textural transform based on spot detectors. Applying a spot detector to an image having two different textures, the spot size matching the coarseness of one of the textures, and then averaging the resultant image, they show that it is easy to segment the image into its two textures.

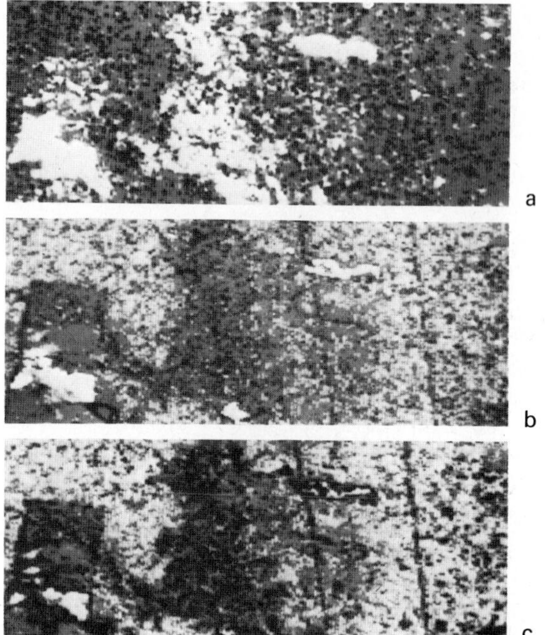

Fig. 2.10a–c. Shows the respective textural transforms of the three images of Fig. 2.9. (a) 1×1 before and 1×1 after $T \times T$, SAMH3 Q19; (b) 2×2 before and 1×1 after $T \times T$, SAMH3 Q29; (c) 3×3 before and 1×1 after $T \times T$, SAMH3 Q39

2.6 Near Real Time Hardware Processing

The ease with which remote sensors can inundate the investigator with data strongly indicates the virtue of a processing system which can process the data quickly. Hardware processing, although limited to performing rather simple basic functions, has the great advantage of near-real-time operation. In this section we survey the typical kinds of operations easily done in a near-real-time hardware system.

Basically, there are two types of image format inputs that a hardware system can have: film or print, and video tape. Systems which have black and white film or print inputs have the advantage of being usable with the most common form of image format data. However, they have the associated problem of registration. The analog tape input usually comes from a multispectral scanner sensor so that the individual images (channels) are already registered on the tape.

Hardware system outputs usually include forms of black and white and color displays. The black and white display can be used for examining the images, one by one, or for showing simple two-category discriminations. The color display may be used to show a processed false color enhanced image or a color map indicating multi-category discriminations.

Those hardware systems having registration problems usually input the images using independently controllable flying spot scanners or vidicons. The operator manually registers the images by adjusting the rotation, translation, and scale controls of the input devices while quickly flickering between the displayed images. Images which are not in registration show on the flickered display a displeasing interference movement which indicates in what way the images have to be rotated, translated or scaled to be put in registration.

There are at least eight basic processing operations which a near-real time hardware device can perform either singly or in various sequences depending on the required processing function. They are:

1) level slicing or thresholding to produce binary images,
2) Boolean operations on binary images to produce new binary images,
3) differentiation for boundary detection,
4) linearly combining images for enhancement or calculation of discriminant functions,
5) quadratically combining images for enhancement or calculation of discriminant functions,
6) determining which of a set of signals is the maximum for category assignments,
7) analog to digital conversion or quantizing,
8) table look-ups for category assignments.

A level slicer or thresholder produces a binary {0, 1} output which is 1 whenever the video signal it is operating on is between two adjustable thresholds, and 0 at all other times. Two or more level slicers may operate simultaneously on different signals and their binary outputs may be ANDed together, thereby producing category decision boundaries which are rectangular parallelepipeds. However, should the images be linearly combined in two or more linear combiners and then level sliced and ANDed, the category decision boundaries would take on the more general form of parallelepipeds; should they be quadratically combined in two or more quadratic combiners, then level sliced and ANDed together, the category decision boundaries would be piecewise quadratic surfaces. In general, the combiner/level slicer combination implements a decision rule for a two-category problem.

If an image is differentiated and displayed, the boundaries can be enhanced. If it is first differentiated and then thresholded, its boundaries can be detected. If two or more images are differentiated, and then the thresholded differentiated outputs are ORed together, all the boundaries on the set of images can be detected. In this case, the differentiator/thresholder combination implements a two category decision rule: boundary, no boundary.

The linear or quadratic combiners may be used by themselves to produce a false color image enhancement. For example, if the images are combined with one combiner driving each primary color in a color display, false color image enhancement results.

The linear or quadratic combiners may be used in conjunction with the maximum selector to implement a multi-category decision rule. In this rule, the combiners act as discriminant functions and the maximum selector determines which combiner has the largest signal. The decision rule then assigns the category associated with the combiner having the largest signal being processed.

The papers by MARSHALL and KRIEGLER [2.124] and ANDERSON et al. [2.54] describe the hardware processing systems at the Universities of Michigan and Kansas, respectively.

2.7 Software Processing

Although software processing of image data on a general purpose digital computer is more costly and definitely slower than near-real time, it has the advantage of great versatility. In fact, software can be written to implement any well defined processing task of finite length. LILLESTRAND and HOYT [2.125] reviewed the nature of the rapidly evolving digital systems for image processing. They characterize the system by the rate at which digitized imagery can be processed, by the complexity of the processing algorithms, and by the type of output produced. In this section we discuss the different formats in which multi-images are stored and the necessity of checking the A/D conversion process, and describe some of the basic library programs which any software processing system must have.

There are relatively few published descriptions of software systems. JOSEPH et al. [2.126] discussed a pattern recognition system for reconnaissance applications. In a continuing series of papers, JOSEPH and co-workers [2.127–129] described a system for the interactive preprocessing of image data. They discussed both the computer hardware configuration and the required image processing functions. FRIEDEN [2.129a] discussed the VICAR system developed at the Jet Propulsion Laboratory. BEBB et al. [2.130] presented an overview of image processing systems. HOFFER [2.131] gave a brief user's description of LARSYAA, a software system designed to analyze multispectral scanner data at Purdue University. JARVIS [2.132] discussed interactive image processing and pattern recognition systems based on minicomputers. GAMBINO and CROMBIE [2.133] described the DIMES software system (Digital Image Manipulation and Enhancement System). TURINETTI and HOFFMANN [2.134] explained how OLPARS (On-Line Pattern Analysis and Recognition System) can be used to process remotely sensed image data. An Electromagnetic Sensing Laboratory report [2.135] describes the IDIM system (Interactive Digital Image Manipulation).

2.7.1 Multi-Image Formats

Since image processing systems must handle such large amounts of data, they are usually input-output bound. This makes the format in which the multi-image data are stored quite important. There are two basic kinds of formats for storing multi-image data. In what we shall call photo format, all the grey tones from the first image are stored as integers in a row by row logical record format on the first file. This file is followed by all the grey tone integers from the second image, also stored in a row by row logical record format, and so on. In what we shall call the corresponding point form, the grey tone integer from the first image's first resolution cell, first line, is followed by the grey tone integer from the second image's first resolution cell, first line, ..., followed by the grey tone integer from the last image's first resolution cell, first line, ..., followed by the grey tone integer from the first image's last resolution cell, first line, followed by the grey tone integer from the second image's last resolution cell, first line, and so on. Each such multi-image line of integers is stored as a logical record.

A format perhaps more versatile than the photo or corresponding point format is one which stores line 1 of image 1 followed by line 1 of image 2, ..., followed by line 1 of image N, and so on. Noticing that logical records are subimages, we see that the line format can be generalized to a subimage format where each logical record is a subimage of say K_1 rows by K_2 columns.

Editing, congruencing, registering, digital printer displaying and texture analysis are most quickly done with data in photo form, while most other operations such as feature extraction, clustering or pattern identification are done more conveniently in corresponding point format.

To conserve magnetic tape or disc storage space, speed up I/O time, and make more effective use of buffer space, the grey tone integers are often packed two, three or four integers per computer word. The packing is usually done by special machine language FORTRAN callable subroutines which convert a line or subimage of compacted integers at a time.

2.7.2 Checking A/D Conversion of Image Data

The first problem any image processing software system has to handle is the checking of the digital image conversion process. The digital tapes must be checked to verify that the A/D conversion was done successfully. Preliminary checking can be done by dumping the first few records on the tape; however, this is by no means a complete check. The image display program can make a complete check by outputting the tape in picture format on the digital printer, creating the grey tones by overprinting. If the number of resolution cells on the image is large enough to make the printing of the digital image awkward, or display of it impossible, a program may be utilized which reduces the image size by averaging blocks of $N \times N$ resolution cells or by selecting every N-th row and every N-th column.

Examination of the digital picture output should indicate what kind of editing needs to be done on the sides and top and bottom of the image, as well as indicating skewing and A/D conversion distortion. (Skewing can occur because it may be impossible to start digitizing each line of the image in exactly the same place. A/D conversion distortion can occur when jitter or noise external to the A/D conversion makes the conversion go awry.) If necessary, a deskewing program may have to be used to remove skew, and a special smoothing-replacement program may have to be used to operate on those resolution cells which were improperly converted. Histograms are helpful in identifying and isolating failure of individual A/D conversions.

2.7.3 User Oriented Commands and Library Programs

The heart of any image processing software system is its utility library programs, written in modular form, which perform each kind of special processing task. For ease of writing the library programs, transferring them to various computers, or understanding them, the programs are usually written in a universally available high level language such as FORTRAN, rather than in machine assembly language. Each library program or set of programs implements a specific user oriented command which can be input in a free format form. The library typically contains programs for image displaying, editing, scaling, registering, congruencing, mosaicking, selecting, quantizing, filtering, clustering, feature extraction, texture analysis, and pattern recognition.

The display commands can display an image either on a video CRT screen [2.136] or by using the digital printer. The printer can overprint characters to obtain shades of grey and use different color ribbons to obtain color map displays.

The editing commands can print out the grey tone values for any subimage, replace all occurrences of one grey tone value by another, change the grey tone in any one resolution cell from one value to another, and locate the next occurrence of any specified grey tone. The image scaling commands can either decrease the size of the image by sampling grey tones at every K_1-th pixel vertically and K_2-th pixel horizontally, or increase the size of the image by duplicating each pixel K_1 times vertically and K_2 times horizontally.

The registering command can translate any image vertically or horizontally so that it is aligned with another image; then it can combine the aligned images into a multi-image. The congruencing command can change the geometry of any image to the geometry of another image and then combine the images into a multi-image by the registering operation. The mosaicking command can paste together images in a side by side manner. The selecting or subimage command can construct a new image consisting of any specified size subimage from any set of bands of a multi-image.

The quantizing command can quantize or contrast stretch an image by an equal interval or equal probability quantizing procedure. The filtering command performs a spatial filtering of the image by a specified point spread function.

The clustering, feature extracting, texture, and pattern recognition operations are much more complex than the utility operations just described. However, one processing operation universally included in this set has the capability to compute histograms and scattergrams of any portion of any image or for any ground truth category of an image. Histograms or scattergrams are usually displayed on the line printer, or more conveniently on a CRT with alphanumeric as well as line generating abilities.

Ideally, all user oriented programs should accept image data in a standard format and accept data parameters and control information in a free-format manner. System routines should provide for error processing and dynamic storage allocation.

2.8 Conclusion

We have briefly discussed aspects of the automatic processing of remotely sensed imagery. We have indicated the importance of various normalizing, feature extraction, decision rule determination and clustering algorithms. We have discussed processing from a hardware and software perspective. Hopefully, we have provided enough insight about the vocabulary and concepts of the automatic data processing world to enable a scientist to read and understand the remote sensing papers in his field which deal with automatic data processing.

References

2.1 R. N. COLWELL: American Scientist **49**, 9–36 (1961)

2.2 R. N. COLWELL: "Uses and Limitations of Multispectral Remote Sensing", *Proc. 4th Intern. Symp. on Remote Sensing of Environment*, Univ. of Mich., Ann Arbor, (1966)

2.3 W. A. MALILA: Photogrammetric Engineering **34**, 566–575 (1968)

2.4 *NASA MSC 3rd Annual Earth Resources Program Review*, Vols. I–III, NASA MSC-03742, Manned Spacecraft Center, Tex. (1970)

2.5 *NASA MSC 4th Annual Earth Resources Program Review*, Vols. I–V, NASA MSC-05937, Manned Spacecraft Center, Tex. (1972)

2.6 J. D. LENT, G. A. THORLEY: Remote Sensing of Environment **1**, 31–45 (1969)

2.7 C. M. HAY: Photogrammetric Engineering **40**, 1283–1293 (1974)

2.8 N. R. NUNNALLY: Remote Sensing of Environment **1**, 1–7 (1969)

2.8a J. D. BARR, R. D. MILES: Photogrammetric Engineering **36**, 1155–1165 (1970)

2.9 S. A. MORAIN, D. S. SIMONETT: Photogrammetric Engineering **33**, 730–740 (1967)

2.10 H. C. MACDONALD, P. A. BRENNAN, L. F. DELLWIG: IEEE Trans. GE-**5**, 72–78 (1967)

2.11 F. T. ULABY: IEEE Trans. AP-**22**, 257–265 (1974)

2.12 P. M. MERIFIELD, J. CRONIN, L. L. FOSHEE, S. J. GAWARECKI, T. J. NEAL, R. E. STEVENSON, R. O. STONE, R. S. WILLIAMS, JR.: Photogrammetric Engineering **35**, 654–668 (1969)

2.13 T. PHILLIPS: "Corn Blight Data Processing Analysis and Interpretation", *4th Annual Earth Resources Program Review*, NASA/MSC (1972)

2.14 K. S. FU, D. A. LANDGREBE, T. L. PHILLIPS: Proc. IEEE **57**, 639–653 (1969)

2.15 R. M. HARALICK, F. CASPALL, D. S. SIMONETT: Remote Sensing of Environment **1**, 131–142 (1969)

2.16 R. M. CENTNER, E. D. HIETANEN: Photogrammetric Engineering **37**, 177–186 (1971)

2.17 P. E. ANUTA, R. B. MACDONALD: Remote Sensing of Environment **2**, 53–67 (1971)

2.18 R. M. HOFFER, P. E. ANUTA, T. L. PHILLIPS: Photogrammetric Engineering **37**, 989–1001 (1972)

2.19 J. D. TURINETTI, O. W. MINTZER: Photogrammetric Engineering **39**, 501–505 (1973)

2.20 P. H. SWAIN: "Pattern Recognition: A Basis of Remote Sensing Data Analysis", LARS Information Note 111572. The Laboratory for Applications of Remote Sensing, Purdue Univ., Lafayette, Ind. (1973)

2.21 D. A. LANDGREBE: "Machine Processing for Remotely Acquired Data", LARS Information Note 031573. The Laboratory for Applications of Remote Sensing, Purdue Univ., Lafayette, Ind. (1973)

2.22 G. NAGY: Proc. IEEE **60**, 1177–1199 (1972)

2.23 D. EGBERT, F. T. ULABY: 'Effect of Angular Variation on Terrain and Spectral Reflectivity", *Proc. of 17th Symp. of the AGARD Electromagnetic Wave Propagation Panel on Propagation Limitations in Remote Sensing* (1971)

2.24 D. STEINER, H. HAEFNER: Photogrammetric Engineering **31**, 269–280 (1965)

2.25 R. M. HOFFER, R. A. HOLMES, J. R. SHAY: "Vegetative Soil and Photographic Factors Affecting Tone in Agricultural Remote Multispectral Sensing", *Proc. 4th Intern. Symp. on Remote Sensing of the Environment*. Univ. of Mich., Ann Arbor, Mich. (1966)

2.26 R. M. HARALICK: Pattern Recognition **5**, 391–403 (1973)

2.27 L. R. Pettinger: "Analysis of Earth Resources on Sequential High Altitude Multiband Photography", Special Report, Forestry and Conservation, Univ. of Calif. (1969)

2.28 F. J. Kriegler, W. A. Malila, R. F. Nalepka, J. Richardson: "Preprocessing Transformations and Their Effects on Multispectral Recognition", *Proc. 6th Intern. Symp. on Remote Sensing of the Environment*, Univ. of Mich., Ann Arbor, Mich. (1969) pp. 97–131

2.29 R. B. Crane: "Preprocessing Techniques to Reduce Atmospheric and Sensor Variability in Multispectral Scanner Data", *Proc. 7th Intern. Symp. on Remote Sensing of Environment*, Univ. of Mich., Ann Arbor, Mich. (1971) pp. 1345–1350

2.30 H. W. Smedes, M. M. Spencer, F. J. Thomson: "Preprocessing of Multispectral Data and Simulation of Earth Resources Technology Satellite Data Channels to Make Computer Terrain Maps of a Yellowstone National Park Test Site", *3rd Ann. Earth Resources Program Review*, NASA MSC-03742 (1970) pp. 10–1 to 10–25

2.31 R. F. Nalepka, J. P. Morgenstern: "Signature Extension Techniques Applied to Multispectral Scanner Data", *Proc. 8th Intern. Symp. on Remote Sensing of Environment*, Environmental Research Institute of Michigan (1972) pp. 881–893

2.32 J. S. Crabtree, J. O. McLaurin: Photogrammetric Engineering **36**, 70–76 (1970)

2.32a R. B. McEwen: "An Evaluation of Analog Techniques for Image Registration", U.S. Geological Survey Professional Paper 700-D (1970) pp. D305–D311

2.33 F. G. Peet: "Affine Transformations from Aerial Photos to Computer Compatible Tapes", *3rd Earth Resources Technology Satellite Symp.*, NASA SP-351, Goddard Space Flight Center (1973) pp. 1719–1724

2.34 R. A. Emmert, C. D. McGillem: "Multitemporal Geometric Distortion Correction Utilizing the Affine Transformation", *Machine Processing of Remotely Sensed Data*, Purdue Univ. (1973)

2.35 M. Szeto: "An Approximation for Projective Transformation for Satellite Imagery", Report RC-3320, IBM T.J. Watson Research Center, Yorktown Heights, New York (1971)

2.36 R. O. Duda, P. E. Hart: *Pattern Classification and Scene Analysis* (John Wiley & Sons, New York 1973) pp. 405–424

2.37 H. Markarian, R. Bernstein, D. G. Ferneyhough, L. E. Gregg, F. S. Sharp: "Implementation of Digital Techniques for Correcting High Resolution Images", Paper No. 71-326, AIAA (1971) pp. 285–304

2.38 P. F. Anuta: IEEE Trans. GE-**8**, 353–368 (1970)

2.39 D. I. Barnea, H. F. Silverman: IEEE Trans. C-**21**, 179–186 (1972)

2.40 D. F. Webber: "Techniques for Image Registration", *Machine Processing of Remotely Sensed Data*, Purdue Univ. (1973)

2.41 S. S. Yao: "A Method for Digital Image Registration Using a Mathematical Programming Technique", *Machine Processing of Remotely Sensed Data*, Purdue Univ. (1973)

2.42 S. S. Riffman: "Digital Rectification of ERTS Multispectral Imagery", *Symp. on Significant Results Obtained from the Earth Resources Technology Satellite*, NASA SP-327, Goddard Space Flight Center (1973) pp. 1131–1142
S. S. Riffman: "Evaluation of Digitally Corrected ERTS Imagery", *Amer. Society of Programmetry Symp. Proc. on the Management and Utilization of Remote Sensing Data*, Sioux Falls, SD (1973) pp. 206–221

2.43 E. Diday, J.-C. Simon: In *Pattern and Speech Recognition*, ed. by K. S. Fu (Springer, Berlin, Heidelberg, New York 1975) Chapter 3

2.44 B. J. Davis, P. H. Swain: "An Automated and Repeatable Data Analysis Procedure for Remote Sensing Application", *Proc. 9th Intern. Symp. on Remote Sensing of Environment*, Environmental Research Institute of Michigan, Ann Arbor, Mich. (1974) pp. 771–774

2.45 A. G. WACKER: "The Effect of Subclass Numbers on Maximum Likelihood Gaussian Classification", *Proc. 8th Intern. Symp. on Remote Sensing of Environment*, Environmental Research Institute of Michigan, Ann Arbor, Mich. (1972) pp. 851–859
2.46 D. FOLEY: IEEE Trans. IT-**18**, 618–626 (1972)
2.47 L. KANAL, B. CHANDRASEKARAN: Pattern Recognition **3**, 225–234 (1971)
2.47a W. G. BROONER, R. M. HARALICK, I. DINSTEIN: "Spectral Parameters Affecting Automated Image Interpretation Using Bayesian Probability Techniques", *Proc. 7th Intern. Symp. on Remote Sensing of Environment*, Univ. of Mich., Ann Arbor, Mich. (1971) pp. 1929–1949
2.48 W. G. EPPLER, C. A. HEMKE, R. H. EVANS: "Table Look-up Approach to Pattern Recognition", *Proc. 7th Intern. Symp. on Remote Sensing of Environment*, Univ. of Mich., Ann Arbor, Mich. (1971) pp. 1415–1425
2.49 W. G. EPPLER: "An Improved Version of the Table Look-up Algorithm for Pattern Recognition", *Proc. 9th Intern. Symp. on Remote Sensing of Environment*, Environmental Research Institute of Michigan, Ann Arbor, Mich. (1974) pp. 793–812
2.50 R. W. ASHBY: "Constraint Analysis of Many-Dimensional Relations", *Yearbook of Soc. General Systems Research*, Vol. 9, 99–105 (1964)
2.51 P. W. COOPER: Cybernetica **5**, 215–238 (1962)
2.52 G. NAGY: Proc. IEEE **56**, 836–862 (1968)
2.53 G. S. SEBESTYEN: *Decision-Making Processes in Pattern Recognition* (The MacMillan Co., New York 1962)
2.54 P. N. ANDERSON, G. W. DALKE, R. M. HARALICK, G. L. KELLY, R. K. MOORE: "Electronic Multi-Image Analog-Digital Processor and Color Display", *1972 IEEE Convention Digest* (1972)
2.55 D. G. LAINIOTIS: IEEE Trans. IT-**15**, 730–731 (1969)
2.56 W. A. WHITNEY: IEEE Trans. C-**20**, 1100–1103 (1971)
2.57 S. KULLBACK: *Information Theory and Statistics* (John Wiley & Sons, New York 1959)
2.58 T. MARILL, D. M. GREEN: IEEE Trans. IT-**19**, 11–17 (1963)
2.59 T. KAILATH: IEEE Trans. CT-**15**, 52–60 (1967)
2.60 N. J. NILSSON: *Learning Machines* (McGraw-Hill, New York 1965)
2.61 T. L. GRETTENBERG: IEEE Trans. IT-**19**, 11–17 (1963)
2.62 K. S. FU, C. H. CHEN: "Sequential Decision, Pattern Recognition and Machine Learning", Techn. Rept. TR-EE 65-6, School of Electrical Engineering, Purdue Univ., Lafayette, Ind. (1965)
2.63 P. H. SWAIN, T. V. ROBERTSON, A. G. WACKER: "Comparison of the Divergence and B-distance in Feature Selection". LARS Information Note 020871, The Laboratory for Applications of Remote Sensing, Purdue Univ., Lafayette, Ind. (1971)
2.63a P. H. SWAIN, R. C. KING: "Two Effective Feature Selection Criteria for Multispectral Remote Sensing", *1st Intern. Joint Conf. on Pattern Recognition*, Washington, D.C. (1973)
2.64 H. JEFFREYS: Proc. of the Roy. Soc. A **186**, 454–461 (1946)
2.65 K. MATUSITA: Ann. Inst. Stat. Math. (Tokyo) **3**, 17–35 (1951)
2.66 S. KULLBACK, R. A. LIEBLER: Ann. Math. Stat. **22**, 79–86 (1951)
2.67 P. C. MAHALANOBIS: Proc. Nat. Inst. of Science (India) **27**, 49–55 (1936)
2.68 C. C. BABU: IEEE Trans. IT-**18**, 428 (1972)
2.69 H. HOTELLING: J. Educ. Psych. **24**, 417–441 (1933)
2.70 S. S. VIGLIONE: "Application of Pattern Recognition Technology", *in Adaptive Learning Systems*, ed. by J. M. MENDEL and K. S. FU (Academic Press, New York 1970)
2.71 P. J. READY, P. A. WINTZ, S. J. WHITSITT, D. A. LANDGREBE: "Effects of Data Compression and Random Noise on Multispectral Data", *Proc. 7th Intern. Symp. on Remote Sensing of Environment*, Univ. of Mich., Ann Arbor, Mich. (1971) pp. 1321–1342

2.72 R. M. HARALICK, I. DINSTEIN: IEEE Trans. CAS-**22**, 440–450 (1975)

2.73 P. J. READY, P. A. WINTZ: IEEE Trans. COM-**21**, 1123–1131 (1973)

2.74 M. M. TAYLOR: "Principal Components Color Display of ERTS Imagery", *3rd Earth Resources Technology Satellite-1 Symp.*, NASA SP-351, Goddard Space Flight Center (1973) pp. 1877–1887

2.75 E. DIDAY, J. C. SIMON: Clustering Analysis, in: *Communication and Cybernetics*, Vol. 10, ed. by K. S. FU (Springer, Berlin, Heidelberg, New York 1975)

2.76 R. M. HARALICK: "Adaptive Pattern Recognition of Agriculture in Western Kansas by Using a Predictive Model in Construction of Similarity Sets", *Proc. 5th Intern. Symp. on Remote Sensing of Environment*, Univ. of Mich., Ann Arbor, Mich. (1967)

2.77 E. M. DARLING, J. G. RAUDSEPS: Pattern Recognition **2**, 313–335 (1970)

2.78 R. M. HARALICK, I. DINSTEIN: IEEE Trans. SMC-**1**, 275–289 (1971)

2.79 L. BORRIELLO, F. CAPOZZA: "A Clustering Algorithm for Unsupervised Crop Classification", *Proc. 9th Intern. Symp. on Remote Sensing of Environment*, Environmental Research Institute of Michigan, Ann Arbor, Mich. (1974) pp. 181–188

2.80 G. H. BALL, D. J. HALL: "ISODATA, A Novel Method of Data Analysis and Pattern Classification", Techn. Rept., Stanford Research Institute, Menlo Park, Calif. (1965)

2.81 A. G. WACKER, A. LANDGREBE: "Boundaries in Multi-Spectral Imagery by Clustering", *Proc. 1970 IEEE Symp. on Adaptive Processes (9th) Decision and Control*, Univ. of Texas, Austin, Tex. (1970)

2.82 H. W. SMEDES, H. J. LINNERUD, L. B. WOOLOVER, M. Y. SU, R. R. JAYROE: "Mapping of Terrain by Computer Clustering Techniques Using Multi-spectral Scanner Data and Using Color Aerial Film", *NASA 4th Ann. Earth Resources Progress Review*, MSC-05937 **3**, Houston, Tex. (1972) pp. 61–1 to 61–30

2.83 M. Y. SU, R. E. CUMMINGS: "An Unsupervised Classification Technique for Multispectral Remote Sensing Data", *Proc. 8th Intern. Symp. on Remote Sensing of Environment*, Environmental Research Institute of Michigan, Ann Arbor, Mich. (1972) pp. 861–879

2.84 C. R. BRICE, C. L. FENNEMA: Artificial Intelligence **1**, 205–226 (1970)

2.85 L. G. ROBERTS: "Machine Perception of Three-Dimensional Solids", in *Optical and Electro-Optical Processing of Information* (MIT Press, Cambridge, Mass. 1965) pp. 159–197

2.86 J. PREWITT: "Object Enhancement and Extraction", in *Picture Processing and Psychopictorics*, ed. by B. S. LIPKIN and A. ROSENFELD (Academic Press, New York 1970) pp. 75–149

2.87 M. H. HUECKEL: J. Asso. Computing Machinery **18**, 113–125 (1971)

2.88 A. ROSENFELD, M. THURSTON: IEEE Trans. C-**20**, 562–569 (1971)

2.89 R. M. HARALICK, G. L. KELLY: Proc. IEEE **57**, 654–665 (1969)

2.90 G. NAGY, G. SHELTON, J. TOLABA: "Procedural Questions in Signature Analysis", *Proc. 7th Intern. Symp. on Remote Sensing of Environment*, Univ. of Mich., Ann Arbor, Mich. (1971) pp. 1387–1401

2.91 R. R. JAYROE: "Unsupervised Spatial Clustering with Spectral Discrimination", NASA Techn. Note TN D-7312, George C. Marshall Space Flight-Center, Ala. (1973)

2.92 T. V. ROBERTSON, K. S. FU, P. H. SWAIN: Multispectral Image Partitioning", LARS Information Note 171373, Purdue Univ., Lafayette, Ind. (1973)

2.93 A. J. LEWIS: "Geomorphic Evaluation of Radar Imagery of Southeastern Panama and Northwestern Columbia", CRES Techn. Report 133–18, Univ. of Kansas Center for Research, Inc., Lawrence, Kan. (1970)

2.94 H. C. MACDONALD: "Geologic Evaluation of Radar Imagery from Darien Province, Panama", CRES Techn. Report 133–6, Univ. of Kansas Center for Research, Inc., Lawrence, Kan. (1970)

2.95 R.M.HARALICK, D.E.ANDERSON: "Texture-Tone Study with Application to Digitized Imagery", CRES Techn. Report 182-2, Univ. of Kansas Center for Research, Inc., Lawrence, Kan. (1971)

2.96 H.KAIZER: "A Quantification of Textures on Aerial Photographs", Techn. Note 121, Boston Univ. Research Laboratories (1955) AD 69484

2.97 G.G.LENDARIS, G.L.STANLEY: Proc. IEEE **58**, 198–216 (1970)

2.98 N.GRAMENOPOULOS: "Terrain Type Recognition Using ERTS-1 MSS Images", *Symp. on Significant Results Obtained from the Earth Resources Technology Satellite*, NASA SP-327 (1973) pp. 1229–1241

2.99 R.J.HORNUNG, J.A.SMITH: "Application of Fourier Analysis to Multispectral/Spatial Recognition", *Management and Utilization of Remote Sensing Data ASP Symp.*, Sioux Falls, South Dakota (1973)

2.100 L.KIRVIDA, G.JOHNSON: "Automatic Interpretation of Earth Resources Technology Satellite Data for Forest Management", *Symp. on Signification Results Obtained from the Earth Resources Technology Satellite*, NASA SP-327 (1973) pp. 1076–1082

2.101 G.MATHERON: *Elements Pour Une Theorie des Milieux Poreux* (Masson, Paris 1967)

2.102 J.SERRA, G.VERCHERY: Film Science and Technology **6**, 141–158 (1973)

2.103 R.M.HARALICK, K.SHANUMGAM: *2nd Symp. on Significant Results Obtained from Earth Resources Technology Satellite – 1*, Goddard Space Flight Center (1973) pp. 1219–1228

2.104 R.M.HARALICK, K.SHANUMGAM, I.DINSTEIN: IEEE Trans. SMC-**3**, 610–621 (1973)

2.105 E.O'NEILL: IRE Trans. IT-**5**, 56–65 (1956)

2.106 L.J.CUTRONA, E.N.LEITH, C.J.PALERMO, L.J.PORCELLO: IRE Trans. IT-**6**, 386–400 (1960)

2.107 J.W.GOODMAN: *Introduction to Fourier Optics* (McGraw-Hill, New York 1968)

2.108 A.R.SHULMAN: *Optical Data Processing* (John Wiley & Sons, Inc., New York 1970)

2.109 K.PRESTON: *Coherent Optical Computers* (McGraw-Hill, New York 1972)

2.110 T.S.HUANG (Ed.): *Topics in Applied Physics*, Vol. 6: Picture Processing and Digital Filtering (Springer, Berlin, Heidelberg, New York 1975)

2.111 D.EGBERT, J.MCCAULEY, J.MCNAUGHTON: "Ground Pattern Analysis in the Great Plains", *Semi-annual Earth Resources Technology Satellite A Investigation Report*, Remote Sensing Laboratory, Univ. of Kansas, Lawrence, Kan. (1973)

2.112 G.SWANLUND: "Honeywell's Automatic Tree Species Classifier", Report 9D-G-24, Honeywell Systems and Research Division (1969)

2.113 N.JENSEN: Photogrammetric Engineering **39**, 1321–1328 (1973)

2.114 E.E.TRIENDL: "Automatic Terrain Mapping by Texture Recognition", *Proc. 8th Intern. Symp. on Remote Sensing of Environment*, Environmental Research Institute of Michigan, Ann Arbor, Mich. (1972) pp. 771–776

2.115 R.SUTTON, E.HALL: IEEE Trans. C-**21**, 667–676 (1972)

2.115a H.MAURER: "Texture Analysis with Fourier Series", *Proc. 9th Intern. Symp. on Remote Sensing of Environment*, Environmental Research Institute of Michigan, Ann Arbor, Mich. (1974) pp. 1411–1420

2.116 R.BIXBY, G.ELERDING, V.FISH, J.HAWKINS, R.LOEWE: "Natural Image Computer Final Report, Vol. 1, C-4035, Philco-Ford Corp. Aeronutronic Division, Newport Beach, Calif. (1967)

2.117 E.M.DARLING, R.D.JOSEPH: IEEE Trans. SSC-**4**, 38–47 (1968)

2.118 J.S.READ, S.N.JAYARAMAMURTHY: IEEE Trans. C-**21**, 803–812 (1972)

2.119 R.M.HARALICK: "A Texture-Context Feature Extraction Algorithm for Remotely Sensed Imagery", *Proc 1971 IEEE Decision and Control Conference*, Gainesville, Fla. (1971) pp. 650–657

2.120 R. M. HARALICK, K. SHANUMGAM, I. DINSTEIN: "On Some Quickly Computable Features for Texture", *Proc 1972 Symp. on Computer Image Processing and Recognition*, Vol. 2, Univ. of Missouri (1972) pp. 12-2-1 to 12-2-10

2.121 B. JULESZ: Scientific American **232**, 34–43 (1975)

2.122 R. M. HARALICK: "A Resolution Preserving Textural Transform for Images", *Computer Graphics Pattern Recognition and Data Structure Conference*, IEEE Computer Society and ACM Group on Computer Graphics, Beverly Hills, Calif. (1975)

2.123 S. W. ZUCKER, A. ROSENFELD, L. S. DAVIS: "Picture Segmentation by Texture Discrimination", Computer Science Techn. Rept. 356, Univ. of Maryland, College Park (1975)

2.124 R. E. MARSHALL, F. J. KRIEGLER: "An Operational Multispectral Survey System", *Proc. 7th Intern. Symp. on Remote Sensing of Environment*, Univ. of Michigan, Ann Arbor, Mich. (1971) pp. 2169–2192

2.125 R. L. LILLESTRAND, R. R. HOYT: Photogrammetric Engineering **40**, 1201–1218 (1974)

2.126 R. D. JOSEPH, R. G. RUNGE, S. S. VIGLIONE: "A Compact Programmable Pattern Recognition System", MDAC Paper WD 1409, McDonnell Douglas Astronautics Co., Newport Beach, Calif. (1970) [presented at *The Symp. on Applications of Reconnaisance Technology to Monitoring and Planning Environmental Change*, Rome Air Development Center, Rome, New York (1970)]

2.127 R. D. JOSEPH, S. S. VIGLIONE: "Preprocessing for Interactive Imagery Analysis", *Two-Dimensional Digital Signal Processing Conf.*, Univ. of Missouri, Columbia, Missouri (1971)

2.128 R. D. JOSEPH, E. E. NELSON: "Orbital Information Systems—Image Processing and Analysis", Report MDC G2593, McDonnell Douglas Co., Huntington Beach, Calif. (1972)

2.129 S. S. VIGLIONE: "Digital Imagery Processing and Analysis", *Symp. on Operational Remote Sensing*, American Society of Photogrammetry, Houston, Tex. (1972) (McDonnell Douglas Report WD 1845, Huntington, Calif.)

2.129a H. FRIEDEN: "Image Processing System VICAR Guide to System Use", Report 71–135, Jet Propulsion Laboratory, Pasadena, Calif. (1971)

2.130 J. BEBB, W. D. STROMBERG, R. E. NIMENSKY: "An Overview of Image Processing", TM-5068/001/00, System Development Corp., Santa Monica, Calif. (1973) AD 757 119

2.131 R. M. HOFFER: "ADP of Multispectral Scanner Data for Land Use Mapping", LARS Information Note 080372, The Laboratory for Applications of Remote Sensing, Purdue Univ., Lafayette, Ind. (1973)

2.132 R. A. JARVIS: Computer **7** (10), 49–59 (1974)

2.133 L. A. GAMBINO, M. A. CROMBIE: Photogrammetric Engineering **40**, 1295–1302 (1974)

2.134 J. D. TURINETTI, R. J. HOFFMAN: Photogrammetric Engineering **40**, 1323–1330 (1974)

2.135 ESL: "PECOS II Users Manual", ESL-TM-503, Electromagnetic Systems Laboratory, Inc., Sunnyvale, Calif. (1974)

2.136 N. H. KREITZER, W. J. FITZGERALD: IEEE Trans. C-**22**, 128–134 (1973)

3. On Radiographic Image Analysis*

C. A. HARLOW, S. J. DWYER III, and G. LODWICK

With 38 Figures

When the initial work on radiographic image analysis was begun, there was concern about several factors involved in the computer analysis of radiographs. This concern focused on three points. First, the known difficulty of analyzing images of a simple nature such as characters and blocks seemed to indicate that it would be unwise to undertake such a difficult project as the analysis of radiographic images. The second point of concern was the tremendous amount of information present in a radiograph, possibly requiring a digitization of 1024×1024 or in excess of one million data points [3.1]. This fact would seem to indicate the necessity of utilizing a large computer which would be impractical in a clinical environment. The third point was the lack of enthusiasm of radiologists for using enhanced or modified images for diagnoses. This means that some of the best developed methodology for image analysis would be of uncertain value. These problems are somewhat mitigated by the advantages that might accrue from having a system for the analysis of radiographic images. That is, a computer system for analyzing radiographs offers the potential advantages of allowing a radiologist to handle more patients, to give improved diagnosis, and to make his diagnosis more repeatable or standardized. Ideally, the computer system should be a diagnostic tool used by the radiologist to give him quantitative and diagnostic information on items of interest in the radiograph.

One difficulty at present in the analysis of radiographic images is the large volume of film to be analyzed. For example, in 1964 there were 506 million radiographs taken of humans, and thus increased to 660 million per year in 1970 [3.2]. In 1970 alone there were 71 million radiographic chest examinations (one or more films) [3.3]. Also, over a fifteen-year period, it was determined that the demand for radiological services increased at a rate of 7.1% while the number of radiologists increased by only 5.9%. It has been estimated that the need for radiologists is twice the number available [3.4]. Another difficulty in current methods of analyzing radiographic images is the variation among different radiologists seeing the same film at different times. For example, some of

* Supported by U.S. Public Health Services, Grant No. GM 17729, NSF Equipment.

our studies show that radiologists' interpretations of heart disease on chest X-rays run under 40% accuracy [3.5]. Similar results will be given in more detail in Section 3.4.

In addition to the efforts in our laboratory, work on radiographic image analysis is being conducted at other institutions. For information on these institutions, consult the Appendix. This paper will emphasize the work done at our laboratory since it is representative of the field, and we are the most familiar with it.

We decided upon the following strategy in our research on radiographic image analysis. First, we sould concentrate our efforts on developing a system to automatically diagnose a film rather than develop an interactive system or systems to enhance images for viewing by radiologists. Enhancement was not emphasized because of the high resolution required, the general lack of high quality displays, and most importantly, lack of enthusiasm by the radiologists with whom we worked. We believed that automatic analysis could mitigate the resolution problem because some important radiographic problems might be handled at a low resolution while very high resolution might only be required on small parts of the film. Also, by concentrating on an automatic system we would more quickly determine which functions could be performed automatically. There is no denying the potential usefulness of an interactive system, but studying the automatic system helps to develop the interactive approach by identifying those tasks that cannot be done by machine.

Secondly, we decided to implement a digital system and actively develop and evaluate hardware and software in our laboratory. A digital approach was chosen because of the flexibility of such a system for research and also because of the tremendous advances being made in the cost/performance ratios of such systems. Within our laboratory it was decided to concentrate upon the development, interfacing and evaluation of displays and scanners since this area of hardware would most critically influence our research. It was decided not to develop processors, operating systems, and programming languages because this effort would dilute our resources unnecessarily for any benefits that might accrue.

Thirdly, it was decided to concentrate on analyzing the more important diseases. For example, we intend to determine if one could predict heart disease from a chest X-ray rather than to determine if a person is normal or abnormal over all possible diseases. This influences the development of libraries, as discussed later. Even when analysis is limited to PA chest radiographs, a tremendous variety of disease patterns can occur. If a master algorithm could be developed to detect every possible diagnostic finding in PA chest radiographs, it would probably be difficult to find the necessary radiographs to test the algorithm. Since different diseases manifest themselves in varied ways on the radiograph, it is doubtful whether one algorithm could detect all the variations. It is, of course, hoped that a useful analysis of important diseases such as heart disease would justify the development of the system.

In discussing radiographic image analysis we have found it useful to consider the following areas. The first is library development, which requires defining the problem to be considered and collecting radiographs showing the diseases. These radiographs are used to develop and evaluate the programs generated for analyzing the films. The second area is the system hardware consisting of scanners, displays and processors necessary for digitizing, displaying and processing the radiographic data. The third area is image analysis software, where the main concern is the development of the appropriate algorithms and techniques for radiographic image analysis. The fourth area is applications, which consists of the specific programs and results obtained on well-defined problems. There might be another area called evaluation, but this is included in the discussion on applications, since the applications programs are constantly being evaluated.

3.1 Library Development

The development of suitable libraries of radiographic images is important in work on radiographic image analysis. This work involves scanning, displaying, processing, transmitting and otherwise modifying the data. The purpose of such manipulation is to improve diagnostic services. The ultimate evaluation of a system to process or transmit radiographs will be done by radiologists. In evaluating any new system the radiologist will want to know how the new system compares with the present system in regard to diagnostic information. For example, in transmitting images, how is image quality affected by the transmission? However, it is unlikely that adequate means for establishing image quality as it relates to diagnosing disease will be deduced. Instead, there will probably develop different measures for different disease classes.

It seems reasonable, therefore, to determine error rates over different disease classes for any proposed diagnostic system. One method for determining the error rate of a diagnostic system would be to establish a film library containing selected radiological cases of the disease of interest. This library can then form the standard upon which to test this system.

Several test libraries have been established at the University of Missouri-Columbia. Our largest library represents all branches of radiological diagnosis and contains normal cases as well as cases representing different degrees of perceptual and diagnostic difficulty. Examples of perceptual difficulty would be small gallstones in a gallbladder or small osteolytic or osteoblastic bone changes. Diagnostic difficulties are provided by groups such as heart disease, bone tumors, and pulmonary disease. The library will ultimately consist of 5000 cases [3.6]. The cases are collected using the MARS radiology reporting system [3.7], and are classified by anatomical site as follows:

Anatomical site	Normal	Abnormal
Abdomen	92	149
Chest	289	740
Genitourinary	30	207
Lower extremities		392
Lower GI		96
Pelvis	5	
Skull	77	272
Spine		201
Thoracic cage		138
Upper extremities	37	318
Upper GI		380
Vessels		251
	Total 530	3144

The cases may be retrieved by patient number, anatomical site, normality versus abnormality, difficulty of perception, difficulty of diagnosis, and the diagnosis itself, as well as by certain other attributes such as which radiologist has or has not seen the case.

In addition to the large library, special libraries have been developed for disease classes of particular interest in our work. One example is the library on congenital heart disease [3.8] which consists of chest radiographs. For each patient both a lateral and a PA (posteroanterior) view are included. In the initial selection of patients, each radiograph and the medical history of the patient were reviewed and examined by a radiologist to insure that every radiograph included was technically acceptable, and no secondary disease was imposed in the radiograph to mask the radiologic finding pertaining to congenital heart disease.

Table 3.1. Disease classes in the congenital heart training library-CDS1

	Disease class	Number of cases
A	Normal	85
B	Atrial septal defect—Secundum type	74
C	Ventricular septal defect	73
D	Pulmonary stenosis	38
E	Patent ductus arteriosus	35
F	Tetralogy of Fallot	20
G	Endocardial cushion defect (ASD primum, etc.)	16
H	Transposition	11
I	Truncus arteriosus	2
J	Coarctation	7
K	Dextrocardia	0
L	Endocardial fibroelastosis	0
M	Ebstein's disease	4
N	Total anomalous venous return	1
O	Tricuspid atresia	5
Z	Abnormal—This includes the following:	
	1. Multiple primary problems	
	2. Pink tetralogy of Fallot	
	3. Very rare diseases	
	4. Abnormals for which a specific diagnosis has not been made	16
	Total	387

Next, patient data were collected and validated. Clinical information such as name, date of birth, date of film, date of catheterization, radiologist's precatheterization diagnosis (if available), clinician's precatheterization diagnosis (if available), and state of the patient as to cyanosis was collected. A summary diagnosis was carefully determined after a review of the patient's medical history, all cardiac catheterization reports, any surgical reports relating to the congenital heart disease, and an autopsy report when applicable. Both the preceding information and a photocopy of the pertinent medical reports are compiled and available with the original PA and lateral chest films.

A final check was made on those patients with congenital heart disease who were to be included in the Congenital Heart Library. If a patient had a secondary diagnosis which might mask the primary diagnosis, the medical history and the radiographic films were reviewed together, and questionable cases were either put into the abnormal class or were removed from the library. Any congenital heart case placed in classes B through O in Table 3.1, therefore, has a definite primary diagnosis corresponding to the class in which it was placed, and any secondary diagnoses are definitely subordinate to the primary one.

For the normal patients the following clinical information was compiled: name, date of birth, and date of film. Normal patients were assumed to be non-cyanotic. Since considerable risk is associated with cardiac catheterization, normal patients do not usually undergo this procedure. Hence, some other means of validating the diagnosis had to be used. The technique selected was a review of the radiographic films of each normal patient by two radiologists. If they both agreed that films appeared normal, the patient was included in the library as normal.

The disease classes presently in the Congenital Heart Training Library and the number of patients in each class are shown in Table 3.1. At the present time two classes do not contain any patients. These classes were included for completeness and to allow for expected future expansion of the library. It is also important to note the definition of class Z, called "abnormal". This class contains those cases which do not fit into any of the other categories. Cases with multiple primary problems of such magnitude that the cases do not fit any one category are included. Pink Tetralogy of Fallot, which radiographically often looks like a ventricular septal defect and does not fall into the Tetralogy of Fallot class, is placed this category. Very rare diseases are grouped into this class for convenience. Also contained in this category are those few cases which could not be definitely diagnosed from the cardiac catheterization into a particular class.

3.2 System Hardware

The primary deficiencies in image processing hardware have been in display technology, although suitable commercial units are beginning to appear on the market. Although there is some question as to the appropriateness of the serial architecture present in most commercial units for image processing applications, it does not seem to hinder progress. The continued price/performance improvement in commercial machines is beginning to make the development of clinical systems feasible. There are many different types of commercial scanners available. The selection of the primary input device is of critical importance in any design.

Radiographic image analysis requires the digitization of regularsize radiographs. However, there are systems commercially available which produce a high-quality miniature film that have been proved to lose little significant information.

A resolution of 256×256 digitized to eight bits (or one byte) has been found satisfactory. A higher resolution would be desirable in some circumstances; however, this tends to place a significant burden on the processor. For example, consider the simple problem of displaying the data without any processing. For a flicker-free display on a CRT, at lest 30 frames per second are required [3.9]. A conventional television frame composed of 525 lines presented at 30 frames per second can be used. To present a 256×256 pixel image each line is repeated in an interlaced line format; that is, 256 lines are written at a 60 frame-per-second rate. A data rate of 65536 bytes per 1/60 s implies 131072 bytes per 1/30 s interlaced or approximately 4 mbytes/s. A typical 32-bit computer operating at a memory cycle time of 1 μs can transfer four bytes in one μs or one byte in 250 ns. This rate almost achieves the four mbyte data rate required (i.e., one byte being required in 250 ns), but at a high cost. The I/O channel of the computer is locked up, and the memory is completely held by the I/O channel. The processor cannot get a memory cycle.

The system used in our laboratory for image analysis research is shown diagrammatically in Fig. 3.1 This facility was designed to be a research facility in radiographic image analysis [3.10]. In our work on radiographic image analysis, the image dissector camera has been the primary image scanning device. Since the image dissector camera is plagued by shot noise typical of photomultipliers, the current is sampled and integrated over a selected time interval during which the x-y deflection angles remain constant for each point. The output of the integrator is held constant after the integration interval by switching the input to zero. This voltage is input to the analog-to-digital converter to generate a form of data suitable for transfer to the computer. The output of the integrator is then reset tc zero, and the deflection angles of the beam are varied to center the next point of the image within the aperture.

Integration time is a critical parameter since longer integration times yield less noise and less error in the digitization of the image but result in longer times required for digitization of the entire image. Typical integration intervals range from 20 μs to 1280 μs per point with signal-to-noise ratios ranging from 23 dB to 45 dB. Maximum possible transfer rates for the digital quantity representing each point range from 50000 points per second to 781 points per second, inversely proportional to the integration interval. For an image consisting of 256×256 points this limits scan times to values of one second to one minute 40 s, directly proportional to integration time. The best choice of integration time is determined by system speed requirements and noise tolerance.

Variations in conversion efficiency across the face of the tube caused by aging and flaws in the photosensitive coating limit the gray level accuracy to approximately 45 dB. The tube and deflection components are subject to stray electromagnetic and some electrostatic sources and are mechanically quite fragile and sensitive to mechanical misalignment

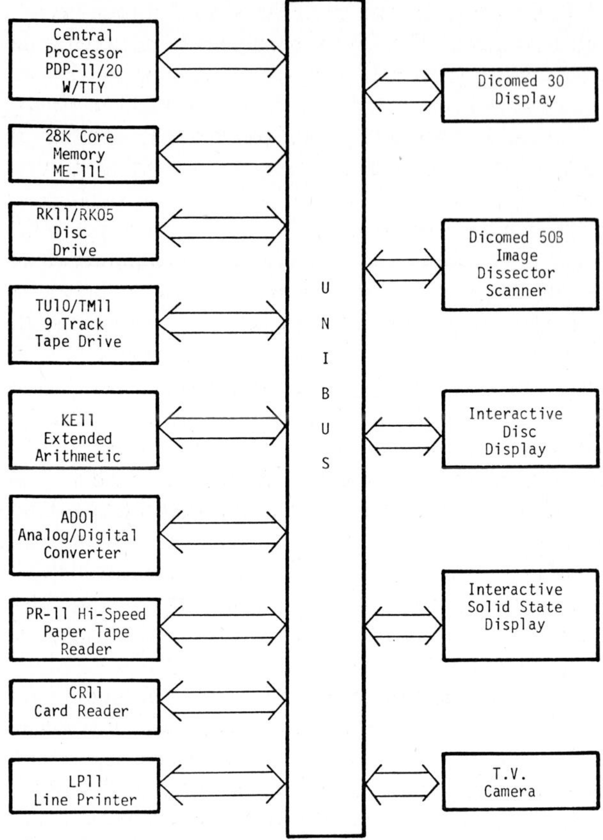

Fig. 3.1. Image analysis hardware system

among components. All of these effects can slightly degrade dot resolution and gray level accuracy from the optimum.

Because of its economy and flexibility in capturing an image the standard television camera has long been used for transmitting and displaying video images. However, digitizing this video for input to a computer has not been so common since the TV camera video output is contaminated with noise. Also, each TV line is generated at the camera at a rate of one line per 63.5 μs; at 500 samples per line, each picture point is allowed only 126 ns for processing. Hence, a very fast and expensive analog-to-digital converter is required.

In recent years, state-of-the-art vidicon tubes and other TV camera developments have produced a camera video output with signal-to-noise ratios of over 50 dB. One system for digital-conversion electronics scans the image in the vertical direction. This requires a conversion of at most 63.8 μs. Since commercially available analog-to-digital converters can convert a 10-V signal to an 8-bit number in 20 μs, the cost for this speed is not prohibitive. This speed then requires 33 ms for each scan to be digitized at full (525 line) resolution. Most computers can keep up with this data rate, leaving time for analysis.

The primary advantages of this system are low cost, reliability, and versatility, coupled with performance suitable for most applications. Since TV cameras are widely used as analog imaging devices, their cost is low, and a variety of optical and mechanical configurations have been developed.

As mentioned previously, the display of picture data is not a trivial problem. Four different display devices have been built or purchased at University of Missouri-Columbia for the purpose of adequately satisfying the image output function. Two of the displays incorporate an analog storage medium, and the others store the image digitally. The first display device uses a direct view storage tube. The contrast and brightness of the resultant image is low, and it requires over 90 s to transfer and store one image for display. The storage tube has been replaced and at present, 18 months later, it shows a great amount of degradation again.

The second display uses a scan converter tube for image storage with a high resolution CRT monitor for viewing the image. In the scan converter system, the picture information is written in an analog fashion on a storage tube screen. Precision sweep circuits position the beam to a location on the target storage screen. The beam is then intensified according to the digital value of the pixel, and the stored charge on the screen is, in turn, read off by another electron beam being scanned in synchronism with the television sweep circuits. The problems associated with this system were beyond the capability of the Image Analysis Laboratory IAL group to solve. There were inherent limitations on the precision of the beam positioning circuits – i.e., linearity of the deflection system. The focusing of the beam was extremely difficult especially over the full screen (dynamic focusing). The extraction of the video signal from the storage tube was also very difficult. After a year of effort this project was discontinued, and a commercial unit was borrowed for evaluation, consisted of a single-gun tube which used the same beam for writing and reading. The unit, while useful for frame stealing in television, was not capable of producing desirable images. The device was not suited for interactive use with readout of the image back to the computer because of reconversion problems associated with this type of device. Industry claims of 32 linear gray shades were not realized.

After these experiences it was decided to store the data digitally. Also, a digital storage technique is more amenable to interactive processing since each picture point is precisely defined as to location and gray shade. However, using a digital medium presents problem as regards both storage capacity and data rate.

Enough storage locations must be provided to store each gray shade code for each picture element of the image. For images of 256 × 256 resolution this requires 65 536 picture point storage locations. The gray shade code length must be between four and eight bits to present the digitized image satisfactorily, depending on how the digital image is converted for video generation. Consequently, between 32 768 bytes and 65 536 bytes of storage are needed for one image. This approaches or exceeds the limit of available core on most minicomputers. A dedicated form of digital storage is, therefore, required for the display system. At this point at least four options exist for storing this image: core memory, solid state memory using Random Access Memory (RAM) elements, solid state memory using a circulation shift register configuration, or a dedicated digital disc. Since our application required the ability to display images of resolution in excess of 256 × 256 up to 1024 × 1024, our necessary storage space increased to 1 048 576 picture elements or over eight million bits.

For 256 × 256 interlaced images the data rate required is 3.9 megabytes per second when one picture element is coded in one byte; a 512 × 512 interlaced pattern requires 7.8 megabytes per second. At this rate the byte access time would be less than 128 ns. As a result, core memory is too slow, and even solid state memory must be strictly specified to achieve this access time. A fixed head-per-track, parallel access digital disc does provide the data rate required to refresh the monitor even up to the 1024 × 1024 raster. Consequently, this device was used as the storage medium for one digital display. However, there are some disadvantages to this type of display. The initial cost of the disc memory and associated circuitry was high ($ 40000). Two massive mechanical failures occurred, involving head crashes and freezing of the heads to the disc surface. One of these cost $ 2000 to repair. Because of these costs one would be reluctant to use the disc display in a production system.

Another display is based on solid-state memory storage. This system should give a superior image because there should be less jitter in the image, since there are no moving parts, and it should also be less expensive for a 256 × 256 resolution [3.11].

MOS shift registers of reasonably long length, 1024 bits, have recently become available. Use of these devices began in 1972. Referring to the numbers previously discussed for the raster required, one finds that 65 536 bytes of memory are still needed for storage of the raster. A simple, ideal architecture for display purposes would be eight 65536-bit long registers composed of sixty-four 1024-bit MOS shift registers per line, each register holding one bit of the pixel. The output of each register could be input to a digital-to-analog converter to give a 256 linear gray shade image. One problem would be to maintain bit line synchronization between all eight lines.

The high data rate requires a clocking frequency of 5 MHz for the registers. Available registers do not have this clock rate. The selected registers have a guaranteed clock rate of 1.5 MHz; a certain amount of demultiplexing and multiplexing is therefore required to get an effective clock rate of 5 MHz. The selected organization has eight registers, each register composed of eight 1024 bit shift registers. The data are entered into the registers via a one-to-eight-line demultiplexor and extracted via an eight-line to single line multiplexor.

A simple raster of 256 lines repeated at a rate of 60 frames per second was chosen to eliminate the necessity for a 256 point buffer. For a viewer who wants a 512 line raster a 256 point buffer can be added. It was decided to give full random access capability to the display, enabling the user to selectively address one point out of 65536 and write into it or read out of it.

At the present time these type of displays are becoming available in the commercial market at moderate prices and should be strongly considered in any system design.

3.3 Image Analysis

The emphasis of our work in image analysis has been to develop techniques for automatically analyzing images, and not enhancing of images for viewing by radiologists. The reasons for this approach are discussed in the introduction to this chapter. A block diagram of an image analysis system that is useful in discussing image analysis is shown in Fig. 3.2.

The function of the monitor is to evaluate the results of any stage of processing and decide what processing should be done next. This implies that the processing does not necessarily proceed from left to right in the diagram. In particular, the recognition and segmentation phases will usually interact very closely, because the initial segmentation may be incorrect. The recognition stage should detect any errors which would involve another attempt at segmentation. Also, it is often useful to perform initial analysis at a low resolution, perhaps to locate the lungs in a chest radiograph. Subsequent processing and scanning at a higher resolution would be used to analyze, for example, the details of the lung.

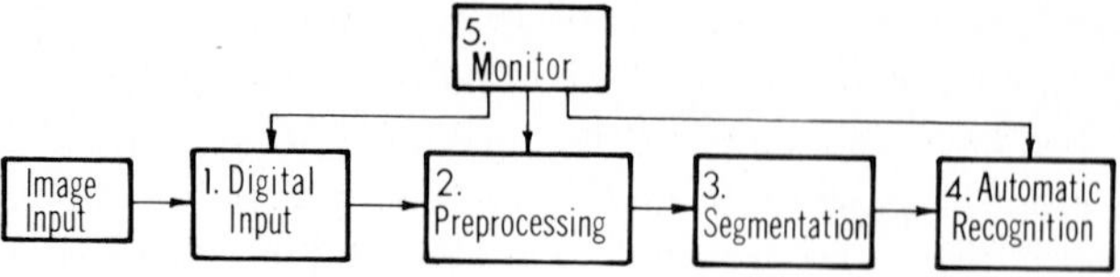

1. Digital input: The camera and associated analogue to digital conversion equipment and drivers required to convert the image into an $M \times N$ array of points whose entries give the brightness values of the image.
2. Preprocessing: Programs used to suppress the noise put into the image data by the digital conversion equipment and also programs used to emphasize certain properties in the image that are more important than others for the later stages of analysis.
3. Segmentation: The location of the significant objects in the scene.
4. Automatic recognition: The automatic recognition of the important objects in the image by the computer.
5. Monitor controls the execution of all programs and evaluates results of all processing steps.

Fig. 3.2. Block diagram of image processing system

Any system for analyzing complex images will likely require an involved interaction of the various stages of processing.

There are two primary problems in developing radiographic image analysis programs. The first is the extreme variety and complexity of radiographic images, and the second is the high resolution present in these images. To a lesser extent, one must also consider the fact that a clinically useful system must be evolved from the research.

From the beginning we considered only certain important classes of radiographic images. The programs were developed by modeling the image class under consideration. This merely takes cognizance of the fact that in order to analyze an image one must have within the program a description of the image class. This approach has long been recognized as important and was first discussed as a linguistic approach to image analysis as reviewed in [3.12–14]. The critical problem is how to formulate these concepts and make them useful in the design of image analysis programs. In other words, one must not only specify linguistic rules, but one must also specify the way they are to be used by the programs in analyzing the images. Some of the later work has emphasized this aspect of the problem. Terms that appear in papers associated with this emphasis are "descriptive" and "semantic" [3.13, 14].

One is ultimately grappling here with the problem of program organization [3.15]. A paradigm is needed for constructing these programs so they are easily understood and modified. Techniques for problem solving such as state-space searching and searching "and/or" graphs, as described in [3.16], can be considered guidelines for constructing programs for solving these problems. It has long been recognized that the paradigm used in classical pattern recognition, sometimes called the receptor/categorizer model (RCM) [3.17], which is a statistical formulation of the problem, is not adequate for constructing scene analysis programs. The RCM model is discussed in [3.14] and in numerous other books. Attempts to discover the correct formulation is an area of much current research activity.

A method we have emphasized could succinctly be called a descriptive approach using top-down analysis and graphical models [3.15, 18, 19]. This approach implies that to analyze an image one will first locate and recognize the large objects in the scene and later analyze the details of these objects. Doing a hierarchical search for objects suggests that the search can begin at a low resolution and later increase the resolution as required. This fact has proved helpful in handling the resolution problem discussed earlier. Also a model can control the processing in the segmentation stage of the analysis. This will be discussed further below.

3.3.1 Pattern Recognition

The proper title of this subsection might be Classical Pattern Recognition. Since this subject area has an extensive history, only the parts relevant to radiographic image analysis will be discussed. Figure 3.3 is a diagram of the structure of a classifier, often called the RCM or receptor categorizer model.

Classical pattern recognition is the means of deciding in which class a given pattern belongs. There are a number of references on the subject including [3.20–22]. The input to the recognizer is the pattern vector $(U_1, U_2, \ldots, U_p)$. The input pattern vector often has far too many components to be used for classification. Therefore, a set of measurements or features, $\{X_1, X_2, \ldots, X_n\}$ is derived from the input pattern, where $n \leq p$. The pattern classifier makes a decision based upon the n measurements and decides to which class the input pattern belongs.

Discriminant Functions

Decisions based on discriminant functions

$$g_i(i=1, \ldots, n) \tag{3.1}$$

assign x to class i if $g_i(x) > g_j(x)$ for all other j. Adding a constant to the g's does not affect the results. Also if f is a monotonically increasing function we can replace every g_i by $f(g_i)$ without changing the results.

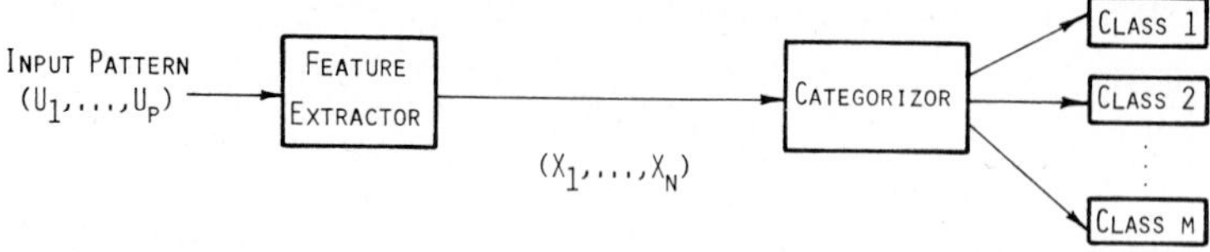

Fig. 3.3. Pattern recognition block diagram

Note that the discriminant functions divide R^n into subsets $R_1, R_2, \ldots, R_m$ which are separated by decision boundaries. In the Bayes case the decision boundaries are

$$g_i(x) = P_i f(x|i) \tag{3.2}$$

or

$$g_i(x) = \log f(x|i) + \log P_i . \tag{3.3}$$

Assuming the distributions are multivariate normal, (3.3) becomes

$$\begin{aligned} g_i(x) = &-(1/2)(x-\mu_i)' \textstyle\sum_i^{-1} (x-\mu_i) - (d/2)\log 2\pi \\ &-(1/2)\log |\textstyle\sum_i| + \log P_i . \end{aligned} \tag{3.4}$$

Modified Maximum Likelihood Decision Rule (MMLDR)

With the assumption of multivariate normal distributions, the Bayes decision rule can be stated as follows: assign x to class k if

$$(x-\mu_i)' \textstyle\sum_i^{-1} (x-\mu_i) - \log (P_i^2/|\textstyle\sum_i|) \tag{3.5}$$

is minimum for $i=k$.

The Bayes classifier does not work well in classifying actual multiclass samples since the *a priori* class probabilities P_i are usually unknown, and the conditional density functions $\{f(x|i)\}$ are not truly multivariate normal distributions. Therefore, the following rule was proposed [3.23]: Assign x to class k if

$$(x-\mu_i)' \textstyle\sum_i^{-1} (x-\mu_i) \tag{3.6}$$

is minimum for $i=k$.

This new decision rule is called the modified maximum likelihood decision rule (MMLDR). This rule will be equivalent to the Bayes decision rule if $P_i^2/|\sum_i|$ is a constant for all i.

Evaluation

The following definitions are useful in comparing the performance of different pattern recognizers. The first definition is as follows: The training result of a pattern recognizer or classifier is the result of classifying the samples used to train the classifier. The 10% jackknife testing result is the sum of the ten different 10% testing results. Each 10% testing result is the result of classifying the unused 10% of the samples from each class while the classifier is trained on the remaining 90% of the samples from each class. The above procedure is repeated ten times, using a different 10% of the samples each time, to obtain ten different 10% testing results.

Measurement Selection

Measurement selection is a well recognized problem in pattern recognition. References [3.23–29] give more information on the subject.

The following are some of the reasons for the growth of interest in measurement selection. First, some of the available measurements may not be useful in distinguishing the classes. These measurements may be redundant because the other measurements convey the same information. Excluding these redundant measurements will, therefore, not deteriorate the performance of the pattern recognizer or classifier.

Secondly, it is possible for the performance of a pattern recognizer to actually decrease when more measurements are used. This effect results from the estimation of the underlying probability structure using a fixed, finite number of training samples. The inclusion of more measurements requires the estimation of more complicated and higher dimensional densities and, in the parametric case, more parameters. The greater number of measurements means that the estimation of the MMLDR or Bayes decision rule is more difficult and hence probably less accurate. If the information provided by the additional measurements does not compensate for the loss of estimation accuracy, then the performance of the pattern recognizer will decrease.

A third reason for utilizing only the subset of available measurements involves the effect of using a large number of measurements relative to the number of training samples. When this is the case, the ensuing decision rule can become overdependent on the particular observations of the random variables. This overtraining of the pattern recognizer creates a very sensitive decision rule, sensitive not to the separation of the underlying distributions, but merely to the separation of the two samples of training vectors.

Since measurement selection is an important part of pattern recognition, it is desirable to find an economical way of choosing a subset of measurements which will give optimum classification results, optimum in the sense that the pattern recognizer will give the best training and 10% jackknife results using the selected subset of measurements. The best solution to this problem is exhaustive search of all possible subsets of measurements. However, this approach is not feasible in practice because its cost is very high.

In our experimental work on radiographic images, forward sequential search has been used to select the subsets of measurements to be used in classification. Let $M=\{x_1, \ldots x_n\}$ be the set of available measurements and $C(M_1)$ the value of the performance criterion evaluated for the subset of measurements M_1. The following is the definition of forward sequential search.

Choose y_1 to be the measurement x for which $C(\{x\}), x\in M$ is optimum. Choose y_2 to be the measurement x for which $C(\{y_1, x\}), x\in(M-\{y_1\})$ is optimum. Having chosen $y_1, y_2, \ldots, y_{k-1}$, choose y_k to be that measurement x for which $C(\{y_1, \ldots, y_{k-1}, x\}), x\in(M-\{y_1, \ldots, y_{k-1}\})$ is optimum. Continue this process until $k=n$. One can stop the search when the performance criterion decreases, or at a predetermined fixed number of variables.

3.3.2 Preprocessing

Preprocessing is the necessary processing of the input image before the subsequent steps (segmentation, feature extraction and recognition) can be applied. This subsection will discuss only those preprocessing techniques that have proved the most useful in radiographic image processing.

One operation that has been found useful is computing projections [3.30] in the horizontal and vertical directions. If $[\mathrm{PIC}(I, J)]_{m,n}$ is the array of picture data then

$$\mathrm{hor}(I)=\sum_{J=1}^{n} \mathrm{PIC}(I, J) \tag{3.7}$$

and

$$\mathrm{ver}(J)=\sum_{I=1}^{m} \mathrm{PIC}(I, J). \tag{3.8}$$

Figure 3.4 shows the form of these curves for a chest X-ray. Projections greatly reduce the data as compared with the original image data. They also have a property similar to that of reduction by averaging, in that the contributions of small objects in the image to the projection are often minimal.

Logarithmic conversion is another useful operation [3.31, 32]. Many images exhibit a gray-level histogram biased on the dark side. This bias tends to give an image with little visible detail. Using logarithmic con-

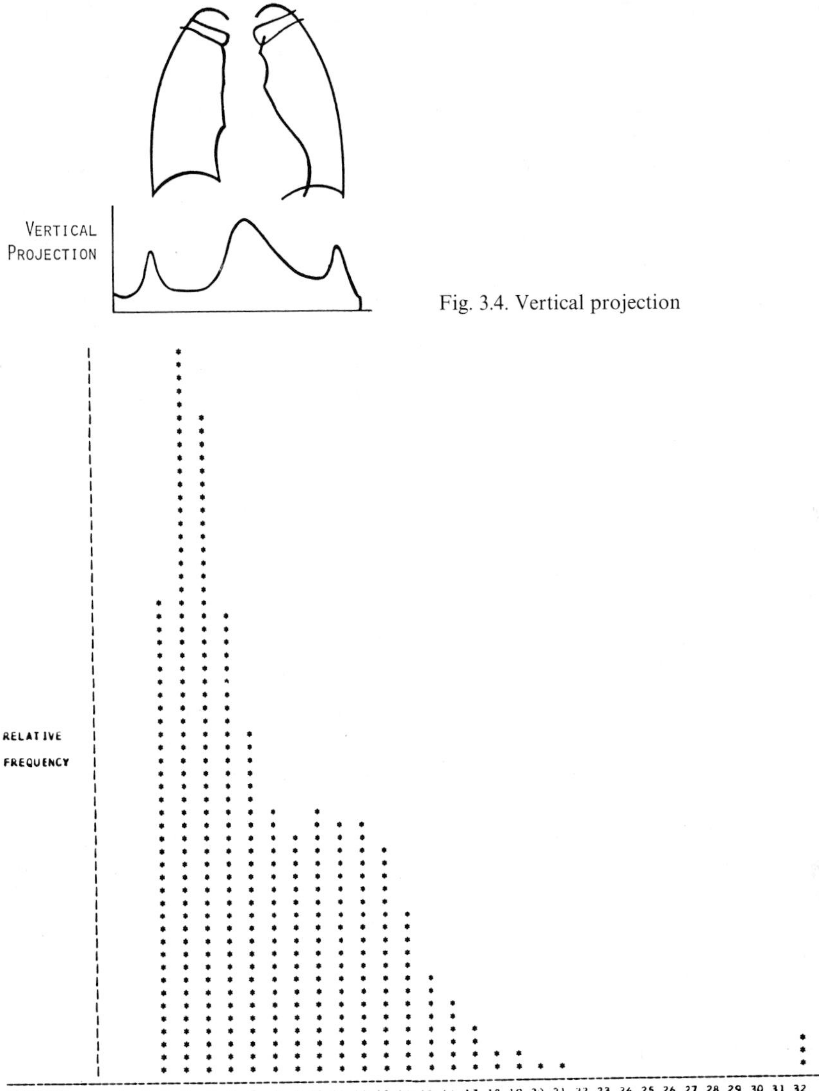

Fig. 3.4. Vertical projection

Fig. 3.5. Histogram of chest image before linear distribution

version to correct the images expands the gray range of the lower brightness levels while compressing the higher brightness levels.

To achieve a uniform distribution, one can use a transformation similar to that employed in statistics. First, the picture is quantized so each intensity value is an integer in the range 0 to $(N_G - 1)$, where N_G

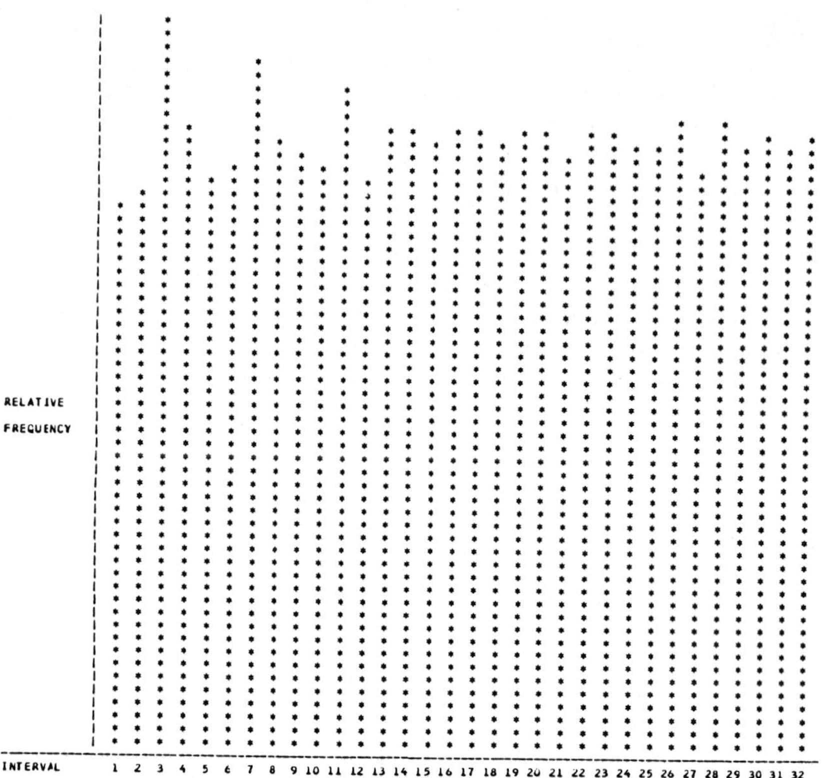

Fig. 3.6. Histogram of chest image after linear distribution

is the number of gray shades. A histogram vector of length N_G is constructed, where the i-th element contains the number of points in the image with intensity $i-1$. Then this vector is integrated so element i becomes the sum of all previous elements $1, 2, \dots i$; i.e., $\text{hist}_i \leftarrow \text{hist}_i + \text{hist}_{i-1}$, $i=2, \dots, N_G-1$. After each element is divided by 2 (to quantize the original intensity range), the vector is used to transform the quantized image; i.e., $\text{pic}_{i,j} \leftarrow \text{hist}(\text{pic}_{i,j})+1$, where $\text{pic}_{i,j}$ is the intensity value of the point in row i, column j of the digitized image matrix. Figure 3.5 shows a histogram before equalization, and Fig. 3.6 exhibits the histogram after equalization. It should be noted that the result is a discrete approximation to the predicted uniform distribution. There are image classes in which this type of a transformation may not be desirable.

Scaling and limiting the gray shades in an image is sometimes useful [3.32]. For example, when one is digitizing with eight bit accuracy it may be useful to limit the range of gray levels to eight. This can be accomplished by ignoring the five least significant bits. It may produce contours in the image, which can be harmful, but it can also help in some situations. For example, if the images are PA chest radiographs a contour can be produced that approximates the heart boundary.

3.3.3 Texture

The problem of textural discrimination involves locating a feature that may be distributed over a wide area of the image rather than being concentrated in one well-defined area. There are a number of properties associated with disease that appear on radiographic images as texture patterns. An example of this, in the case of chest radiographs, is increased or decreased vascular change, which is related to heart disease. Another example is interstitial fibrosis, which is related to several diseases. In relation to bone cancer, a significant problem is to describe the destruction as geographic, permeated or moth-eaten.

References [3.33–39] give more details on the general problem of texture discrimination. We describe here the essential ideas that have been applied to radiographic problems.

The texture problem has been considered as basically a problem in statistical analysis. That is, texture is characterized as a statistical distribution of the gray shades in a given area. In the approach outlined here a random process is considered, where the variable is indexed by (I, J), the picture coordinates. It is also assumed that the process is stationary [3.39]. In this situation, PIC(I, J) is a sample function of the random process. From this model a number of statistical features are determined.

Suppose the sampling is with $N_{\mathrm{G}}=64$ gray shades. Then for total specification of the first-order probability distribution, the entire discrete density function must be specified

$$f(i),\ i=0, 1, \ldots, 63\,. \tag{3.9}$$

The use of 64 features is undesirable; therefore, one must settle for less than total specification. One alternative is to estimate the more important parameters of the distribution.

The r-th moment about zero, m_r, is given by

$$m_r = \textstyle\sum_{i=0}^{63} i^r f(i)\,. \tag{3.10}$$

The mean μ is given by $\mu=m_1$. The r-th central moment (moment about the mean) is given by

$$c_r = \textstyle\sum_{i=0}^{63} (i-\mu)^r f(i) = \sum_{j=0}^{r} \left[\binom{r}{j} m_{r-j} + (-m_1)^j\right]. \tag{3.11}$$

The variance V is given by

$$V = c_2 = m_2 - m_1^2 \,. \tag{3.12}$$

The skewness S is given by

$$S = \frac{\sum_{i=0}^{63} (i-\mu)^3 f(i)}{V^{3/2}} \,. \tag{3.13}$$

While variance is a measure of the dispersion of a distribution, skewness measures the departure from symmetry.

The kurtosis K is given by

$$K = \frac{\sum_{i=0}^{63} (i-\mu)^4 f(i)}{V^2} \,. \tag{3.14}$$

The kurtosis expresses the tendency of a distribution to either cluster about the mean or spread out toward the tails of the distribution.

In addition to first-order statistical properties, several features can be defined based upon second-order statistics. We will consider eight possible orientations around any point, corresponding to its eight neighbors on the raster scan. For each of these orientations I one can define a probability matrix P_I, $I=0, \ldots, 7$, where $\mathrm{dir}(I) = 45\, I°$. If there are N_G shades of gray, then P_I consists of $N_G \times N_G$ elements denoted as follows

$$P_I = \begin{bmatrix} f_I(0,0) & f_I(0,1) & \ldots & f_I(0, N_G-1) \\ f_I(1,0) & f_I(1,1) & \ldots & f_I(1, N_G-1) \\ f_I(N_G-1, 0) & f_I(N_G-1, 1) & \ldots & f_I(N_G-1, N_G-1) \end{bmatrix} .$$

Note that

$$\sum_{k=0}^{N_G-1} \sum_{l=0}^{N_G-1} f_I(k,l) = 1, \text{ for all } I=0, \ldots, 7 \,. \tag{3.15}$$

These matrices are not symmetric. However,

$$\begin{aligned} P_0 &= P_4^T \\ P_1 &= P_5^T \\ P_2 &= P_6^T \\ P_3 &= P_7^T \,. \end{aligned}$$

If there is no need to distinguish between rotations of 180 degrees then the following symmetric matrices can be formed.

$P_H = 0.5(P_0 + P_4)$ [horizontal joint probability matrix],

$P_V = 0.5(P_2 + P_6)$ [vertical joint probability matrix],

$P_{DL} = 0.5(P_3 + P_7)$ [left diagonal joint probability matrix],

$P_{DR} = 0.5(P_1 + P_5)$ [right diagonal joint probability matrix].

Even using symmetric matrices, the number of features required to specify all four joint probability matrices is given by

$$4 \times \frac{N_G(N_G+1)}{2} = 2N_G(N_G+1). \tag{3.16}$$

In the case of first-order probabilities the number of features required for specification was reduced by using the statistics, mean, variance, skewness, and kurtosis. The same type of reduction is also desirable in the case of second-order probabilities, but the quantities used are somewhat more heuristic. Each of the four symmetric joint probability matrices will be represented by the following functional quantities

$$E(P) = \sum_{k=0}^{N_G-1} \sum_{l=0}^{N_G-1} f(k,l)^2, \tag{3.17}$$

the energy;

$$H(P) = \sum_{k=0}^{N_G-1} \sum_{l=0}^{N_G-1} [-f(k,l)\log f(k,l)], \tag{3.18}$$

the entropy;

$$C(P) = \frac{\sum_{k=0}^{N_G-1} \sum_{l=0}^{N_G-1} (k-M)(l-M) f(k,l)}{\sum_{k=0}^{N_G-1} (k-M)^2 \sum_{l=0}^{N_G-1} f(k,l)}, \tag{3.19}$$

the correlation, where

$$M = \sum_{k=0}^{N_G-1} k \sum_{l=0}^{N_G-1} f(k,l); \tag{3.20}$$

$$L(P) = \sum_{k=0}^{N_G-1} \sum_{l=0}^{N_G-1} \frac{1}{1+(k-l)^2} f(k,l), \tag{3.21}$$

the local homogeneity: and finally

$$I(P) = \sum_{k=0}^{N_G-1} \sum_{l=0}^{N_G-1} (k-l)^2 f(k,l), \tag{3.22}$$

Table 3.2. Texture features

I.D.	Symbol	Measurement description	I.D.	Symbol	Measurement description
1	M	Mean	14	$I(\hat{P}_V)$	Vertical inertia
2	V	Variance	15	$E(\hat{P}_{DR})$	Diagonal from right energy
3	S	Skewness	16	$H(\hat{P}_{DR})$	Diagonal from right entropy
4	K	Kurtosis	17	$C(\hat{P}_{DR})$	Diagonal from right correlation
5	$E(\hat{P}_H)$	Horizontal energy	18	$L(\hat{P}_{DR})$	Diagonal from right local homogeneity
6	$H(\hat{P}_H)$	Horizontal entropy	19	$I(\hat{P}_{DR})$	Diagonal from right inertia
7	$C(\hat{P}_H)$	Horizontal correlation	20	$E(\hat{P}_{DL})$	Diagonal from left energy
8	$L(\hat{P}_H)$	Horizontal local homogeneity	21	$H(\hat{P}_{DL})$	Diagonal from left entropy
9	$I(\hat{P}_H)$	Horizontal inertia	22	$C(\hat{P}_{DL})$	Diagonal from left correlation
10	$E(\hat{P}_V)$	Vertical energy	23	$L(\hat{P}_{DL})$	Diagonal from left local homogeneity
11	$H(\hat{P}_V)$	Vertical entropy			
12	$C(\hat{P}_V)$	Vertical correlation	24	$I(\hat{P}_{DL})$	Diagonal from left inertia
13	$L(\hat{P}_V)$	Vertical local homogeneity			

the moment of inertia about the main diagonal. Thus, 24 statistical features are extracted as representing the textural information. They are summarized in Table 3.2. These features can then be the input features to a classical pattern recognition system for the purpose of discriminating textures.

It should be observed that one could choose other features. In fact, the set of features most appropriate is still an open question. However, the features described are commonly used and typify the method of texture discrimination.

3.3.4 Picture Segmentation

Picture segmentation is the process of locating objects of interest in the picture array PIC(I, J). Writing programs to do segmentation is perhaps the most difficult task in image analysis. This is particularly true in radiographic image analysis where often one needs accurate outlines of an object which is not too clearly defined in the original image. An example of this problem would be locating cardiac silhouettes in a PA chest radiograph. Also, the class of all radiographic films is very complex with a variety of views of the chest, pelvis, skull, upper and lower GI, abdomen, etc. Each radiograph in itself is very complex with much detail and a large number of superimposed objects recorded on the film. Many subtle findings in the radiograph can yield significant diagnostic information, and it is difficult for the non-radiologist (and non-radiologists are usually involved in the system development) to distinguish these findings from noise.

We have tried to approach the analysis of radiographic images in two ways. The first problem considered is the development of analysis techniques that have general applicability to radiographic image processing. It is hoped this will facilitate the transfer of picture processing techniques from one type of image to another type; for example, from chest radiographs to knee radiographs.

The second problem considered in picture segmentation was the design and development of programs to solve well-defined problems in radiographic image analysis. This is necessary because it is important to demonstrate working programs that can be used to evaluate the automated approach to radiographic image analysis. These efforts are described in detail in Section 3.4. Some concepts of general value in radiographic image processing are indicated below.

As Fig. 3.2. indicates, the segmentation stage of processing must be integrated with the recognition stage. This consideration takes formal cognizance of the fact that in a complex scene one must use the knowledge about an object to extract the boundary of the object, for example. In a complex scene there will be many overlapping objects to confuse primitive local algorithms. In order to successfully locate objects, the search must be context-sensitive, and a model of the object must be used in the search for the object. Moreover, the recognition stage must be utilized to confirm every step in segmentation so that the results conform to the model of the object.

The analysis of the scene should be top down, in the sense that one first attempts to locate large objects, and later to locate the smaller details of the objects [3.19]. It seems natural to utilize a tree structure to describe the hierarchical structure of the image class; however, there is no unique tree to describe the image class. Figure 3.7 shows several for PA chest X-rays. The tree utilized dependes on the details of the processing. For example, consider the trees in Fig. 3.7b and c. The difference is that there is only one node for the heart and mediastinum in Fig. 3.7c. This might be a better tree in cases where there is no significant difference in gray shade between the mediastinum and heart. One should also observe that there is no well-defined part of the image that corresponds to the physical object, the heart. Certain parts may not be visible at all on the radiograph, such as the top of the heart, but rather must be computed. The tree in Fig. 3.7d is much more refined at the first level than the others. This tree might require processing at a higher level of resolution to determine the objects on the first level of the tree than would be required for the other two trees.

Another example is given in Fig. 3.8a, which is a diagram of some of the anatomical features that would be present in a lateral view of the brain. The objects of interest in brain scans are lesions, which appear as

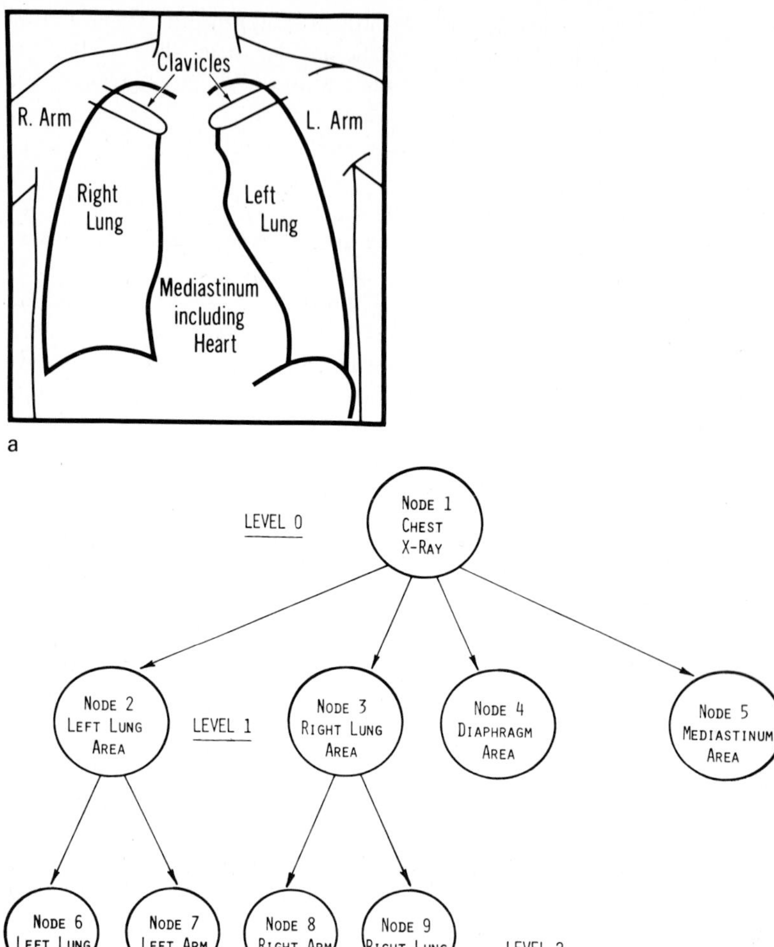

Fig. 3.7. (a) Schematic view of PA chest radiogram showing the large anatomical areas within the scene. (b) Tree structure currently used to describe the top down ordering of PA chest radiogram. (c) Tree structure combining the heart and mediastinum nodes from the previous tree. (d) Tree structure calling for greater differentiation of areas at low level

dark spots within the lighter brain region. Figure 3.8b gives a hierarchical structure for analyzing these images [3.40]. The images have rather simple structure, and this is reflected in the tree used for analyzing them.

A hierarchical structure is important in analyzing images, but missing from this part of the model is any description of the computer processing to be done. The tree describes the objects in the scene. Choosing larger

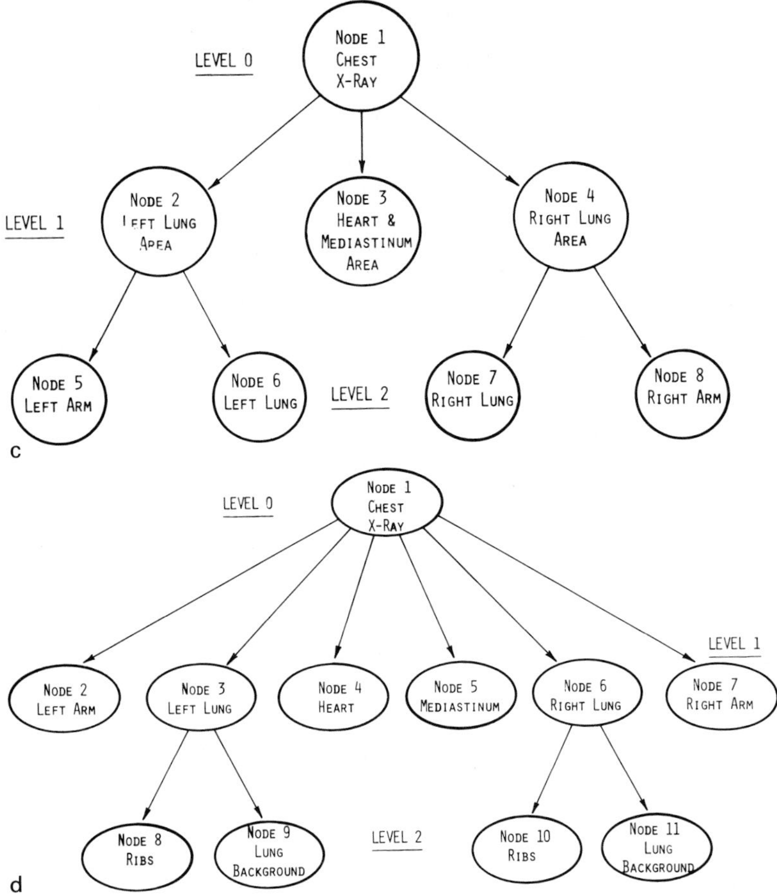

Fig. 3.7c, d

objects closer to the root of the tree implies that they can be analyzed at a lower resolution. This is one of the advantages of using top-down analysis. In looking for large objects at a low resolution, the small objects or details disappear from the image, leaving a relatively clutter-free image for analysis of the objects of current interest.

When referring to the tree structure we use common terminology from graph theory. Each node in the tree represents an object that may occur in the image. Therefore, it is convenient to refer to node(n) and object(n), where node(n) is the node in the tree representing object(n). If the model is to control the analysis of the image, then more than just hierarchical information is needed. One way to include additional information is to attach an attribute list to each node of the tree. The

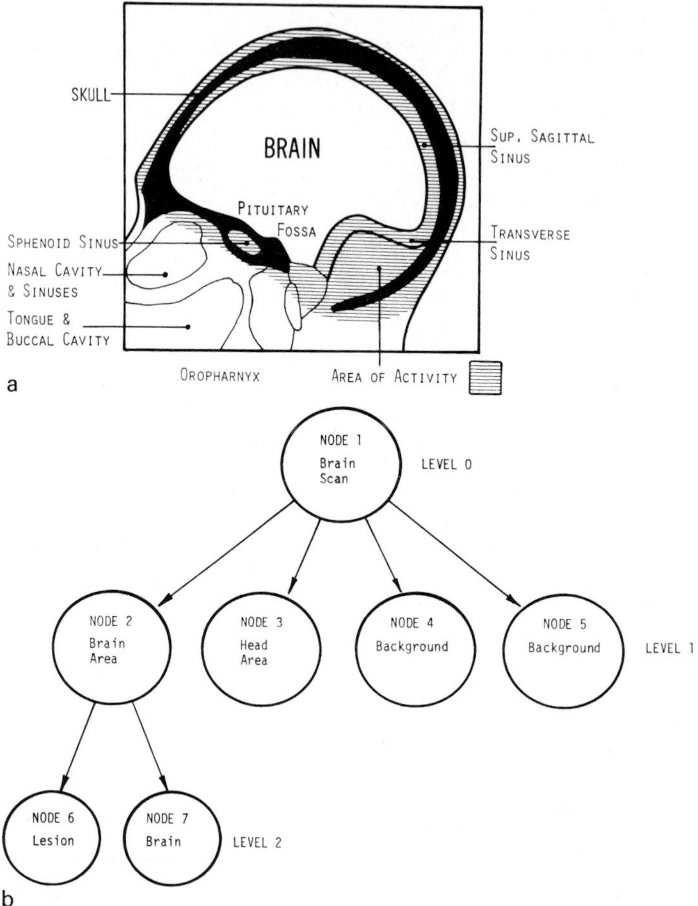

Fig. 3.8. (a) Lateral view of the brain. (b) Tree for lateral view of the brain

attribute list can be as long as desired, but should be fairly simple in order to be useful in the segmentation phase. This is because one can structure the processing so that only objects on one level of the tree need be distinguished. It is often the case that objects on the same level are dissimilar – for example, the heart and lungs in Fig. 3.7, in which the heart is distinguished by its gray level, and the lungs are distinguished by their spatial locations. If the objects are specified too closely, one must allow for all possible variations, which may be very difficult. Some attributes used include AREA (described as the area in the scene in which the object may be located), DELTA (a parameter used by a program to enumerate regions), INT (expected average gray shade of the object), and RES (the resolution at which the object is to be identified). The reader can easily identify other potentially useful attributes such as higher-order moments [3.41]. These ideas are discussed in more detail in [3.19]. Similar ideas are mentioned in [3.42] in relation to other image analysis problems.

Additional information that may be useful in the model is spatial information about the objects. For example, does one object lie to the left or right of another? There is no doubt that this would be valuable information, because in some images a boundary is not always given by a gray shade gradient but may be defined by the spatial relations of the objects. For example, consider the brain scan in Fig. 3.8. According to the geometry of the image, the brain area is to be totally enclosed by what we call the head area. In nuclear medicine images, there are a large number of blood vessels which contain radioactivity and totally enclose the brain (Fig. 3.8). In an actual image the radioactive uptake may be low, and the vessels may not be visible in the image even though the proper area for the brain is still the one enclosed by the vessels. Therefore, edge information is of little help in completing the brain region. It is tempting to use a predicate in locating the brain region that describes the fact that region 2 is to be enclosed by region 3. Some methods using this type of spatial information are described in [3.19, 43], where they are called structure tests. In [3.44] spatial information is used in segmentation, where region merging is used to locate objects.

In our original study a standard program for enumerating regions was used to examine the picture data to locate object(n) associated with node(n) in the tree. This implies that the computer processing was standard, and only the tree was to be changed when changing from one class of images to another. Even though we had some success with this effort, we concluded that, in general, one needed more flexibility to analyze complex radiographic images. For example, there are instances when it is better to locate an object by thresholding or contour tracing than by region enumeration. Also, one often discovers a better method of analysis to use after a more thorough investigation of the problem. This implies that developing image analysis programs is an experimental activity, and one is faced with continually modifying the program to achieve the desired result.

Approaches to programming concerned with criteria of flexibility, readability and portability are current popular subjects; see [3.15] for references on these topics.

One basic problem is how to manage complexity in the development of programs for image analysis. One approach to this problem is to incorporate the program control in a graph [3.15]. The supervisor which interprets the graph is very simple. The graphs themselves may become fairly complex, but it has been possible to construct them so the states of the graph reflect fairly clearly the current status of the processing of the image. In order to avoid confusion with the tree structures discussed earlier, we use the notation arc and state in referring to the control graph, which we call an abstract image analysis graph (IAG). It may be defined as follows:

I.a) A finite directed graph
$[S, L, E, l, q_0, H]$, where
S is a finite set called the states of the graph
L is a finite set of labels of the states
$E \subseteqq S \times S$ is the set of arcs of the graph
$l: S \to L$ is the function that labels the graph
q_0 is the start state
$H \subseteqq F$ is the set of halt states.

b) A non-empty domain D

c) A set of $m \geqq 0$ distinct input variables $\boldsymbol{x} = (x_1, x_2, \ldots, x_m)$.

d) A set of $n \geqq 1$ distinct variables $\boldsymbol{y} = (y_1, y_2, \ldots, y_n)$, called program variables.

II.a) With each pair $(q, l(q))$, where $q \in S$, is associated a function f: $D^{n+m} \to D^n$.

b) With each arc set $\{e_1, e_2, \ldots, e_k\}$, where each $e \in E$, emanating from state q there exist predicates $p_1, p_2, \ldots, p_k$ which are complete, $(\forall \boldsymbol{x})(\forall \boldsymbol{y})(p_1 \vee \ldots \vee p_k)$, and mutually exclusive, $(\forall \boldsymbol{x})(\forall \boldsymbol{y})(\sim p_i \vee \sim p_j)$ where $1 \leqq i \neq j \leqq k$. Each predicate maps $D^{m+n} \to \{T, F\}$, i.e., each predicate is of the form $p_i(\boldsymbol{x}, \boldsymbol{y})$.

References [3.45–46] give similar concepts relating to programming theory. The above definition seems a little forbidding but in an actual example it takes on more meaning. One should note that the graph control structure separates the decision making or control part of the program from the computational part, or from primitive programs such as region enumeration or contour trace programs. The graph also indicates when the primitive programs are to be called and evaluates the results of the processing by the primitive programs.

As an example, consider Fig. 3.9. Let REG indicate a program to enumerate regions, CONT a program to trace contours, and ADJ a program to adjust the parameters used by the primitive programs. DELTA is a parameter for the program REG that influences the size of the region determined. SUCC and NODE are program variables; SUCC assumes values T and F. The operation of this IAG might be as follows:

Start in state 1. The label on this state indicates the execution of the program REG to search for object 1. Node 1 of the tree is interrogated

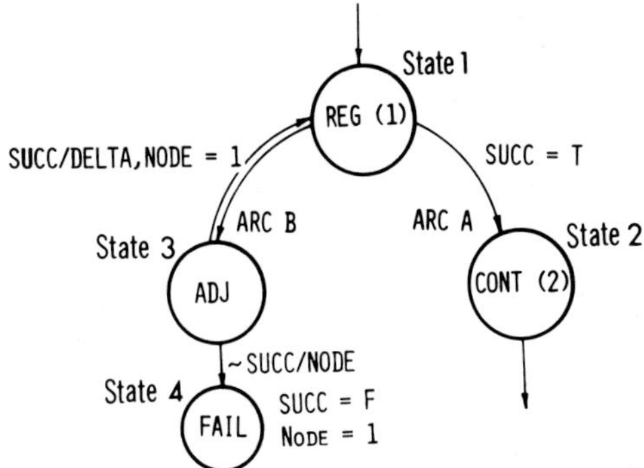

Fig. 3.9. Sample IAG

for the characteristics of object 1. REG is executed and the variable SUCC is set to *T* if a region is located suitable for representing object 1, otherwise SUCC=*F*. If SUCC=*T*, then arc *A* is traversed to state 2. This state label indicates a search for object 2 utilizing the primitive program CONT. If SUCC=*F*, arc *B* is traversed to state 3; NODE=1 indicates object 1 was not found. NODE=1 might be considered to be the output associated with the arc. The state label of state 3 is ADJ which calls the program ADJ. The ADJ program tends to adjust the parameter DELTA for NODE=1 so a suitable region can be found for object 1. If an adjustment of DELTA can be made, then SUCC=*T*; otherwise, SUCC=*F*. If SUCC=*F*, then state 4 with NODE=1 is entered to indicate a failure to locate object 1. The program FAIL might print a suitable message on the console indicating the reason for failure.

A useful IAG would be more involved, but it is hoped that the above example demonstrates the construction process and the manner in which the state of the graph reflects the state of the analysis. It is rather easy to adapt the system to different machines. In fact, the most difficult task usually is modifying the primitive programs. Since the primitive programs are computational programs, they could be implemented in Fortran to aid portability.

3.4 Applications

In this section, some of the work that has been done on radiographic image analysis is described. Normally the work in our laboratory is done on a project basis. Each project is concerned with a specific class of images and has fixed goals. The projects have varied comprehensiveness due to financial limitations, and to some extent reflect the priorities and interests of our laboratory. Our most intensive work has been on chest radiographs, which will be discussed last. Other institutions have also emphasized the analysis of chest radiographs; this discussion thus summarizes the current state of radiographic image processing.

3.4.1 Nuclear Medicine

Because of its technical simplicity, brain scanning is one of the most widely used procedures in nuclear medicine. It is useful in detecting abnormalities such as arteriovenous malformations, inflammatory diseases, subdural hematoma, cerebral infarcts, tumors, and focal diseases of the skull. The normal brain is relatively impermeable to most radiopharmaceuticals injected into the bloodstream. Tumors and other lesions of the brain have increased radioactivity on a brain scan because these lesions are more permeable to radiopharmaceuticals than normal brain tissue.

An anterior, posterior, and two lateral views are obtained in a routine brain scan. The right and left lateral views are mirror images. The scalp, skull, superior sagittal sinus vein and vessels of the meninges form a band of activity around the cerebral hemispheres in the lateral views, Fig. 3.8a. This band of activity is narrow anteriorly and thickens posteriorly toward the confluence of the venous sinuses at the base of the skull. Increased activity below the orbits represents the nasal cavity, oropharynx, mouth, maxillary sinuses, and salivary glands.

In an initial study, images containing 600000 counts per view were obtained on a pho-Gamma III scintillation camera. Pictorial representations of the emitted radiation were recorded both on Polaroid film and on 35×43 [cm] X-ray film. The back-illuminated scintigram (X-ray film) image is focused by a simple optical system onto the front surface of the dissector camera. These are low resolution images, so they are scanned in the standard 256×256 format and then reduced by averaging to a suitable resolution for processing. Usually 32×32 is adequate. There is available commercial equipment for directly digitizing the count data.

The initial task for computer processing is to locate the brain. A technique of variable thresholding and region enumeration was initially developed to accomplish this [3.47]. The array of points digitized to 100 gray levels can be transformed into a simpler, binary array of 0's and 1's by assigning to each point whose value is at or above a given value or threshold a value of 1, and to each point below the threshold a value of 0. Consider a typical point PIC(i,j) and its eight neighboring points.

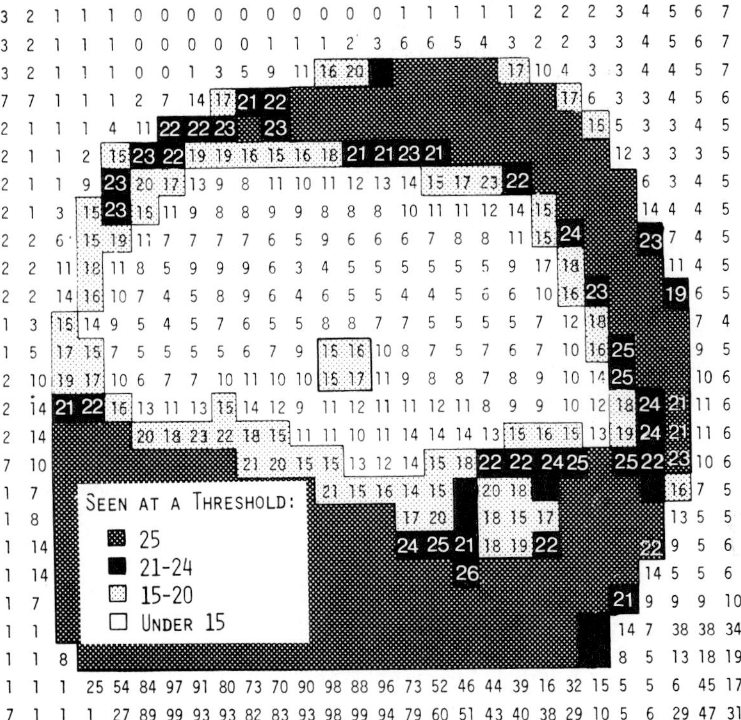

Fig. 3.10. Thresholding on a brain scan

It is possible to divide the binary array into sets of regions by the following rules:

(a) Every point must belong to one and only one region.

(b) If any point has the same value (0 or 1) as a neighboring point, then the two points must belong to the same region.

In this manner, all points below any threshold are collected into a set of "cold" regions, and all points at or above the threshold are collected into another set of "hot" regions. Figure 3.10 shows an example of the regions obtained when the brain is examined at decreasing thesholds of 25, 21, 15.

In the terminology of nuclear medicine, the brain is described as a cold region contained within a hot region (the surrounding tissues). Regarding the original image as a series of binary images at all thresholds, the brain is the largest cold region at any threshold which is completely enclosed within a hot region at that same threshold. If an interior region is one which contains no point on the edge of the array, then the brain

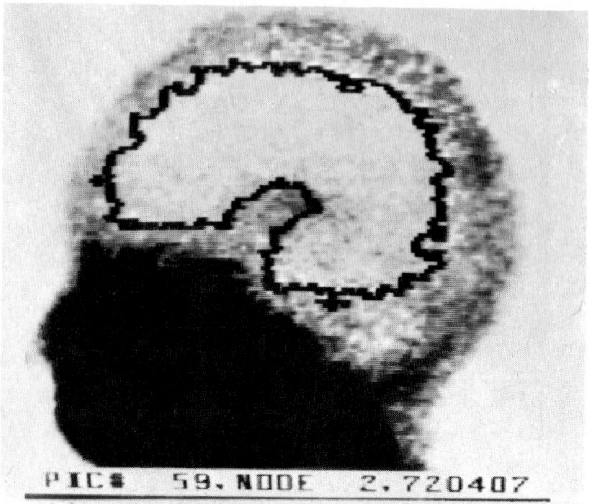

Fig. 3.11. Abnormal brain scan—Lesion lying along outer edge

may be defined as the largest cold interior region. Therefore, the problem is to find the threshold at which the largest interior region exists.

One approach varies a threshold T from high to low and determines the regions below T. Since the brain region has the characteristic of being a region below the threshold, or light, and is enclosed by a dark region above the threshold, we have a mechanism for locating the brain—we use the first threshold at which a cold interior region is found. This approach is exhaustive and tends to be time-consuming. Some other methods will be discussed later.

Once the brain has been identified by the computer, the next step is to identify any abnormalities which appear within it. Many lesions can be described as hot regions contained within a cold region (the brain). This problem is very similar to the one stated for finding the brain except that the words "hot region" and "cold region" are interchanged. Therefore, with some modifications, one can use similar techniques for finding hot spots in the brain.

This technique will suffice for detecting lesions which when projected appear as though they are completely surrounded by the low-activity brain region. Frequently, however, a lesion is projected in such a manner that it blends into the hot region surrounding the brain (Fig. 3.11) and cannot be detected by the methods described above. In these circumstances, the lesion appears as an indentation distorting the outline of the brain. To detect this type of abnormality, a method using chain codes can be employed. After defining the brain region, it is fairly simple to join the points on the outer boundary of that region. The resulting

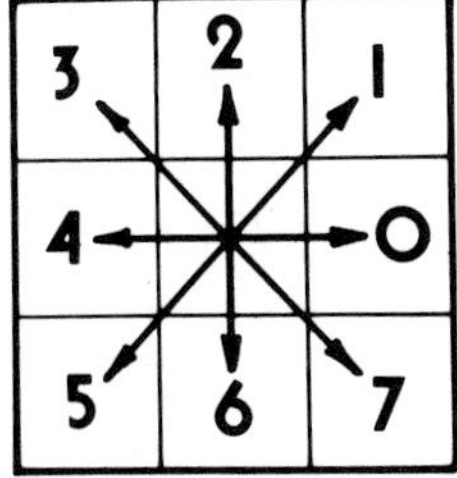

Fig. 3.12. Chain code

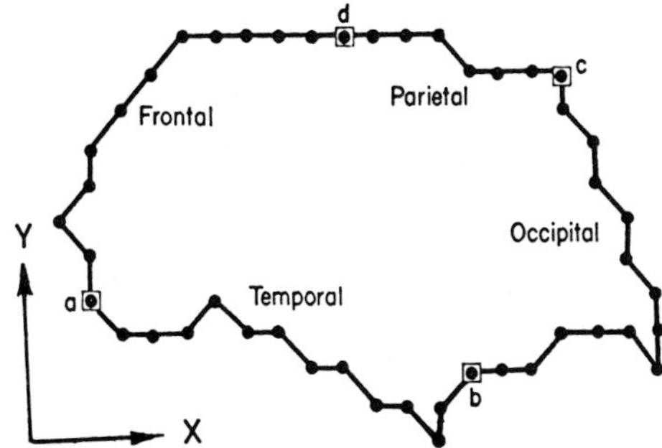

Fig. 3.13. Brain outline

closed curve consists of a sequence of line segments, adjoining adjacent elements of the picture array. Each segment is assigned an integer from 0 to 7 according to its orientation (Fig. 3.12), and the resulting sequence of integers forms the chain code representation of the brain outline (Fig. 3.13).

In order to have the outline correspond more closely to the anatomical features, the boundary line is subdivided into four parts. This has the advantage of allowing one to work with smaller chain codes and also helps to pinpoint the area of the abnormality. On the lateral view this outline is subdivided by points selected according to the following rules (Fig. 3.13):

Point a: Consider a subset of edge points such that $(x+y)$ is a minimum. Point a is chosen to be the point in that subset with a minimum value of x.

Point b: Let x_{max} and x_{min} be the greatest and least x coordinates of the edge points. Define a coordinate $x_{2/3}$, two thirds of the way from anterior to posterior, as follows:

$$x_{2/3} = (x_{min} + 2x_{max})/3 .$$

Then b is selected to be the edge point of the form $(x_{2/3}, y)$, with the smallest value of y.

Point c: This is the point in the subset of edge points such that $(x+y)$ is a maximum, with the largest value of y.

Point d: With use of x_{max} and x_{min} as defined for point b, let x_{av} be their mean. The d is the edge point of the form (x_{av}, y), with the largest value of y. These four points divide the brain outline into four segments corresponding roughly to the borders of the frontal, parietal, occipital, and temporal lobes.

The problem of detecting tumors that lie along the boundary now involves comparing the boundary segments of normal patients to those of patients with tumors. This is basically a classical pattern recognition problem which extracts features characterizing the shape of the segment to be used by the classifier. Features used in this application are based on the chain codes, using sequences of fixed length [3.47].

Because the variable thresholding technique is slow and at times does not give an accurate boundary, other algorithms were developed for obtaining the brain region. One method used the descriptive approach discussed previously. Figures 3.8a and b give an indication of the hierarchical analysis [3.40].

Applying the descriptive approach to brain scans presented some difficulties when locating every object using region enumeration. Nodes 2 and 4 (Fig. 3.8b) tend to grow together before Node 3 can encircle the brain. This occurs because the activity in the superior sagittal sinus becomes a very narrow band on the superior aspect of the brain scan, which causes this area of the image to have a very low contrast with respect to the background. Thus, a portion of the background will sometimes be included in the brain area.

When a thresholding method is used a similar problem occurs. Usually a threshold is picked which is low enough to determine the cut-off of the head and the background, but often with low quality scans this threshold is set so low that only a portion in the upper left part of the brain is found. For these reasons a combined approach was used which locates an initial region of the brain region, Node 2, using region enumeration. Node 3 is determined by placing every point in the picture which is above a threshold value into the head region. This situation shows the utility of using different primitive programs in picture segmentation.

A program utilizing projections was incorporated into the algorithm to determine the threshold value. This program also sets the start of region coordinates (STARTI, STARTJ) for the brain region and determines the value DELTA which is the input parameter to the region enumeration program for the brain region. Consider the function

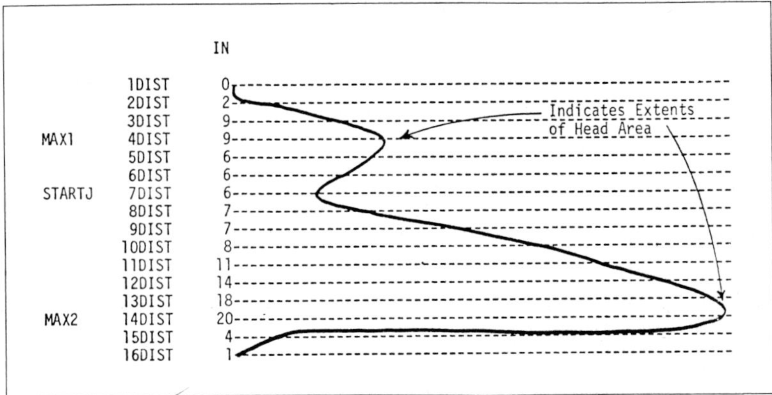

Fig. 3.14. Horizontal projection of a brain scan

shown in Fig. 3.14. As the brain is always lighter, its intensities are lower; therefore, the first minimum point (indicating the brain region) following a local maximum point (indicating the upper portion of the head region) is chosen as the STARTI coordinate of Node 2. This row is known to intersect the brain region. Thus, it can be examined in a similar fashion to determine the STARTJ point. The STARTI row is searched for the first MAX1 and last MAX2 local maximum points and the minimum STARTJ between these two. The maxima indicate the horizontal extents of the head region while the minimum indicates a point in the brain region. Therefore, the point (STARTI, STARTJ) is located. One can go further and analyze the STARTJ column in a similar fashion to spot a point MAX3 located in the sagittal sinus, Fig. 3.8a.

The locations of three maximum points, to the left, right and above the brain region (MAX1, MAX2, MAX3), give useful data for analyzing the image. For example, they allow reasonable calculation of DELTA and the computation of a threshold T for locating the head region, Node 3. This calculation is fast and determines the brain region quickly and more reliably than the variable thresholding method. Figure 3.15 displays some processed images.

3.4.2 Bone Disease

Another class of significant images are those pertaining to bone diseases. There are several reasons for considering these images. Radiographs of pathologically proven bone tumors are available through the UMC Medical Center and the Mid-America Bone Evaluation Center. Much effort has been expended in developing a protocol for analyzing these images [3.48]. The existence of a protocol of demonstrated value for

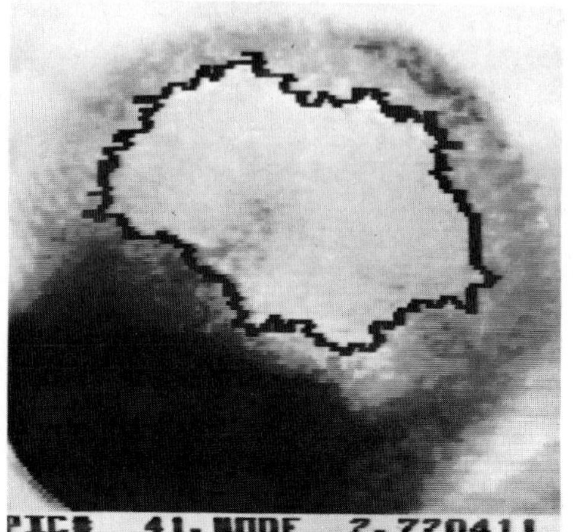

a

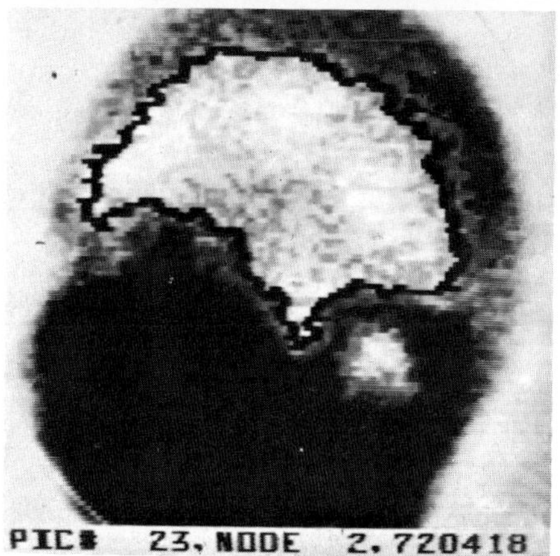

b

Fig. 3.15 a and b. Processed brain scans. (a) Lesion in center of brain. (b) Scan of normal brain

predicting disease is a valuable asset in developing an image analysis system for the images under consideration.

A protocol gives important information about the strategy for analyzing the images and about the variables that must be measured. For example, some variables of importance in diagnosing bone diseases are:

a) age of the patient,
b) location, size and shape of the tumor,
c) density of the tumor,
d) bone destruction visibility,
e) type of bone destruction; geographic, moth-eaten, permeated,
f) penetration of cortex,
g) fracture sign.

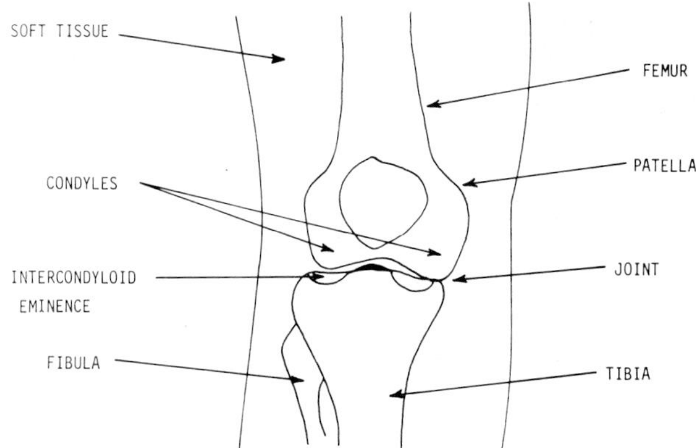

Fig. 3.16. Anteroposterior view of knee

In order to begin the study of this class of images, knee radiographs were considered. This selection was made because the bones of the knee are the most frequent sites for bone tumors. Figure 3.16 shows the structure of these images.

A program for locating the basic anatomical features of the PA view of the knee was implemented. Top down analysis with the scene modeled as a tree was again utilized, and the program was implemented using an IAG [3.43].

A degree of *a priori* knowledge about the knee was used in the algorithm to construct a graphical representation, or tree structure, of the knee X-ray (Fig. 3.17). As the segmentation progresses, regions or objects located by the algorithm are compared with the nodes of the tree. If the attributes match closely, the region is linked to the corresponding node. If the region intersects a previous region (one or more points satisfy the criteria for membership in both regions), either or both regions may be in error, and a link will not be made. A feedback loop is provided to adjust parameters so that the correct regions are located.

The program to analyze knee X-rays uses an IAG as the control. Several considerations prompted the use of a graphical program organization. It seemed inappropriate to analyze a knee X-ray using only one particular primitive. Referring to Fig. 3.16, the different structures can be described and located by several different methods. The joint is a dark line separating the femur and tibia, and a line following technique is appropriate for tracing the joint. There is generally a large variation in intensities within the bones, but the edges are usually well defined. This suggests the use of some form of edge detection or contour trace for defining the bones. The areas of soft tissue are bounded by bone on

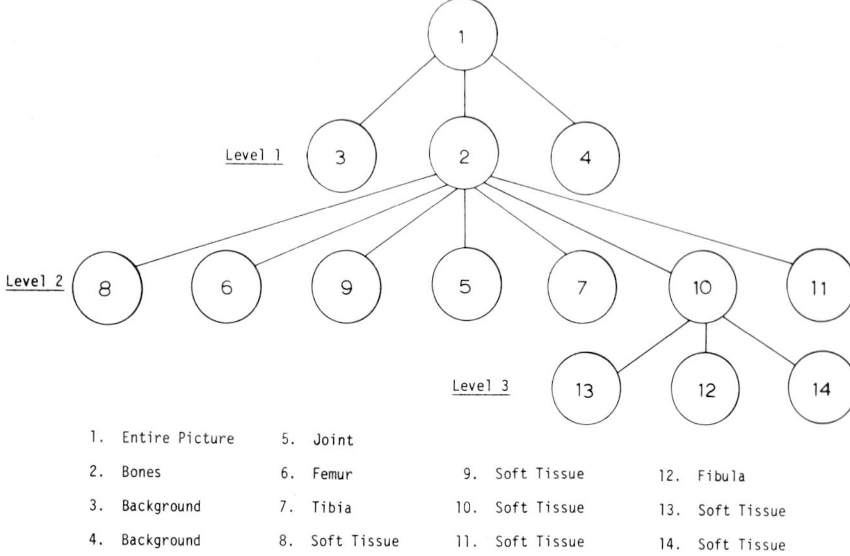

Fig. 3.17. Tree description of the knee

one side and darker background on the other side, forming a "region" or set of connected points which share similar gray shades.

Variations in the quality and technique used to produce X-ray films result in a wide range of normal data. Variations in size and position must be allowed for. Also, variations in intensity of the original radiographs may cause one or more of the soft tissue regions to be obscure or to appear missing. Even a slight rotation in a standard anteroposterior X-ray will cause the fibula to be either partially hidden by the tibia, or placed in such a prominent position that its edges coincide with the edges of the tibia.

In the tree description of the knee (Fig. 3.17), nodes are numbered consecutively beginning with zero at the root of the tree. Node 0 represents the entire picture. Since the descendants of a node completely fill the area represented by that node, measurements made on the node can be used to set the parameters for its descendants.

Level 1 of the tree (nodes 1–3) is designed to separate the useful bone information from the background. Very little picture information is needed at this stage. The original 256×256 picture array is reduced by averaging to a 16×16 array. Intensities are linearly rescaled into 64 gray shades. The approximate locations of the bone and background regions are needed for the region enumeration. These three structures always extend from the top of the picture to the bottom, so row 8, the center

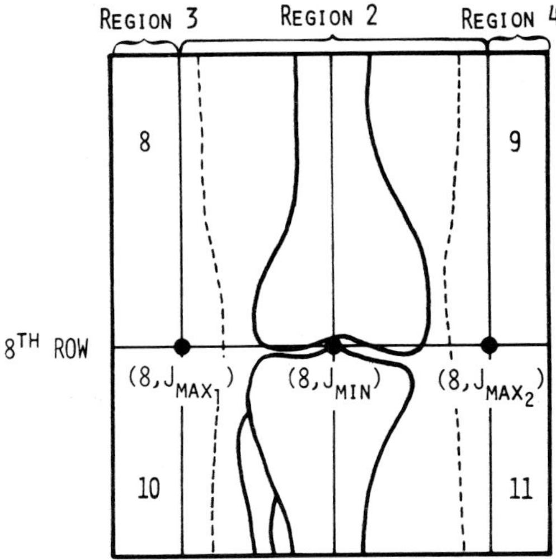

Fig. 3.18. Regions of the knee

row, approximates the "I" coordinates. Examining row 8, the absolute minimum value J_{MIN} in this row approximates the "J" coordinate for the bone region. The absolute maxima to either side of the minimum, $J_{\text{MAX}\,1}$ and $J_{\text{MAX}\,2}$, approximate the "J" coordinates of the background regions (see Fig. 3.18). This information yields the starting points for the three regions. They are completed by a region growing process. These regions are then linked to the respective Nodes 1–3 of the tree. This method works because the objects are always in approximately the same places.

Level 2 of the tree (Nodes 4 through 12) presents some problems. Level 2 is designed to define accurately the joint space, the femur, and the tibia and fibula combination. Node 4, the joint space, is the main problem. The femur and tibia can easily be recognized at low resolution (i.e., a 32×32 picture matrix), but at this resolution the joint is just a dark line, not really a region. Hence, a horizontal projection technique was developed to locate and trace the joint space at this resolution. Figure 3.19 is a plot of the horizontal projection of the region linked to Node 2. The horizontal projection is computed by averaging the points in region 2 in each row across the picture. The projection is standardized by linear rescaling between 0 and 63. The joint is located by finding the greatest maximum near the center.

Once a row has been selected, the midpoint (midway across region 2, not across the entire picture) of that row is employed as an initial point for the joint. A contour trace technique is used to follow the edges of the joint, femur and tibia (Nodes 5–7). The joint and the vertical center

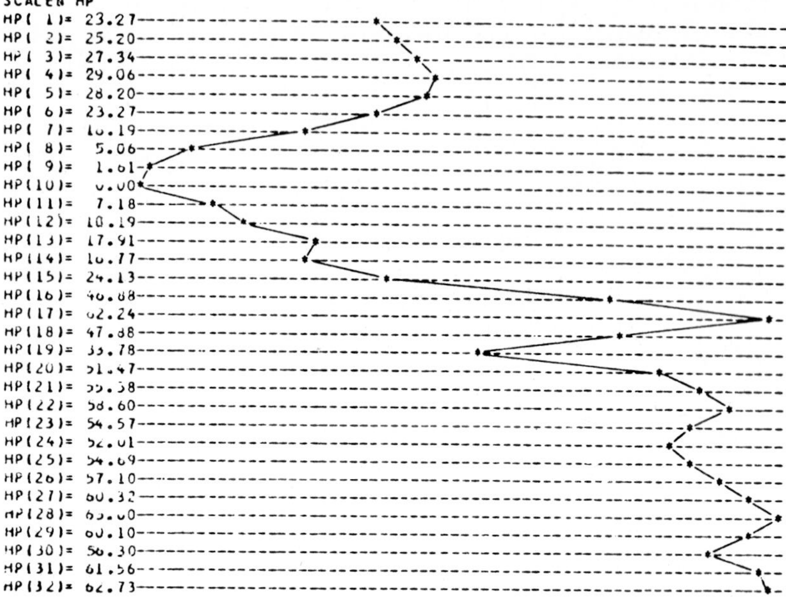

Fig. 3.19. Horizontal projection of joint space

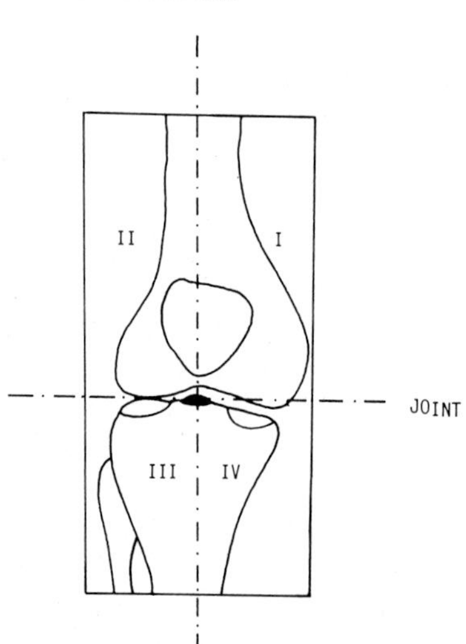

Fig. 3.20. Four quadrants of knee radiograph

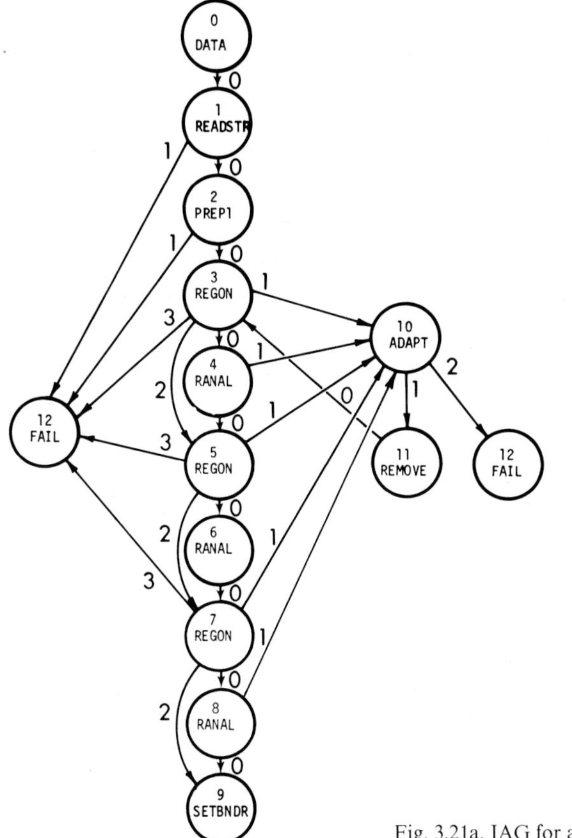

Fig. 3.21a. IAG for analyzing knee radiograph

line serve to partition the picture into four quadrants (Fig. 3.20). Each quadrant contains a section of the edge of either the femur or tibia. The quadrants are approximately symmetric, so an edge following technique can be used with only minor modifications in all four quadrants.

The soft tissue regions (Nodes 8 through 11) are all very similar and lie at the corners of region 1. These regions can be located by choosing either the top row (for Nodes 8 and 9) or bottom row (Nodes 10 and 11) and searching out from the center to the extreme points of region 3 (for Nodes 8 and 10) or region 4 (Nodes 9 and 11). The only problem that may arise is if one of the soft tissue regions is missing. A soft tissue region may be missing for two reasons. It may not show up on the original radiograph, or it may have been mistakenly included in the background. Neither of these cases is significant, but they must be allowed for. If a soft tissue region cannot be found after two tries, it is assumed missing.

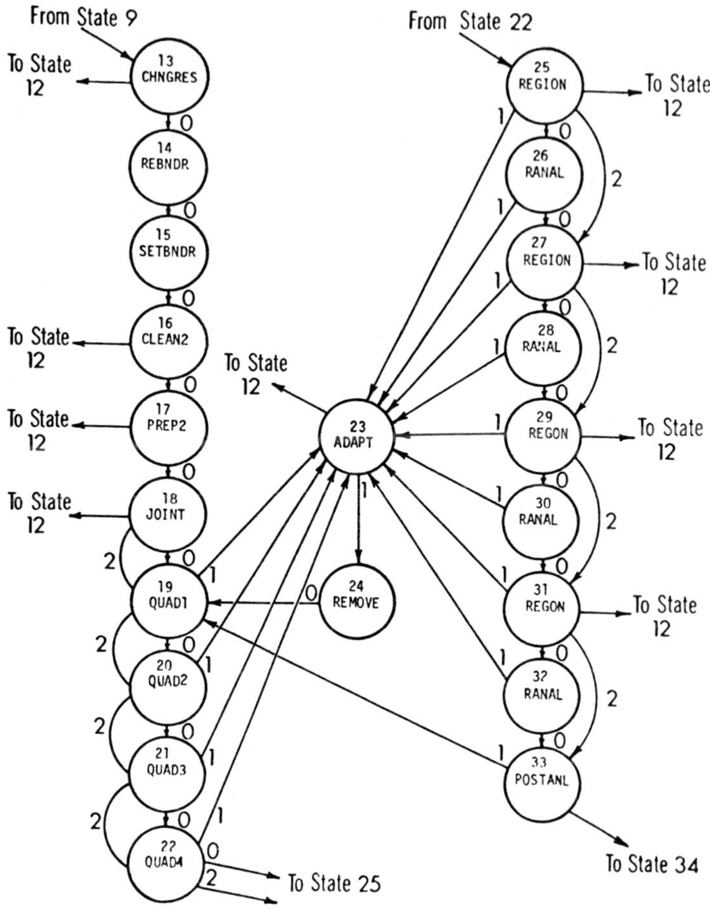

Fig. 3.21b. Part of graph control for knee radiograph

Level 3 is included to separate the tibula from the soft tissue region. For a right knee, Nodes 12–14 are descendants of Node 10. For a left knee, they are descendants of Node 11.

An IAG, shown in Fig. 3.21, was used to implement this strategy. Each circle represents a state which contains a relative state number and the name of the function associated with that state. Arcs leaving the states are numbered from zero, where zero is considered the success branch. The variables are not included on the arc labels; therefore, the figures primarily show the interactions of the modules. The meanings of the other arc numbers are relative to the states that they leave. Table 3.3 is a brief explanation of the action performed by each routine. Table 3.4 summarizes the possible arcs leaving each routine and their meaning.

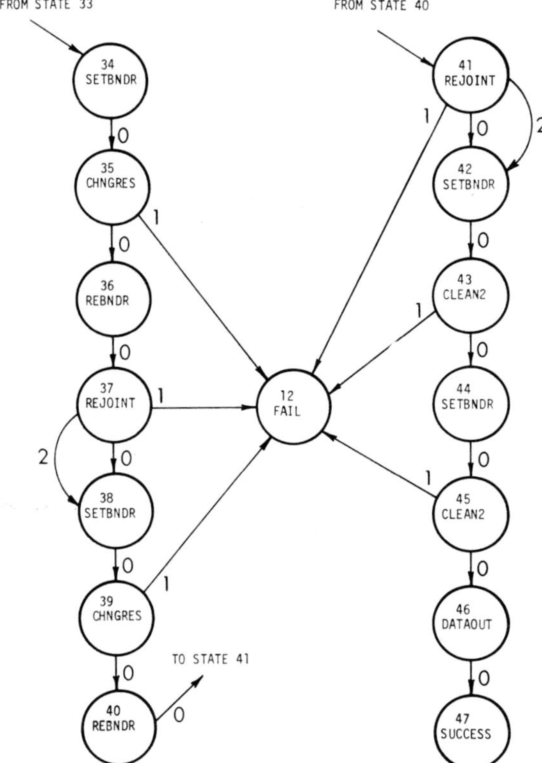

Fig. 3.21c. Part of graph control for knee radiograph

The graph contains information which may not be obvious. First, the tree description of the X-ray is contained implicitly in the graph. Second, the graph is designed so that features are searched for in the order in which they are numbered in the tree. Referring to the ADAPT routine, if exit zero is taken, control goes to the state corresponding to the lowest numbered feature in the tree whose parameters were adjusted by ADAPT. Third, there is a definite grouping of graph states 3 through 9 corresponding to level 1 of the tree, and of graph states 17 through 34 corresponding to level 2 of the tree. Figure 3.22 shows a processed image.

3.4.3 Lungs

Radiographic examination of the chest usually begins with the exposure of a standardized view, the PA erect 14×17 film; a lateral projection is commonly obtained at the same visit. After examination of this pre-

Table 3.3. Actions performed by basic functions

Action number	Name	Explanation
0	DATA	Reads graph structure into memory.
1	FAIL	Terminal action to indicate failure.
2	SUCCESS	Terminal action to indicate success.
3	READSTR	Reads structural description of picture into memory.
4	REGON	Enumerates a region in the picture array $P(x, y)$, flags it in the work array $W(x, y)$. Depending on a parameter, different region properties will be used.
5	NO-OP	No operation. Dummy action.
6	RANAL	Analyzes the region found by REGON, to see if its attributes match the expected attributes.
7	PREP1	Initialization routine for nodes on level 1 of the tree. This routine is picture class dependent.
8	ADAPT	Implement automatic parameter adjustment for primitive feature extraction routines. Parameters are adjusted by several methods, corresponding to the particular feature extraction primitive.
9	REMOVE	Modifies the structural description of the picture if a feature cannot be found.
10	SETBNDR	Allows the initial approximations to the features to grow together, under the constraint of the structure tests, to accurately set the boundaries.
11	REBNDR	Resets boundary points so that they will be reclassified in the presence of additional edge information which may be present at a higher resolution.
12	CHNGRES	Increases the resolution of the picture array $P(x, y)$ and the work array $W(x, y)$ to the next higher power of 2.
13	CLEAN2	Removes groups of isolated points which may have been classified incorrectly so they may be reclassified.
14	PREP2	Initialization routine for nodes on level 2 of the tree (see 8, PREP1)
15	JOINT	Working from the start point, traces the darkest line across the picture.
16	QUAD1	Locates the S.O.C. point and traces a contour for quadrant *I*.
17	QUAD2	See 16, QUAD1.
18	QUAD3	See 16, QUAD1.
19	QUAD4	See 16, QUAD1.
20	POSTANL	If a soft tissue region is missing, POSTANL resets the parameters for the corresponding contours, so that the contours may be retraced using the original parameter settings.

Table 3.3 (continued)

Action number	Name	Explanation
21	REJOINT	Removes and the retraces the joint. Takes advantage of additional edge information which may be present at a higher resolution.
22	DATAOUT	Extracts chain code representation for the boundaries of the femur, tibia and joint, and superimposes the boundary representation on the original data to produce a pictorial display of results.
23	SIGOUT	Calculates the horizontal projection of the picture $[H(x) = \Sigma P(x, y)]$ and writes it out for further analysis.

Table 3.4. Summary of actions

Action number	Function	
1	DATA	0 SUCCESS 1 E.O.F. on SYSIN
2	FAIL	No returns, exit
3	SUCCESS	No returns, exit
4	READSTR	0 SUCCESS 1 bad structure card
5	REGION	0 SUCCESS 1 S.O.R. point covered 2 STATUS, region already found 3 workstack overflow
6	NO-OP	0 SUCCESS
7	RANAL	0 region OK 1 mismatch or intersection
8	PREP1	0 SUCCESS 1 no picture
9	ADAPT	0 SUCCESS 1 Remove a feature 2 Processing cannot continue if feature to remove is removed.
10	REMOVE	0 SUCCESS
11	SETBNDR	0 SUCCESS
12	REBNDR	0 SUCCESS
13	CHNGRES	0 SUCCESS 1 no picture
14	CLEAN2	0 SUCCESS 1 workstack overflow
15	PREP2	0 SUCCESS 1 no picture

Table 3.4 (continued)

Action number	Function	
16	JOINT	0 SUCCESS 1 Start point covered 2 STATUS
17	QUAD1	0 SUCCESS 1 no S.O.C. point 2 STATUS
18	QUAD2	0 SUCCESS 1 no S.O.C. point 2 STATUS
19	QUAD3	0 SUCCESS 1 no S.O.C. point 2 STATUS
20	QUAD4	0 SUCCESS 1 no S.O.C. point 2 STATUS
21	POSTANL	0 SUCCESS 1 refind one or more contours
22	REJOINT	See JOINT
23	DATAOUT	0 SUCCESS
24	SIGOUT	0 SUCCESS

liminary film by a medical doctor specially trained in radiologic diagnosis, more views may be obtained. The initial view is taken from a standard projection to simplify the identification and anatomical location of any pathological (disease) process.

The X-ray tube to film distance is 180 cm. The peak voltage across the tube is usually around 140 kV. The patient is positioned with the front of the chest to the film, and the back to the X-ray tube. X-rays then pass from posterior to anterior through the patient. Exposure of the film by X-rays passing through the patient results in blackening of the film in proportion to exposure. The process is complicated by all the nonlinearities experienced in other photographic processes.

At the energy level commonly used, X-ray absorption is primarily by Compton scattering, the important factor being density of electrons in the absorbing medium. Materials of high atomic number are more efficient absorbers than those of low atomic number; liquids and solids more efficient than gases. This is depicted on the chest film by air-containing structures appearing black, more solid tissues appearing light. Bones which contain calcium are more radio-opaque than soft tissues.

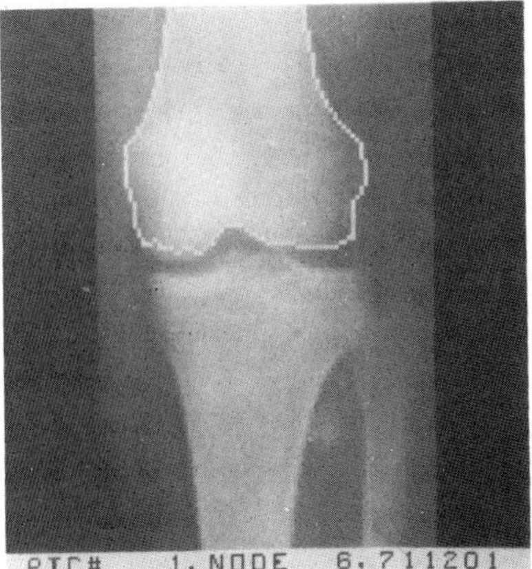

Fig. 3.22. Normal knee X-ray contour traces

Anatomical structures readily noted on the normal chest film are the ribs, the thoracic spine, the heart, and the diaphragm separating the chest cavity from the abdominal cavity.

While all structures in the film are examined for abnormality by the radiologist, primary emphasis is on the lungs and heart. Originating in the center of the film radiating laterally in a more or less fan-shaped manner are the shadows corresponding to the bronchi or large air passages and the pulmonary arteries which carry blood from the heart to the lungs.

There are a number of disease processes with diagnostic signs that can be identified on a PA chest radiograph. An important indicator of heart disease is the vascular pattern within the lung field. The numerous disease processes categorized by the term "interstitial fibrosis" cause an increase in the amount of tissue adjacent to the small bronchi and blood vessels. These changes result in a "reticular" network pattern in the lung field of a chest radiograph. If the alveoli, or air-containing structures where gas exchange takes place, become filled with fluid, a characteristic pattern of alveolar infiltrate is seen. A great number of unrelated disease processes produce interstitial patterns, alveolar pattern, and various mixtures of both. Some of these are the pneumoconioses including anthracosilicosis, asbestosis, and beryllosis. These diffuse abnormalities are to be contrasted with localized abnormalities such as small tumors, rib fractures, and localized pneumonia.

Several studies have recently been conducted on these problems. In one study, the area of consideration in the PA chest radiograph was restricted to a 76.8 [mm] square area in the region of the right apex, bounded by the projection of the thorax laterally, the clavicle superiorly,

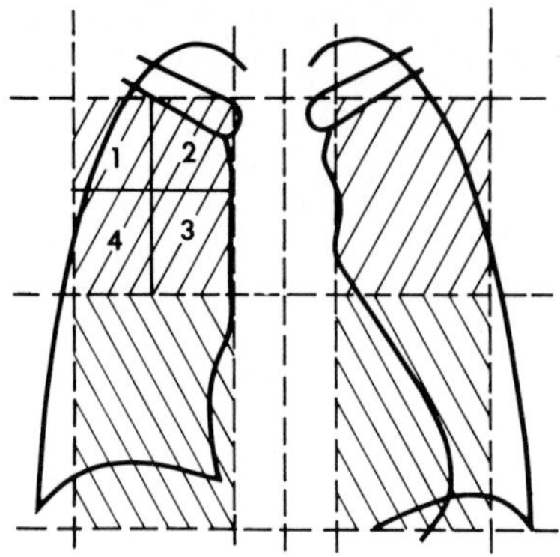

Fig. 3.23. Quadrants of the lung

and the superior mediastinum medially [3.49]. The readily identifiable structures in the normal film are: two or three ribs, a clavicle projecting more or less laterally, and the structures of the hilum at the lower right with normal vessels and bronchi projecting out in a fan-like fashion. A measure was constructed by subdividing the area of interest into quadrants, calculating an edge measure in each quadrant, then comparing the values of the measure in adjacent quadrants. The edge measure calculated in each quadrant was the total "edge" in the x direction, or

$$\begin{aligned} &E_k = \sum_i \sum_j |\mathrm{Pic}(I, J) - \mathrm{PIC}(I+1, J)| \\ &\text{Region } k; \quad k = 1, 2, 3, 4, \end{aligned} \tag{3.23}$$

where k is the quadrant number. Figure 3.23 shows the quadrants.

A consistent finding is that in normal cases more x-direction edge points are present in the area corresponding to the anatomical structure, the hilum (quadrant 3), than in the area corresponding to peripheral lung (quadrant 4). The ratio $M = E_3/E_4$ expresses a relationship between the edge content of the regions. This ratio was found in the test series to be almost always greater than 1 for normal cases and less than 1 for abnormal cases. The results are given in Table 3.5 for a training set of 9 normal and 15 abnormal cases. The ratio was computed at three different resolutions. Thus, $M_{256 \times 256}$ corresponds to the computation for the full 256×256 picture; $M_{128 \times 128}$ corresponds to the computation on a 128×128 array produced by averaging each four points in the original array, and $M_{64 \times 64}$ corresponds to a similar computation on the image, again reduced. The ratio was also computed after linear redistributing was applied to the input picture. These results are given in the $M'_{256\times256}$ row. A simple two class classifier was used, based on the single ratio measurement.

Table 3.5. Summary of classification results using ratio of directional edge measurement described in test at various resolutions

Measurement	Classification percent correct	
	Training	Testing
$M_{256\times256}$	92	75
$M_{128\times128}$	92	85
$M_{64\times64}$	71	50
$M'_{256\times256}$	92	75

Other studies have been directed specifically toward the pneumoconioses or black lung disease prevalent in coal miners [3.50]. These studies used essentially the same approach for measuring texture described in the section on texture. A square area in the upper lung field was scanned. The inter-rib spaces were considered as separate zones for purposes of classification. Linear redistribution was utilized on the input picture data before the texture features were extracted. The testing results from a library of 100 films using a 50% jackknife test procedure were

		Actual class	
		1	2
Assigned class	1	90.0	5.2
	2	10.0	94.8

When 846 observations by radiologists on 141 films were taken the results were

		Actual class	
		1	2
Assigned class	1	82.5	2.6
	2	17.5	97.4

The studies on detecting texture patterns in lungs are encouraging, but the subject is still in a state of development. More work needs to be done on developing measurements and making certain that variations in film technique, with no relation to disease processes, have no effect on the measurements.

3.4.4 Heart Disease

Heart disease is very common, which makes diagnosis of this disease very important. Because it is known that quantitative measures such as heart size and shape are important diagnostic signs for heart disease, computer analysis offers hope of improved diagnosis. The fact that the ordinary PA chest radiograph is the primary diagnostic medium justifies attempts to automate the diagnosis using this medium. Before discussing the computer analysis of the films, some facts about the operation of the heart will be explained.

In the operation of the heart, blood is forced through the heart by concentration of muscles in the heart wall. The direction of blood flow is governed by two atrioventricular and two semilunar valves. One of the atrioventricular valves is the tricuspid valve located between the right atrium and ventricle. The other is the mitral valve located between the left atrium and ventricle. The semilunar valves are termed aortic and pulmonary; they are located at the junctures of the left ventricle—ascending aorta, and right ventricle—pulmonary artery, respectively. Properly functioning valves serve mainly to prevent the backflow of blood into their respective heart chambers during diastole or chamber filling.

Several definitions of frequently used medical terms will also be useful.

Stenosis is a narrowing or stricture of a duct, canal, or valve. If the valve leaflets become scarred and adhere to each other or fail to open completely the valve opening will be smaller than normal, resulting in valvular stenosis.

Insufficiency refers to the condition of being insufficient or inadequate to perform an allotted function. Valvular insufficiency is a condition in which the valves do not close perfectly so that the blood passes back through the orifices during each diastolic phase.

Atresia represents the absence or closure of a normal body orifice or passage. For example, tricuspid atresia is the absence of the tricuspid valve.

Hypertrophy of the heart muscle is the natural long-term method of increasing the strength of cardiac contraction. The heart fibers increase in length and bulk resulting in an increase in chamber wall thickness. Cardiomegaly is a term for general enlargement of the heart and often includes hypertrophy.

Cyanosis is a bluish discoloration, usually applied to such discoloration of the skin or mucous membranes, caused by excessive concentration of reduced hemoglobin in the blood. Cyanosis results from a right-to-left shunt where unoxygenated blood from the right side of the heart is shunted to the left side and pumped out into the systemic circulation.

3.4.5 Rheumatic Heart Disease

Rheumatic heart disease affects the mitral, aortic, and tricuspid valves in descending frequency of occurrence. Inflammation caused by the primary disease may cause scarring of the valve leaflets. The result is either stenosis, a reduction in the valvular area which produces obstruction to blood flow across the involved valve; or insufficiency, continual blood flow across the involved valve throughout the cardiac cycle. Stenosis or insufficiency may occur with any valve or valves, and any valve may be involved simultaneously with both stenosis and

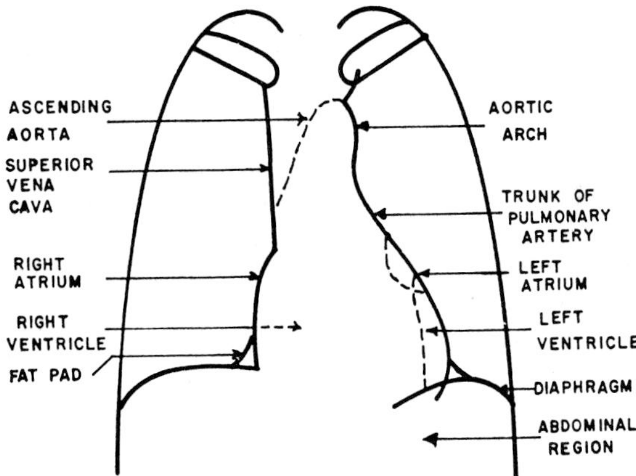

Fig. 3.24. Heart anatomy

insufficiency. The involved valve or valves produce anatomical changes in the heart and great vessels which often can be detected in chest radiographs. Figure 3.24 shows some of the features of interest.

In most cases of mitral stenosis (MS), the radiograph changes are characterized by a filling of the region immediately above the heart and lateral to the pulmonary artery and left hilar region. A second characteristic of mitral stenosis is a reduction or disappearance of the aortic knob. A third characteristic associated with this lesion is a slight to severe protrusion of the left atrial appendage and a steep descent of the left ventricular arch to the diaphragm.

Pure mitral insufficiency (MI) is uncommon; it is usually combined with mitral stenosis. Mitral insufficiency is characterized by enlargement of the cardiac projection with elongation and greater rounding of the left ventricular arch. The left ventricle enlarges both in systole and diastole.

Aortic stenosis (AS) is characterized by elongation, slight to marked outward bulging, and rounded angulation of the right border of the vascular band above the heart. This bulging is due to elongation or dilation of the ascending aorta. In some cases there also may be increased prominence of the aortic knob.

Aortic insufficiency (AI) produces an overall cardiac enlargement that is often greater than the enlargement caused by mitral stenosis. This enlargement, principally left ventricular in nature, is caused by blood regurgitation in diastole through the incompetent semilunar valve, producing a retrograde filling dilation. While aortic stenosis and insufficiency commonly occur together, the dependence is not as strong as in the case of mitral stenosis.

Tricuspid stenosis and insufficiency are not considered as separate classes because they are usually accompanied by aortic or mitral involvement.

A goal for one study on rheumatic heart disease was to develop a computer program for diagnosing heart disease from the standard PA chest X-ray and to compare the performance of the program with that of radiologists [3.51]. A library of 281 cases was created for the study.

The diagnoses were divided into five classes, and the 16 possible combinations of aortic and mitral valve disease were divided into four separate groups. The classes considered for the study were:

Class 1. Normal.
Class 2. Mitral stenosis only.
Class 3. Other mitral valve lesions—MSMI; MI only.
Class 4. Aortic and mitral involvement.
Class 5. Aortic involvement—ASAI, AS only, AI only.

All patients in the abnormal group had right and left heart catheterization and appropriate cineangiography to establish the correct diagnosis. Eighty-eight normal chest examinations were also selected from patients with no history of cardiac disease, no murmurs, and no cardiac symptoms. The normals were generally in the age group 20–45. To test further the normal-abnormal classification, 38 adult patients with other types of cardiac disease were included. These were proven examples of coarctation, ventricular septal defect, atrial septal defect, and pericardial effusions. The radiologist test set library included normals and all categories of rheumatic heart disease and was used to compare the diagnostic accuracy of the computer and the radiologist.

The first step in the computer processing was to extract size, contour, and shape of the heart from the standard PA film. This was accomplished

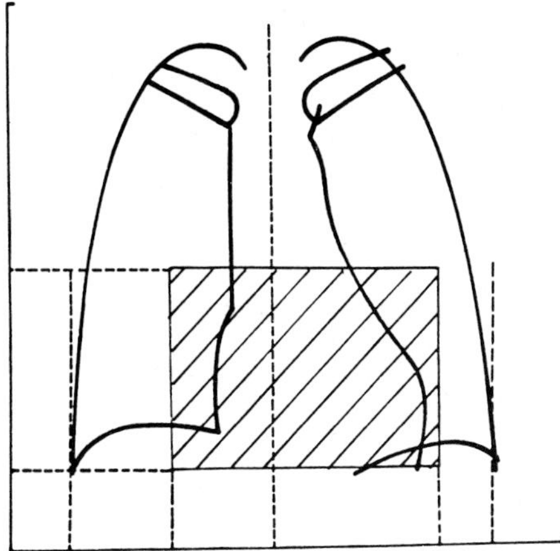

Fig. 3.25. Cardiac rectangle

by an algorithm that constructs a cardiac rectangle around the heart. The cardiac rectangle varies in size depending on the heart size (Fig. 3.25).

The extent of the rectangle was determined from features present in the horizontal and vertical projections of the original image. Once the cardiac outline is obtained, it is straightforward to obtain measurements characterizing the shape of the heart. The measurements used in this study are similar to those described in medical practice. They are shown in Fig. 3.26.

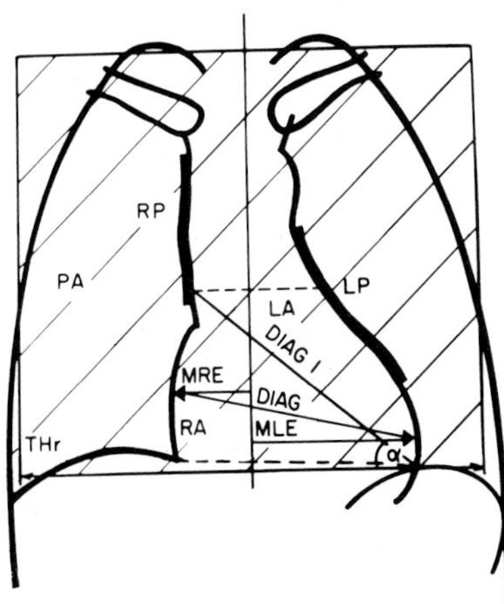

Fig. 3.26. Measurements and two polynomials

Table 3.6. Computer testing results

		θ_1	θ_2	θ_3	θ_4	θ_5	Total cases
Cardiac catheter results	θ'_1	78	1	0	5	4	88
	θ'_2	3	18	1	9	1	32
	θ'_3	0	0	5	14	2	21
	θ'_4	3	6	2	76	4	91
	θ'_5	6	0	1	13	29	49

Table 3.7. Differential diagnostic testing rates

Class θ_i	Testing diagnostic rates [%]
1	89
2	56
3	24
4	83
5	59

All cardiac measurements were normalized so that a ratio figure was obtained. The linear measurements were divided by THr, the thoracic width, and the area measurements were divided by TA, the thoracic area. This allows correction for variation in heart sizes related to the patient's overall size.

Once these heart measurements were extracted from the cardiac rectangle, this information was used as the basis for classifying the case as normal or abnormal. If a particular case was abnormal, the classification went further by placing the case into the correct group of rheumatic heart diseases. Computer classification was accomplished through the use of linear and quadratic discriminant functions.

In order to estimate the unbiased capabilities of the computer diagnostic classification technique for the evaluation of new samples, the 10% jackknifing test procedure was used. The testing results are shown in Tables 3.6 and 3.7. The overall correct classification rate was 73%.

In order to evaluate the system the following study was made. Ten radiologists were asked to individually diagnose 135 representative cases of the 281 rheumatic and normal cases evaluated by the computer. This group of radiologists consisted of seven board certified radiologists and three third-year residents whose training was nearly completed.

Table 3.8. Combined physician results

		θ_1	θ_2	θ_3	θ_4	θ_5	Total observations
Cardiac catheter results	θ'_1	183	3	2	6	27	221
	θ'_2	7	77	41	41	9	155
	θ'_3	4	20	51	13	7	95
	θ'_4	1	20	40	28	6	95
	θ'_5	15	0	0	2	56	73

Table 3.9. Differential diagnostic physician rates

Class θ_i	Physician diagnostic rates [%]
1	83
2	50
3	54
4	29
5	76

Each radiologist was given the PA and lateral views and told that each case was either normal or rheumatic heart disease. He then made a complete radiological diagnosis and recorded his answers on a form designed for the study. Each case diagnosed was counted as one physician observation. Not all of the physicians completed the task of reading all 135 films. The combined results of the physicians' observations are shown in Tables 3.8 and 3.9. Subdividing the results of Table 3.9 into classes, as in the computer study, yields the correct physician diagnosis with respect to each class. The overall correct physician differential diagnostic rate was 62%. These results indicate it is possible in some contexts to perform comparably to radiologists. The results can only be considered tentative and do not demonstrate a clinically applicable program.

A significant problem noticed with this approach was that the segmentation program did not perform well with children. This problem could be somewhat circumvented in the case of rheumatic disease but not in the study of congenital heart disease which must be detected in children.

3.4.6 Congenital Heart Disease

There have been several efforts at utilizing image analysis techniques to detect congenital heart disease from chest radiographs. First, some details of congenital heart disease will be described.

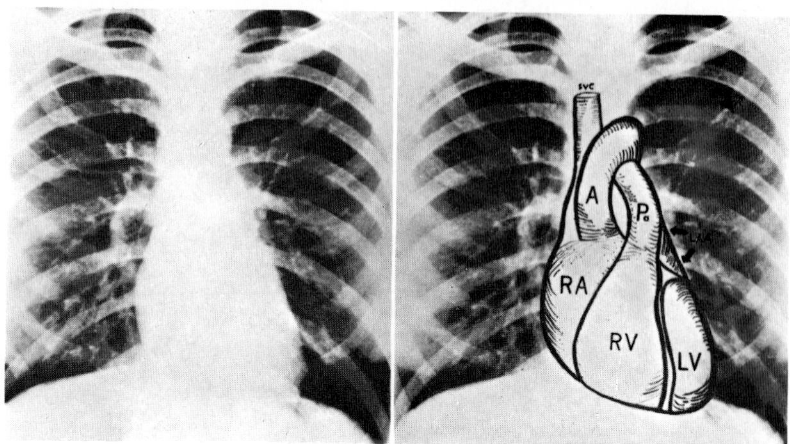

Fig. 3.27. Normal heart, posteroanterior radiograph (Reproduced from L. E. SWISCHUK: *Plain Film Interpretation in Congenital Heart Disease*, Fig. 4. With kind permission from Lea & Febiger, Philadelphia, Pennsylvania, USA)

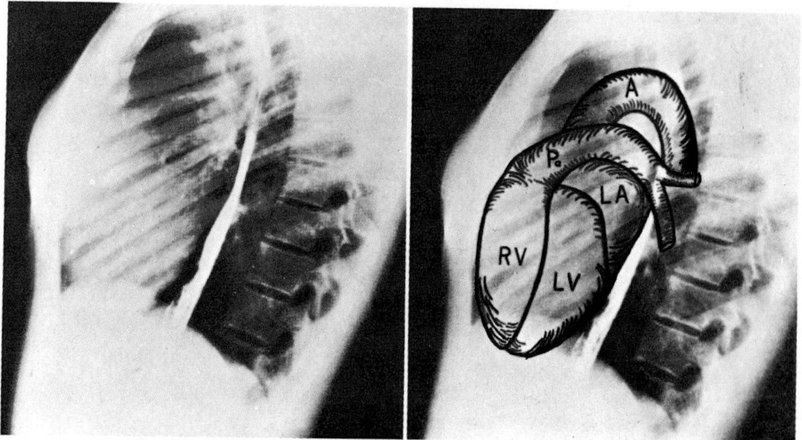

Fig. 3.28. Normal heart, lateral radiograph (Reproduced from L. E. SWISCHUK: *Plain Film Interpretation in Congenital Heart Disease*, Fig. 5. With kind permission from Lea & Febiger, Philadelphia, Pennsylvania, USA)

Congenital heart disease is a result of improper embryologic development of the fetus and is associated with a broad range of structural anomalies of the heart and great vessels. Its incidence is approximately 8 per 1000 total births which includes both live births and stillbirths.

As it is not possible to see all chambers of the heart on any one projection, multiple radiographic views are employed. To make a

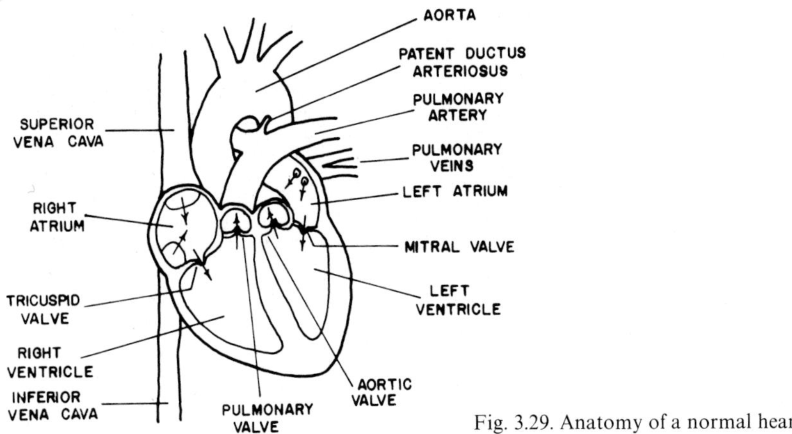

Fig. 3.29. Anatomy of a normal heart

diagnosis of congenital heart disease, the PA view is most important because it contains the most diagnostic information. The lateral view is a complementary view which shows both the right ventricle and the left ventricle. Figure 3.27 is a PA view of a normal heart showing how the cardiac silhouette can be interpreted. Figure 3.28 is a similar example of a normal heart in the lateral view.

Radiologists and cardiologists also use a right anterior oblique radiograph and a left anterior oblique radiograph. These views are part of a cardiac series but are not routinely taken.

Figure 3.29 illustrates the anatomy of a normal heart. The pulmonary vascular pattern is important in the description of each class of congenital heart disease because it is one of the most important assessments a radiologist makes when diagnosing congenital heart disease. Pulmonary vascular patterns are subdivided into increased, decreased, and normal patterns. Increased pulmonary vascularity results from either active or passive congestion. Active congestion is caused by increased blood flow through the pulmonary vessels in left to right shunts. Passive congestion is the result of elevated pulmonary venous pressure. Decreased pulmonary vascularity indicates diminished pulmonary blood flow through the lungs and occurs in right to left shunts and with lesions which obstruct the flow of blood through or out of the right side of the heart. Normal individuals have normal pulmonary vascularity. Normal vascularity also may occur in cases with only left-sided valvular or vascular lesions.

The Congenital Heart Library was based on four major categories of disease pertinent to our study of congenital heart disease. These categories, based on pulmonary vascularity, are 1) congenital heart diseases with increased pulmonary vascularity without cyanosis, 2) those with increased pulmonary vascularity with cyanosis, 3) those with decreased pulmonary vascularity, and 4) those with normal vascularity. Consult the section on libraries for the construction of the Congenital Heart Library.

Small Computer Analysis of Congenital Heart Disease

The goal of this study was to implement a system for diagnosing congenital heart disease on a small computer [3.32]. It was hoped this would yield some insight into how a clinical system could be configured as well as providing insight into the algorithms required.

One preprocessing step is used in the system. Linear redistribution is utilized to standardize the expected contrast values of the films. This tends to compensate for the wide ranges of contrast in chest radiographs due to variations in patient size and position, X-ray exposure techniques, and development times.

The heart contour is extracted in the following manner. First, the vertical midline of the heart is determined from the vertical projections. Having chosen the midline, the algorithm proceeds to trace the left edge of the heart shadow. This edge is determined by a constrained maximum gradient method. Each horizontal line is scanned point by point from the midline to the right edge of the picture. The difference in intensity level between each two successive points is recorded. The location where this difference is maximum is chosen as an edge candidate. If the edge candidate chosen for a line is more than two columns away from the edge candidate of the previous line, the new edge point is adjusted so that the horizontal distance between these two points is equal to two columns. This constraint is added to prevent the algorithm from tracing such light areas as ribs and excessive vascularity.

Determining the right edge of the heart shadow is more difficult because of the overlap of the heart border and the right main pulmonary vessels. Any edge detection based only on gradients would fail because the heart shadow is not very sharp in this area. A thresholding technique is used instead. In computing the threshold, the algorithm examined only the three most significant bits of each intensity value to minimize any differences remaining after linear distribution.

Choosing a threshold at the third level on the three bit scale produced an approximate right edge which often included the vascularity. Since the amount of vascularity varies from picture to picture and the border between the vascularity and the true edge of the heart is often obscured, it was decided to use a second threshold value to produce a second right edge trace further inward. Like the first, this second threshold value is the same for all pictures. The two traces are used to determine

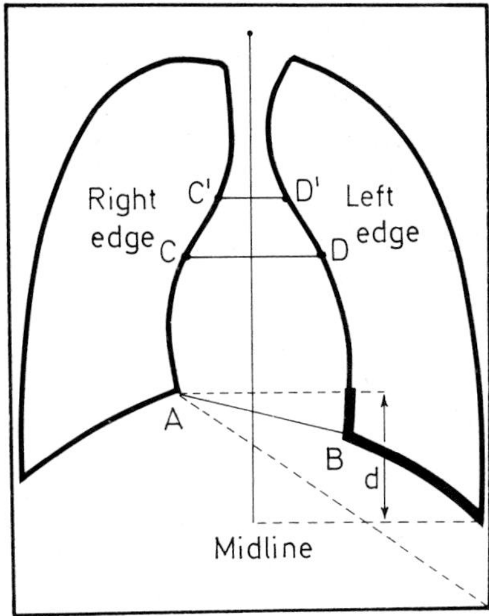

Fig. 3.30. Features of the chest

an approximate right edge using a gradient function between these two lines.

After determining the left and right edges, the next step is to determine the lower-right corner of the heart (point A in Fig. 3.30). As can be seen from Fig. 3.30, this is the point on the right edge which has the shortest distance to the lower-right corner of the picture (point T).

The next step is to find the lower-left corner of the heart. This is found with respect to point A. We know point B cannot be higher than point A. Therefore, a limited search is initiated on the left edge for a vertical distance of d rows. In this search the point that has the minimum distance to the midline and largest y coordinate is chosen as the lower left corner of the heart.

Determining the top of the heart is difficult. Since there is some disagreement as to the location of the top of the heart, we chose an arbitrary criterion suggested by radiologists. This criterion says that the distance between A and C should be equal to the distance between C and D. It was used successfully to produce reasonable outlines. Since the aortic knob has diagnostic significance, it was also included in the outline. This was accomplished by extending the top of the heart by a predetermined amount. (See line C′–D′, Fig. 3.30.)

Figure 3.31 shows chest radiographs after processing which exhibit typical outlines, as well as showing some of the difficulties encountered in certain circumstances.

Once the heart outline has been determined, the next task is to extract a finite number of measurements of the cardiac outline which will

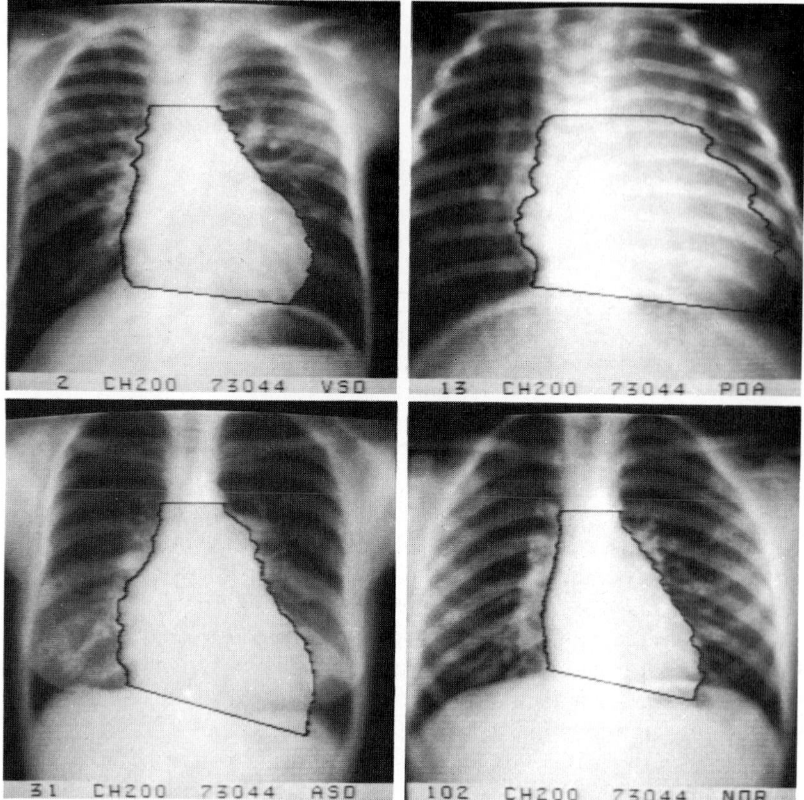

Fig. 3.31. Processed chest radiographs

describe it well and can be used to classify the shape as belonging to one class. For this study the measurements used are Fourier descriptors and the area of the heart in the PA view chest radiograph, normalized by dividing by the total estimated area of the thoracic cavity. Fourier descriptors were chosen because the computational technique is well defined, easily implemented and powerful. It was decided to use the total normalized cardiac area (TCA) because the relative size of the heart is important in diagnosis, and the Fourier descriptors do not contain any information about the size of the object.

The most useful feature of the Fourier descriptors is the invariance of the corresponding amplitudes under translations, rotations, changes in size, reflections, and shifts in the starting point. In other words, if two shapes differ only by one or more of these factors, corresponding amplitudes will be the same. If necessary, reflections and shifts in the starting

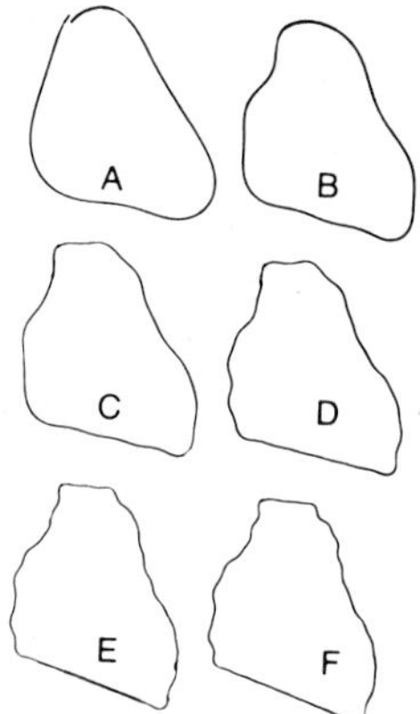

Fig. 3.32. Outline of image reconstructed from: (A) 5 pairs; (B) 10 pairs; (C) 15 pairs; (D) 20 pairs; (E) 25 pairs; (F) 30 pairs of Fourier descriptors

point can be detected by appropriate examination of the phase angles; however, for the present work this is immaterial. The Fourier descriptors also provide for adjustment of the level of shape detail desired by merely adjusting the number of descriptors extracted, as ahown in Fig. 3.32.

Thirty Fourier descriptors and the TCA were extracted for 209 normal and 272 abnormal cases. Measurement selection using forward sequential search was utilized to compress the number of measurement descriptors for each case to 21. The data base for the test data set was then established from these 21 descriptors for each picture.

This program was implemented in real-time on a DEC PDP-11/20. Each radiograph was scanned twice; the first time to compute the linear distribution and the second time to locate the heart boundary, line by line. An optional third scan may be performed for display on the high resolution television monitor.

Approximate processing times are:

1) Compute linear distribution −45 s.
2) Determine heart outline, compute TCA −35 s.
3) Display (optional) −25 s.
4) Compute chain code and F.D.'s −10 s.
5) Perform classification −4 s.

Overall correct classification using 21 descriptors was 87.0% for the training set and 77.0% for 10% jackknife test.

Diagnosis of Congenital Heart Disease from Boundary Data

In the previous study the computer algorithm had some difficulty in tracing correct outlines for very small children. In an incorrect diagnosis there are two sources of error. One is that the measurements are incorrectly taken, and the other is that the classifier may classify the case

Table 3.10. Radiologists' diagnostic accuracy using only the PA and lateral chest radiographs on CDS1

		Radiologists' diagnoses							
		N	ASD	VSD	PS	PDA	ABN	Total	Percentage correct
Verified diagnoses	N	148	35	24	6	19	23	255	58.0
	ASD	16	95	58	10	28	15	222	42.8
	VSD	43	50	63	7	23	33	219	28.8
	PS	29	14	10	23	11	27	114	20.2
	PDA	13	12	32	2	17	29	105	16.2
	ABN	19	37	39	2	17	132	246	53.7
	Total	268	243	226	50	115	259	1161	
Overall percentage correct diagnoses									41.2

Table 3.11. Radiologists' diagnostic accuracy using only the PA and lateral chest radiographs on CDS1

		Radiologists' diagnoses			
		N	ABN	Total	Percentage correct
Verified diagnoses	N	148	107	255	58.0
	ABN	120	786	906	86.8
	Total	268	893	1161	
Overall percentage correct diagnoses					80.4

incorrectly. This section describes a study [3.5] that attempted to remove errors of the first type by having the heart boundaries hand-drawn.

A library was composed for the experiment (see Sect. 3.1). In order to evaluate the computer classification results, the library was analyzed by three radiologists. The results of the radiologists' diagnoses are given in Tables 3.10 and 3.11. In these tables and subsequent ones, a large class called ABN has been created which consists of all the cases listed below item E(PDA) in Table 3.1. The other categories are the same.

The next step in the study was to determine the measurements to be extracted from the PA and lateral radiographs for purposes of computer analysis. The criterion for selecting measurements was that they characterize the disease patterns, correspond to measurements with which physicians are familiar, and can be reliably determined from the radiograph. Figure 3.33 and Table 3.12 describe the measurements for the PA view. Figure 3.34 shows the measurements taken from the lateral view.

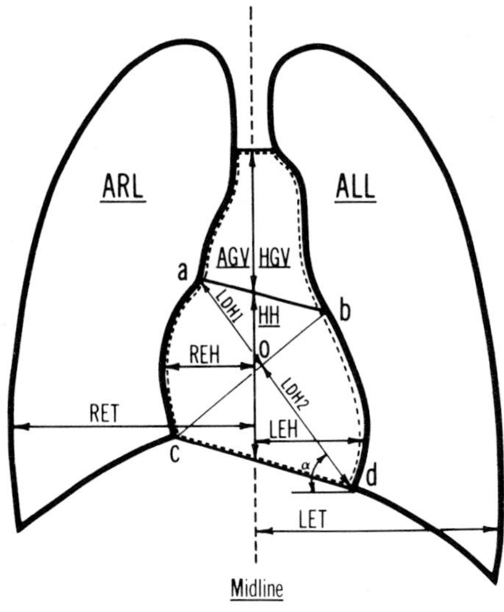

Fig. 3.33. Measurements extracted from the outline of the heart, great vessel, lungs, and thoracic cavity taken from the PA chest radiograph

Table 3.12. Measurements extracted from PA view (see Fig. 3.33)

Definitions of Points a, b, c, d, and o:	
a	The point where the right heart border is lost in either the ascending aorta or the spine.
b	Intersection of the border of the left ventricle with the pulmonary outflow tract or, if this point is not visible, the point where the inferior edge of the left bronchus intersects the border of the left ventricle.
c	Point where the right heart silhouette intersects the right hemidiaphragm.
d	Point where the left heart silhouette intersects the left hemidiaphragm.
o	Point at which lines *ad* and *cb* intersect.
REH	Maximum distance from the midline to the right heart border.
LEH	Maximum distance from the midline to the left heart border.
RET	Maximum distance from the midline to the right edge of the thoracic cavity.
LET	Maximum distance from the midline to the left edge of the thoracic cavity.
LDH1	Length of line *ao*
LDH2	Length of line *od*.
α	The angle between line *ad* and a line perpendicular to the midline.
HH	Height of the heart.
HGV	Height of the great vessels.
AH	Area of the heart.
AGV	Area of the great vessels.
ARL	Area of the right lung.
ALL	Area of the left lung.
FD'S	First 20 amplitude coefficients of the Fourier descriptors which represent a closed curve around the heart and great vessels.

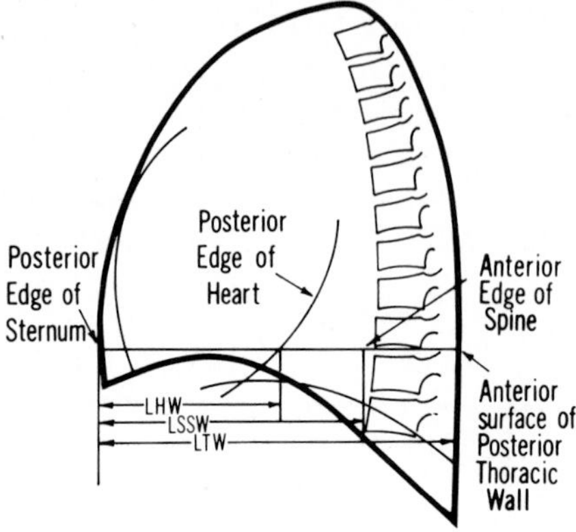

Fig. 3.34. Measurements extracted from the lateral chest radiograph

Table 3.13. Features calculated from measurements extracted from both the PA and lateral chest radiographs

A1	CTR	$= \dfrac{\text{REH}+\text{LEH}}{\text{RET}+\text{LET}}$ cardiac thoracic ratio
A2	RLH	$= \dfrac{\text{REH}}{\text{LEH}}$ right to left heart ratio
A3	AVR	$= \dfrac{\text{LDH1}}{\text{LDH2}}$ atrio-ventricular ratio
A4	HTAR	$= \dfrac{\text{AH}}{\text{ARL}+\text{ALL}+\text{AH}+\text{AGV}}$ heart thoracic cavity area ratio
A5	GHAR	$= \dfrac{\text{AGV}}{\text{AH}}$ great vessel heart area ratio
A6	GHHR	$= \dfrac{\text{HGV}}{\text{HH}}$ great vessel heart height ratio
A7	ALPHA	$= \alpha =$ angle between line *ad* and a line perpendicular to the midline
A8 to A27	FD'S	= First 20 amplitude coefficients of the Fourier descriptors representing a closed curve around the heart and great vessels
A28	LTCTR1	$= \dfrac{\text{LHW}}{\text{LSSW}}$ cardiac thoracic ratio 1 for a lateral radiograph
A29	LTCTR2	$= \dfrac{\text{LHW}}{\text{LTW}}$ cardiac thoracic ratio 2 for a lateral radiograph

The measurements taken from the radiographs determine the features to be used by the pattern classifier. These features were chosen in consultation with the medical literature and with radiologists to correspond to physiological models of the heart and to describe mathematically the shape of the heart. One of the goals of the study was to determine how significant these measurements were. Table 3.13 shows the features.

The six ratios of lengths, one angle measurement, 20 harmonic amplitude measurements from the PA view, and the two ratio features from the lateral view, give a total of 29 features to be evaluated. All length and area ratio features are size normalized to overcome the problem of absolute size discrepancies between cases. This is necessary because the patient cases in the Congenital Heart Library vary from infants to adults.

The measurements for the features were extracted from outlines of the cardiac, greater vessel, and thoracic regions. These acetate outlines were manually traced from the radiographs and digitized at a 256×256 resolution. The needed measurements were then extracted by a computer algorithm.

It is not an altogether simple problem to obtain the measurements from the PA and lateral chest radiographs consistently. In order to be consistent rules were formulated for obtaining these measurements in consultation with radiologists. Table 3.14 shows the rules for the PA view, and Table 3.15 the rules for the lateral view. It was intended that these rules would cover all cases encountered. It was also hoped that this type of protocol analysis for obtaining measurements would aid in the formulation of computer algorithms for the same purpose subsequent to this study.

Using these rules the measurements and features to be input to pattern classification programs were determined. The next part of the study was to determine the significance of these features for classification of congenital heart disease. The decision rule used was the Bayes rule assuming the class conditional density functions are Gaussian. When the class covariance matrices are assumed to be identical, the decision rules are linear; otherwise a quadratic decision rule is obtained. Both of these assumptions were tested using the 10% jackknife procedure. The feature selection was based on minimum probability of error. Forward linear sequential search was the search strategy used for finding the best features.

The first experiment was normal/abnormal classification. The format first studied the ratio measurements, next the Fourier descriptors, then the lateral, and finally combined all the results. The measurements selected as best are listed above the tables. Table 3.16 gives the results for the ratio measurements. Table 3.17 enumerates the testing results for the Fourier descriptors. Table 3.18 gives the results for the lateral features. Table 3.19 lists the results when all the features are combined. Notice that the ratio features are the most significant.

The next experiment performed was one of subclassification. Here the best features and the decision rule, linear or quadratic, were chosen between each two classes. The same set of features was used as in the

Table 3.14. Rules for extracting boundaries in PA view

1. Define midline as the center of the spine.
2. When possible, outline the heart and aorta separately from the thymus. When the heart and aorta cannot be differentiated from the thymus, outline the whole structure.
3. When the aortic knob is visible, locate the top of the aorta by drawing a horizontal line across the spine at the height of the top of the aortic knob. If no aortic knob is visible, locate the top of the aorta at a point one half the distance between the top of the heart at the midline and the top of the lungs.
4. To outline the right border of the heart and aorta, begin at the right diaphragm intercept and follow the cardiac outline upward until the contour is lost in the spine. Follow, the spine upward in a straight line to the level of the top of the aorta. When the aorta projects over the spine onto the lung field, follow the aortic edge to the level of the top of the aorta.
5. Locate the left side of the heart and great vessels by following the silhouette from the top of the aorta to the diaphragm intercept.
6. Omit fat pads when outlining the heart.
7. Locate the left side of the top of the heart by first looking for the pulmonary artery. If the pulmonary artery is visible, use the intercept of the pulmonary artery and left heart border as the top of the heart. If the pulmonary artery is not visible, use the intercept of the left silhouette and the inferior edge of the left bronchus.
8. Draw the top of the heart perpendicular to the midline unless an indentation in the right heart shadow is visible. Always draw the top of the heart so that it is horizontal or slopes down toward the left side of the heart except in one special case. If a patient has a rightsided aorta, the top of the heart may slope down toward the right side of the heart.
9. To locate the bottom of the heart, draw a straight line from the right diaphragm intercept of the cardiac silhouette to the left diaphragm intercept.
10. To outline the lungs, draw each lung separately. Follow the inside of the rib cage. Locate the apex of the lung at the lower surface of the first rib. Follow fluid lines when present.
11. The stomach bubble is located under the left hemidiaphragm unless the radiograph is visibly marked by a marker which denotes the left side.

Table 3.15. Rules for extracting measurements in lateral view

1. Orient the lateral radiograph so the patient is vertical.
2. Locate the top of the right leaf of the diaphragm.
3. Measure all distances along a line horizontal to the top of the right leaf of the diaphragm.
4. Measure the distance from the posterior edge of the sternum to the posterior edge of the heart.
5. Measure the distance from the posterior edge of the sternum to the anterior edge of the spine.
6. Measure the distance from the posterior edge of the sternum to the anterior surface of the posterior thoracic wall.

Table 3.16. Testing results using ratio features
Decision rule: Bayes. Measurements used: A_1, A_3, A_5, A_7
Data set: Congenital heart hand drawn outline—CDS1

		Computer diagnoses			
		N	ABN	Total	Percentage correct
Verified diagnoses	N	59	26	85	69.4
	ABN	37	265	302	87.7
	Total	96	387		
Overall percentage correct diagnoses					83.7

Table 3.17. Testing results using the set of Fourier descriptors
Decision rule: Bayes. Measurements: A_9, A_{11}, A_{16}, A_{20}
Data set: Congenital Heart hand drawn outline—CDS1

		Computer diagnoses			
		N	ABN	Total	Percentage correct
Verified diagnoses	N	21	64	85	24.7
	ABN	21	281	302	93.0
	Total	42	345	387	
Overall percentage correct diagnoses					78.0

Table 3.18. Testing results using the set of lateral features A.28, A.29
Decision rule: Bayes
Data set: CDS1—Hand drawn outline

		Computer diagnoses			
		N	ABN	Total	Percentage correct
Verified diagnoses	N	51	34	85	60.0
	ABN	99	203	302	67.2
	Total	150	237	387	
Overall percentage correct diagnoses					65.6

Table 3.19. Testing results using the set of combination features
Decision rule: Bayes. Measurements used: A_1, A_5, A_7, A_{16}
Data set: Congenital heart hand drwan outline—CDS1

		Computer diagnoses			
		N	ABN	Total	Percentage correct
Verified diagnoses	N	46	39	85	54.1
	ABN	29	273	302	90.4
	Total	75	312	387	
Overall percentage correct diagnoses					82.4

Table 3.20. Feature selection on ratio features

Class pair	Recognizer	Features
1–2	Linear	1, 7, 2
1–3	Linear	1, 5, 7, 4, 3, 6
1–4	Linear	1, 6, 4
1–5	Linear	1, 7, 5, 2
1–6	Linear	1, 7, 2
2–3	Linear	7, 4, 1, 2
2–4	Linear	2, 1, 7, 3
2–5	Linear	4, 2, 1, 3, 5
2–6	Quadratic	4, 2
3–4	Linear	1, 4
3–5	Linear	4, 2, 7
3–6	Linear	4, 3, 2
4–5	Linear	1, 3
4–6	Quadratic	1, 3, 2
5–6	Quadratic	2, 3

Table 3.21. Testing results using ratio features
Decision rule: Bayes
Data set: Congenital heart hand drawn outline—CDS1

		Computer diagnoses							
		N	ASD	VSD	PS	PDA	ABN	Total	Percentage correct
Verified diagnoses	N	62	3	1	6	1	12	85	72.9
	ASD	9	45	1	0	0	19	74	60.8
	VSD	16	20	7	1	2	27	73	9.6
	PS	16	4	5	2	1	10	38	5.3
	PDA	3	7	3	0	0	22	35	0.0
	ABN	7	12	5	1	6	51	82	62.2
	Total	113	91	22	10	10	141	387	
Overall percentage correct diagnoses									43.2

Table 3.22. Feature selection on Fourier descriptors

Class pair	Recognizer	Features
1–2	Linear	16, 11, 18, 19
1–3	Linear	16, 8, 15
1–4	Linear	16, 8
1–5	Linear	8, 10, 14, 16
1–6	Linear	13, 10, 16, 9, 15, 22
2–3	Linear	11, 18, 10, 16, 12
2–4	Linear	11
2–5	Linear	15, 22, 10, 12, 14, 16
2–6	Linear	11, 10, 9, 8, 18, 16
3–4	Quadratic	13, 15, 9
3–5	Linear	15, 14, 24, 22
3–6	Quadratic	12, 13, 8
4–5	Linear	10, 14, 24
4–6	Linear	15, 20, 10
5–6	Linear	15, 14, 8

Table 3.23. Testing results on Fourier descriptors
Decision rule: Bayes
Data set: Congenital heart hand drawn outline—CDS1

		Computer diagnoses							
		N	ASD	VSD	PS	PDA	ABN	Total	Percentage correct
Verified diagnoses	N	48	7	6	1	6	17	35	56.5
	ASD	11	34	5	2	3	19	74	45.9
	VSD	17	7	17	3	4	25	73	23.3
	PS	9	2	5	4	4	14	38	10.5
	PDA	7	6	2	1	10	9	35	28.6
	ABN	14	9	4	1	5	49	82	59.8
	Total	106	65	39	12	32	133	387	
Overall percentage correct diagnoses									41.9

normal-abnormal study. Table 3.20 gives the features selected, and Table 3.21 lists the classification results for the ratio features. Table 3.22 enumerates the features selected, and Table 3.23 gives the classification results for the Fourier descriptors. Table 3.24 lists the classification results using the lateral features; since there are only two measurements, feature selection was not used. Table 3.25 gives the features selected, and Table 3.26 lists the classification results for the set of combination features.

In summary, these results show that the best computer results compare well with the radiologists' results (Tables 3.10, 26, 11, and 19). Also,

Table 3.24. Testing results on lateral features
Decision rule: Bayes. Data set: CDS1

		Computer diagnoses							
		N	ASD	VSD	PS	PDA	ABN	Total	Percentage correct
Verified diagnoses	N	51	17	7	0	0	10	85	60.0
	ASD	23	31	3	0	1	16	74	41.9
	VSD	29	16	2	0	2	24	73	2.7
	PS	17	11	2	0	0	8	38	0.0
	PDA	8	10	2	0	2	13	35	5.7
	ABN	22	18	3	0	4	35	82	42.7
	Total	150	103	19	0	9	106	387	
Overall percentage correct diagnoses									31.3

Table 3.25. Feature selection on combination features

Class pair	Recognizer	Features
1–2	Linear	1, 7, 2, 16, 10
1–3	Linear	1, 7, 4, 10, 16
1–4	Linear	16, 1, 13, 4
1–5	Linear	1, 7, 9, 14, 8, 4
1–6	Linear	1, 2, 13, 10, 8, 7, 15
2–3	Linear	2, 18, 16, 4, 12, 1
2–4	Linear	2, 1, 7, 4, 13, 16
2–5	Linear	4, 15, 24, 1
2–6	Quadratic	4, 2, 8, 11
3–4	Quadratic	15, 13, 1, 9
3–5	Linear	15, 14, 4
3–6	Linear	1, 18, 12, 3, 24, 13
4–5	Linear	1, 3, 14, 9, 7
4–6	Quadratic	1, 8, 2
5–6	Linear	15, 14, 8

ratio measurements are the most valuable measurements, followed by some of the Fourier descriptors. Table 3.27 gives the frequency with which each measurement was selected. These results provide a comparison for subsequent automated approaches. Next we discuss one such approach.

Automated Extraction of Boundary Data for Congenital Heart Disease

The study described in the previous subsection yielded information on the ability to diagnose congenital heart disease from boundary information on the PA and lateral chest radiographs. The boundaries were

Table 3.26. Testing results on combination features
Decision rule: Bayes. Data set: Congenital heart hand drawn outline—CDS1

		Computer diagnoses							
		N	ASD	VSD	PS	PDA	ABN	Total	Percentage correct
Verified diagnoses	N	66	0	3	6	2	8	85	77.6
	ASD	6	44	4	0	2	18	74	59.5
	VSD	13	12	18	4	3	23	73	24.7
	PS	8	4	4	8	2	12	38	21.1
	PDA	3	5	5	0	9	13	35	25.7
	ABN	6	10	7	0	7	52	82	63.4
	Total	102	75	41	18	25	126	387	
Overall percentage correct diagnoses									50.9

Table 3.27. Frequency of selection of features

Feature number	Feature class	Frequency of selection
1	Ratio	12
2	Ratio	6
3	Ratio	2
4	Ratio	8
7	Ratio	6
8	Fourier descriptor	5
9	Fourier descriptor	3
10	Fourier descriptor	3
11	Fourier descriptor	1
12	Fourier descriptor	2
13	Fourier descriptor	5
14	Fourier descriptor	4
15	Fourier descriptor	5
16	Fourier descriptor	5
18	Fourier descriptor	2
22	Fourier descriptor	0
24	Fourier descriptor	2
58	Lateral	0
59	Lateral	0

extracted using rules interpreted by a lay person who drew all the boundaries. It was shown that computer analysis of this information gave results comparable to those obtained by radiologists. In this subsection a study is described that was related to the previous study except that a computer algorithm was employed to extract the boundaries. Also, in this study only the PA view was considered. As previously shown, the

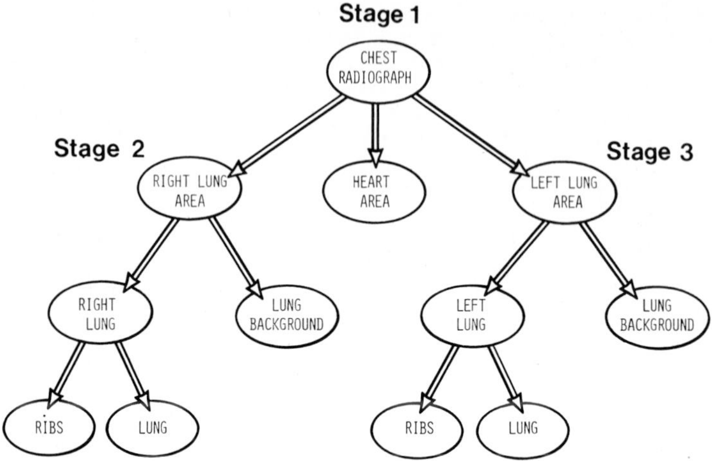

Fig. 3.35. Hierarchical structure of PA chest radiograph

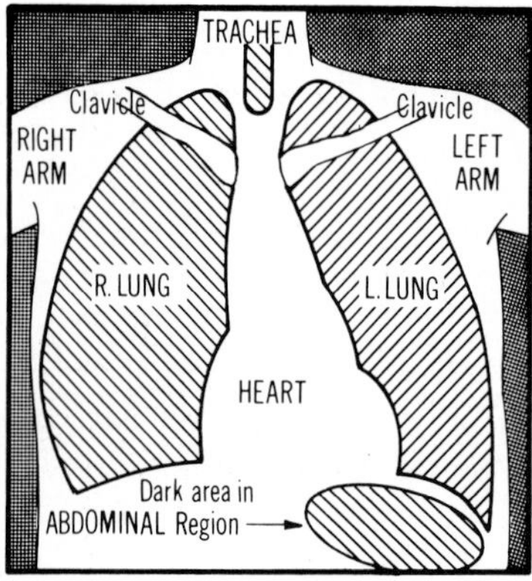

Fig. 3.36. Regions appearing in PA chest radiograph

lateral measurements used were not very significant in detecting congenital heart disease. Classification results were obtained in a fashion similar to those given in the previous study in order to compare the radiologists' results against the computer results from hand-drawn and computer-drawn outlines.

First, the algorithm for determining the boundaries will be discussed [3.52]. The strategy for boundary detection is based on top-down analysis. Because vascularity, ribs, and clavicles are superimposed on the lung field, global information is used to guide local primitive operators. This approach might be called planning [3.53]. Figure 3.35 shows the hierarchial strategy for analyzing the image. Figure 3.36 is a diagram of the requisite anatomy of the chest.

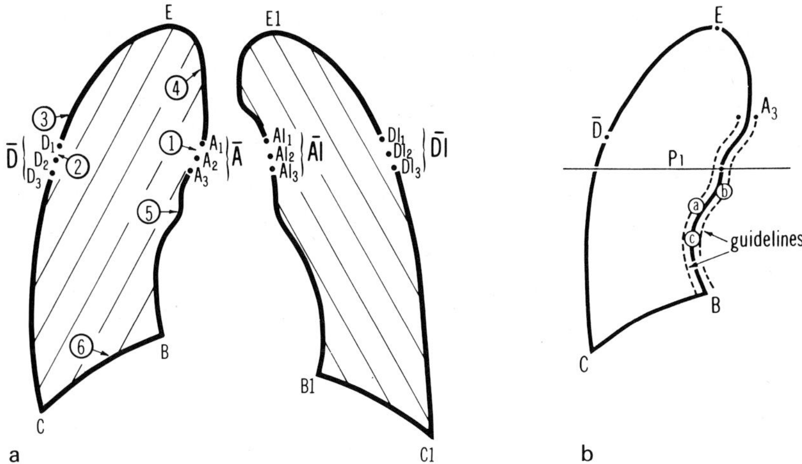

Fig. 3.37. (a) Points and lines detected. (b) Detection of the lower portion of the inside boundary

The procedure for locating the lung boundary is carried out in six steps (consult Fig. 3.37a):

1) Locate the starting vector $\boldsymbol{A}=(A_1, A_2, A_3)$, where A_1, A_2, and A_3 are the starting points of the inside boundary line of the right lung.
2) Locate the starting vector $\boldsymbol{D}=(D_1, D_2, D_3)$, where D_1, D_2 and D_3 are the starting points of the outside boundary line of the right lung.
3) Proceed upward, row by row, from the starting point D_1. Then proceed downward, row by row, from the starting point D_3. Search the points on each row horizontally using a local algorithm to detect the outside boundary point of the right lung.
4) Beginning with the starting point A_1, search the points on each row to detect the upper inside boundary of the right lung. Proceed upward, row by row, until the point E is detected.
5) Locate the lower inside boundary line in three steps:

   a) Beginning with starting point A_3, examine the points on each row using a global algorithm to detect the first approximate boundary point. Proceed downward, row by row, until the first estimated location of the cardiac-diaphragm intercept is detected.

   b) Beginning with starting point A_3, examine the points on each row using a global and a local algorithm to detect the second approximate boundary point. Proceed downward, row by row, until the first estimated location of the cardiac-diaphragm intercept is reached. The second approximate boundary line is then obtained by connecting the boundary points detected at this step; see Fig. 3.37b.

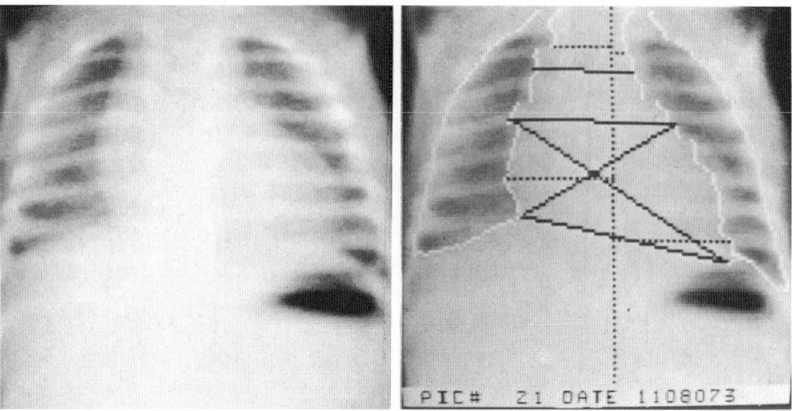

Fig. 3.38. Processed radiographs

c) Use local information between the first and second approximate lines to detect the actual inside boundary of the lower portion of the lung. The cardiac-diaphragm intercept is also detected at this step.

6) Starting from the cardiac-diaphragm intercept, proceed toward the outside boundary, column by column. Examine the points in each column to detect edge points until the point C is detected.

After the starting vector $\boldsymbol{A}$ has been obtained in Step 1, the intensities of the first and last elements of the starting vector are used as thresholds to draw the approximate boundary lines. Guided by these approximate boundary lines, only a small range of neighboring points needs to be searched in Step 5c. Moreover, the approximate boundary lines indicate the approximate location of the cardiac-diaphragm intercept and, thereby, facilitate the detection of the accurate location of the cardiac-diaphragm intercept (B or B1 in Fig. 3.37).

The radiopaque chest walls, appearing as light regions in a chest radiograph, indicate the location of outside lung boundaries. Therefore, a heuristic edge detector [3.15] is implemented in our system to detect this portion of the lung boundary.

The lung outline program is written in PL/1 compiler F, version 5.0 and requires about 0.2 s running time in 168 Kbytes of memory on an IBM 370/165 computer. The program has been run on a test library of chest radiographs of patients of all ages including infants. The algorithm works only moderately well in infant cases because an infant's lungs appear lighter in the chest radiograph, making it more difficult for the radiologist as well as the computer to identify their actual boundaries.

Two radiographs of infants' chests with the processed outlines are shown in Fig. 3.38 (1 month and 6 day old infants). Three radiographs

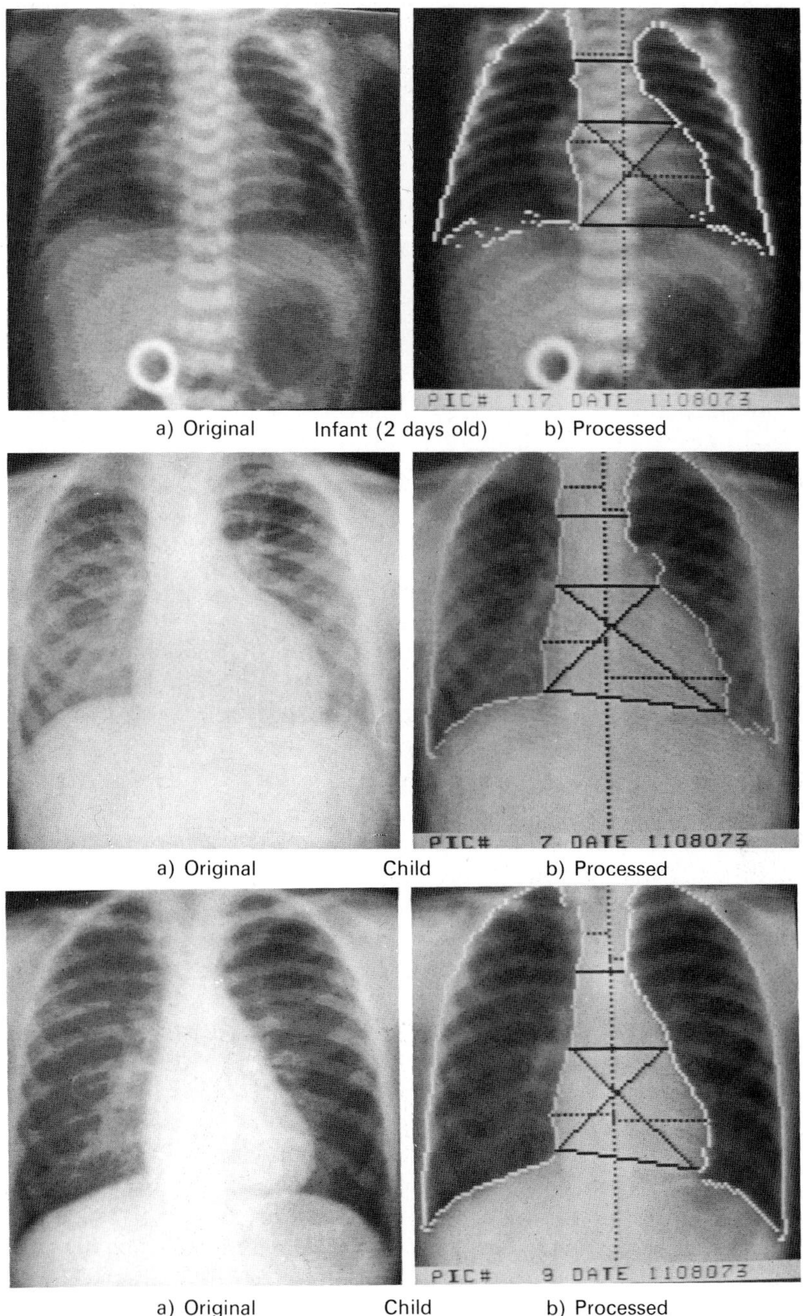

a) Original Infant (2 days old) b) Processed

a) Original Child b) Processed

a) Original Child b) Processed

Fig. 3.38 (continued)

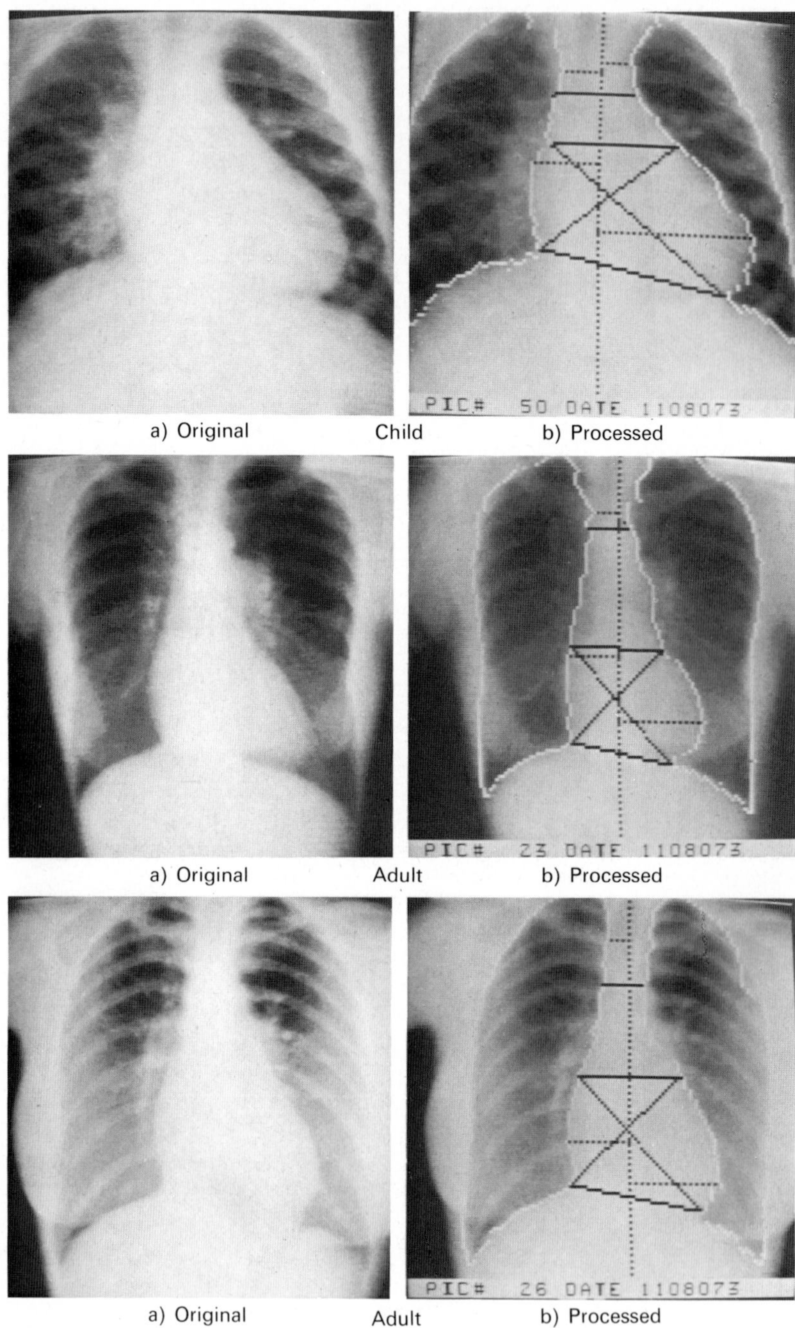

Fig. 3.38 (continued)

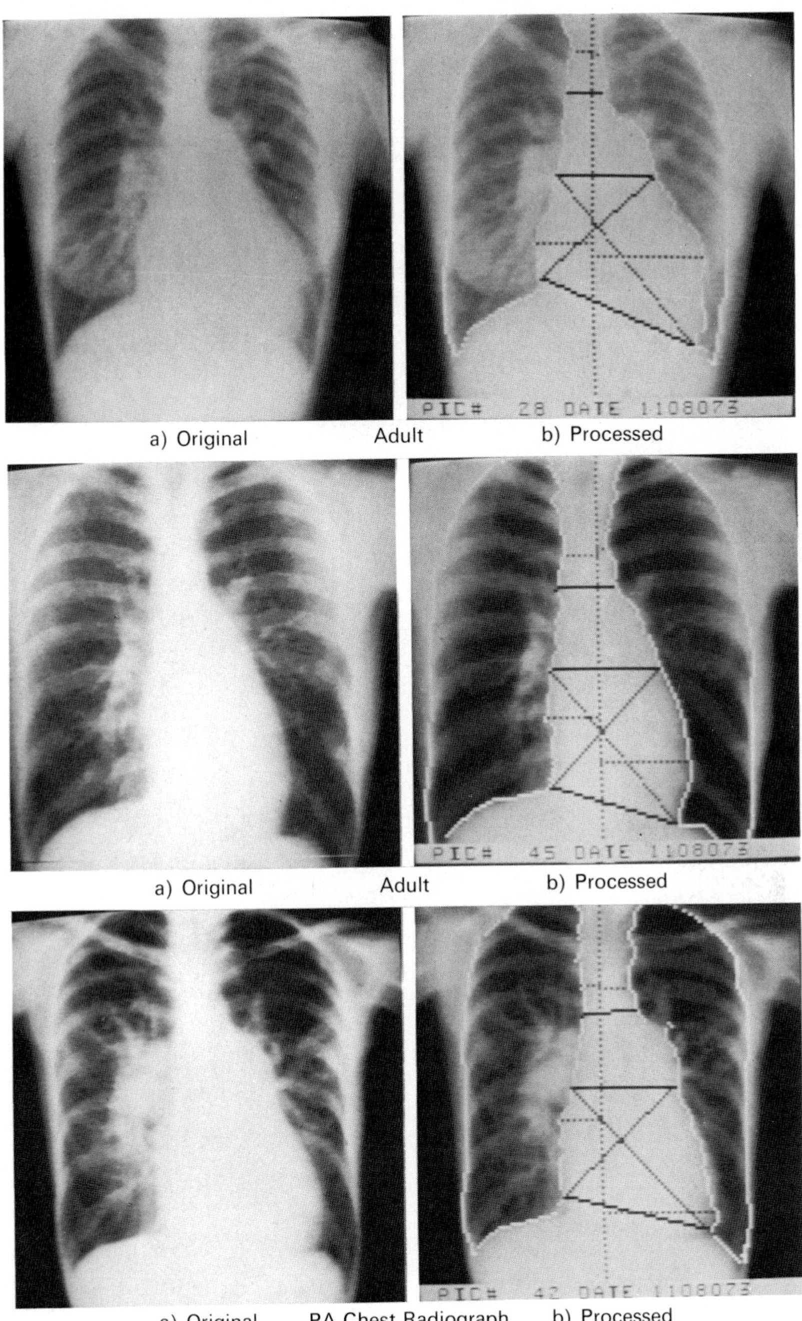

Fig. 3.38 (continued)

of childrens' chests with processed outlines are shown in Fig. 3.38, along with four adult chests and their resultant outlines. Note that these outlines correspond closely to the outlines obtained in the previous study where the outlines were hand drawn. The measurements extracted are the same. They are restated below.

REH = Maximum distance from the midline to the right heart border.
LEH = Maximum distance from the midline to the left heart border.
RET = Minimum distance from the midline to the right lung boundary.
LET = Minimum distance from the midline to the left lung boundary.
HUB = Upper bound of the heart.
HLB = Lower bound of the heart.
UGV = Upper bound of the great vessel.
HH = Height of the heart.
HGV = Height of the great vessel.
AH = Area of the heart.
AGV = Area of the great vessel.
ARL = Area of the right lung.
ALL = Area of the left lung.

The features derived from these measurements are again the same. They are the ratio features:

1) Cardiac-thoracic ratio

$$\mathrm{CTR} = \frac{\mathrm{REH} + \mathrm{LEH}}{\mathrm{RET} + \mathrm{LET}}.$$

2) Right to left heart ratio

$$\mathrm{RLH} = \frac{\mathrm{REH}}{\mathrm{LEH}}.$$

3) Atrioventricular ratio

$$\mathrm{AVR} = \frac{\mathrm{LDH1}}{\mathrm{LDH2}}.$$

4) Great vessel heart area ratio

$$\mathrm{GHAR} = \frac{\mathrm{AGV}}{\mathrm{AH}}.$$

5) Heart thoracic cavity area ratio

$$\mathrm{HLAR} = \frac{\mathrm{AH}}{\mathrm{ARL} + \mathrm{ALL} + \mathrm{AGV} + \mathrm{AH}}.$$

6) Great vessel heart height ratio

$$\mathrm{GHHR} = \frac{\mathrm{HGV}}{\mathrm{HH}}.$$

7) Angle between line AB1 and a line perpendicular to the chest midline:

Let $\mathrm{A} = (i, j)$ and $\mathrm{B1} = (i1, j1)$; then

$$\mathrm{ALPH} = \mathrm{TAN}^{-1}\left(\frac{i1 - i}{j1 - j}\right)$$

The second set of features consists of the first 20 amplitude coefficients of the Fourier descriptors for the outline of the heart and great vessel. The six ratios of lengths, one angle measurement, and 20 amplitude coefficients total 27 measurements selected for classification (Table 3.13).

The classification method used here is called the modified maximum likelihood decision rule (MMLDR), which is described in Subsection 3.3.1. The test library used by our system consisted of 388 chest radiographs shown in Table 3.28. For each of the chest radiographs the 27 measurements were extracted.

Several classification experiments were performed. The decision rule used was the MMLDR. Table 3.29 gives testing results using only the seven ratio measurements, while Table 3.30 gives testing results using only the Fourier descriptors. Measurement selection was performed on

Table 3.28. Disease classes in congenital heart computer drawing outline library

Disease class		Number of cases
A	Normal (N)	159
B	Atrial septal defect (ASD)	72
C	Ventricular septal defect (VSD)	58
D	Pulmonary stenosis (PS)	24
E	Patent ductus arteriosus (PDA)	30
F	Abnormal (ABN)	45
		Total 388

Table 3.29. Testing results using 7 ratio measurements
Decision rule: MMLDR. Measurements used: A_1–A_7
Data set: Congenital heart computer drawing outline

		Computer diagnoses			
		N	ABN	Total	Percentage correct
Verified diagnoses	N	125	34	159	78.62
	ABN	56	173	229	75.55
	Total	181	207	388	
Overall percentage correct diagnoses					76.80

Table 3.30. Testing results using 20 Fourier descriptors
Decision rule: MMLDR. Measurements used: A_8–A_{27}
Data set: Congenital heart computer drawing outline

		Computer diagnoses			
		N	ABN	Total	Percentage correct
Verified diagnoses	N	79	80	159	49.69
	ABN	76	153	229	66.81
	Total	155	233	388	
Overall percentage correct diagnoses					59.79

Table 3.31. Testing results using set of F.D. (13 measurements)
Decision rule: MMLDR. Measurements used: A_9, A_{11}, A_{12}, A_{13}, A_{14}, A_{16}, A_{17}, A_{19}, A_{23}, A_{22}, A_{24}, A_{25}, A_{27}
Data set: Congenital heart computer drawing outline

		Computer diagnoses			
		N	ABN	Total	Percentage correct
Verified diagnoses	N	71	88	159	44.65
	ABN	62	167	229	72.93
	Total	133	255	388	
Overall percentage correct diagnoses					61.34

Table 3.32. Testing results using 27 measurements
Decision rule: MMLDR. Measurements used: A_1–A_{27}
Data set: Congenital heart computer drawing outline

		Computer diagnoses			
		N	ABN	Total	Percentage correct
Verified diagnoses	N	83	76	159	52.20
	ABN	34	195	229	85.15
	Total	117	271	388	
Overall percentage correct diagnoses					71.65

Table 3.33. Testing results using 16 measurements
Decision rule: MMLDR. Measurements used: A_1, A_2, A_3, A_4, A_5, A_6, A_8, A_9, A_{10}, A_{11}, A_{15}, A_{16}, A_{17}, A_{20}, A_{21}, A_{22}
Data set: Congenital heart computer drawing outline

		Computer diagnoses			
		N	ABN	Total	Percentage correct
Verified diagnoses	N	121	38	159	76.10
	ABN	42	187	229	81.66
	Total	225	163	388	
Overall percentage correct diagnoses					79.38

the Fourier descriptors, which resulted in 13 measurements. Table 3.31 gives the testing results obtained using these measurements. All 27 measurements were used to obtain the testing results in Table 3.32. Using measurement selection on the 27 measurements gave the testing results in Table 3.33. Note that the ratio measurements are the most important ones.

Next the multiclass problem was considered. The best testing results obtained and the measurements used are shown in Tables 3.34 and 3.35, using measurement selection on the 27 measurements.

These results, though incomplete, give some indication of the accuracy of the automated approach. For example, one can compare these results with the results obtained from the hand drawn outline and with the radiologists' results (Tables 3.26 and 3.34).

Table 3.34. Testing results using set of combination measurements (6 measurements)
Decision rule: MMLDR. Measurements used: A_1, A_3, A_6, A_{10}, A_{12}, A_{22}
Data set: Congenital heart computer drawing outline

		Computer diagnoses							
		N	ASD	VSD	PS	PDA	ABN	Total	Percent correct
Verified diagnoses	N	107	1	29	9	4	9	159	67.30
	ASD	6	14	27	1	4	20	72	19.44
	VSD	16	5	21	1	5	10	58	36.21
	PS	11	2	2	3	1	5	24	12.50
	PDA	5	4	11	0	2	8	30	6.67
	ABN	10	4	17	2	3	9	45	20.00
	Total	155	30	107	16	19	61	388	
Overall percentage correct diagnoses									40.21

Table 3.35. Testing results using set of combination measurements
Decision rule: MMLDR. Measurements used: A_1, A_3, A_6, A_{10}, A_{12}, A_{22}
Data set: Congenital heart computer drawing outline

		Computer diagnoses			
		N	ABN	Total	Percentage correct
Verified diagnoses	N	107	52	159	67.30
	ABN	48	181	229	79.04
	Total	155	233	388	
Overall percentage correct diagnoses					74.23

3.4.7 Conclusions and Summary

Even though radiographic image processing is still in its infancy, a wide variety of problems have been considered. In addition to the work previously described, work has been done on locating tumors in chest radiographs [3.54], computing lung and heart volumes from PA and lateral chest radiographs [3.55], classifying lung textures [3.56–59], analyzing cineangiograms [3.60, 61], and detecting atherosclerosis [3.62]. There has also been much work in nuclear medicine and CAT (Computer Transverse Axial Tomography) associated with scanning and display system development. The Appendix lists some of the other individuals and institutions involved in this work.

The devices that are the nearest to clinical applications are the basic scanning and display systems. As yet none of the automated diagnosis systems are in use.

It seems likely that in the future there will be advances in scanning and display systems that will have a direct impact on clinical applications. The automated diagnosis tools will also improve. It is likely that their first clinical applications will be as part of a total system that the radiologist would use as a tool to aid his diagnosis. It is easy to envision a radiologist requesting a computer analysis of the shape of a heart. As other proven programs are developed they would be included in the system as sources of diagnostic information. It is thus envisioned that the evaluation of radiographic image processing systems would be an evolutionary process.

Appendix

Other Laboratories Involved in Radiographic Image Analysis

The following is a list of other laboratories involved in Radiographic Image Analysis. Detailed information on their work is usually available in report form and can be obtained by contacting the persons involved.

1. University of Southern California

   W. K. PRATT
   USC University Park
   Olin Hall, Room 330H
   Los Angeles, CA 90007, USA

2. Jet Propulsion Laboratory

   ROBERT SELZER
   JPL, California Institute of Technology
   4800 Oak Grove Drive
   Pasadena, CA 91103, USA

3. Georgetown University Medical Center

   ROBERT S. LEDLEY
   Georgetown University Medical Center
   3900 Reservoir Road, N.W.
   Washington, D.C. 20007, USA

4. University of California-Irvine

   JACK SKLANSKY
   School of Engineering
   University of California
   Irvine, CA 92664, USA

5. Purdue University

   K. S. FU
   School of Electrical Engineering
   West Lafayette, IN 47907, USA

6. Nagoya University
JUN-ICHIRO TORIWAKI
Nagoya University
Department of Electrical Engineering
Furo-Cho, Chikusa-Ku
Nagoya, Japan

7. Medical Research Council
KEITH PATON
MRC Clinical Research Center
Northwick Park Hospital
Watford Road, Harrow, Middlesex
HA1 3UJ, England

Acknowledgements

The authors would like to recognize the contributions made by many of the UMC students and staff, past and present, to the work described in this paper. Dr. DAVID SCHRUNK, Dr. JAMES LEHR, and Dr. CORINNE FARRELL provided much insight into the disease processes and their radiographic characteristics. Dr. BILL MCFARLAND and Dr. TED MCCRACKEN provided valuable hardware support, and Dr. J. K. CHANG gave support in the area of pattern recognition. PATRIC CAUDILL, BOB NESTOR, and SCOTT HENDERSON provided software support. RICHARD BROOKS, PEGGY TSIANG, RICHARD KRUGER, FRANK ROELLINGER, and others worked on the applications. BONNIE KIMBERLING and KAREN GRAINGER prepared the manuscript and graphics. We also gratefully acknowledge the support provided by NIGMS GM 17729, NSF Grant 20401, and the College of Engineering.

References

3.1 R.P. KRUGER, S.J. DWYER III, D.L. HALL, G.S. LODWICK: IAL Technical Report: "Computer Processing of Radiographic Images", Image Analysis Laboratory, Departments of Electrical Engineering and Radiology, University of Missouri-Columbia, Columbia, Missouri (1971)

3.2 Estimates from the U.S. Public Health Service 1970 X-Ray Exposure Study, Department of Health, Education, and Welfare, U.S. Public Health Service, Bureau of Radiological Health, Washington, D.C. (1972)

3.3 Vital and Health Statistics, Series 10, No. 81, DHEW Publication No. (RSM) 73-1507, Department of Health, Education and Welfare, Health Services and Mental Health Administration, U.S. Government Printing Office, Washington, D.C.

3.4 Report to the Surgeon-General on Protecting and Improving Health through the Radiological Sciences, National Advisory Committee on Radiation, U.S. Public Health Service, Washington, D.C. (1966)

3.5 R.C. BROOKS, S.J. DWYER III, G.S. LODWICK: Technical Report IAL-TR 25–73, Image Analysis Laboratory, Bioengineering Program, Departments of Electrical Engineering and Radiology, University of Missouri-Columbia, Columbia, Missouri (1973)

3.6 M. O. Braaten, S. J. Dwyer III, C. Farrell, G. S. Lodwick: Technical Report IAL-TR 3–72, Image Analysis Laboratory, Departments of Industrial Engineering, Electrical Engineering, and Radiology, University of Missouri-Columbia, Columbia, Missouri (1972)
3.7 J. L. Lehr, G. S. Lodwick, L. J. Garrotto, D. J. Manson, B. F. Nicholson: AFIPS Conf. Proc., Spring Joint Computer Conference **40**, 999 (1972)
3.8 M. Elder, R. James, R. C. Brooks: Technical Report IAL-TR 27–74, Image Analysis Laboratory, Bioengineering Program, Departments of Radiology and Electrical Engineering, University of Missouri-Columbia, Columbia, Missouri (1974)
3.9 P. W. Neurath: Medical Res. Eng. **10**, 9 (1971)
3.10 W. D. McFarland, S. J. Dwyer III, G. S. Lodwick: Technical Report IAL-TR 22–73, Image Analysis Laboratory Bioengineering Program, Departments of Electrical Engineering and Radiology, University of Missouri-Columbia, Columbia, Missouri (1973)
3.11 T. E. McCracken: An Economical Tonal Display for Interactive Graphics and Image Analysis, Ph. D. Dissertation, Electrical Engineering Department, University of Missouri-Columbia, Columbia, Missouri (1974)
3.12 A. Rosenfeld: Computing Surveys **5**, 81 (1973)
3.13 A. Rosenfeld: *Picture Processing by Computer* (Academic Press, New York 1969)
3.14 R.O. Duda, P.E. Hart: *Pattern Classification and Scene Analysis*, (Wiley, New York 1973)
3.15 C. A. Harlow: Computer Graphics and Image Processing **2**, 60 (1973)
3.16 N. J. Nilsson: *Problem Solving Methods in Artifical Intelligence*, (McGraw-Hill, New York 1971)
3.17 A. Shaw: The Formal Description and Parsing of Pictures, Ph. D. Dissertation, Computer Science Report **94**, Computer Science Department, Stanford University (1968)
3.18 S. E. Henderson, C. A. Harlow, G. S. Lodwick: Feature Extraction of Knee X-Rays, IEEE Conf. on Decision and Control, Miami, Florida (1971) p. 216
3.19 C. A. Harlow, S. A. Eisenbeis: The Analysis of Radiographic Images, Conf. on Two-Dimensional Signal Processing, University of Missouri-Columbia, Columbia, Missouri (1971); and IEEE Trans. C-**22**, 678 (1973)
3.20 J. B. Thomas: *An Introduction to Applied Probability and Random Processes* (Wiley, New York 1971) pp. 209–215
3.21 G. Nagy: Proc. IEEE **56**, 836 (1968)
3.22 R. G. Leonard: Automated Selection of Measurements for Pattern Recognition, Ph. D. Dissertation, University of Missouri-Columbia (1971)
3.23 J. K. Chang: Decision Rules and Measurement Selection in Pattern Recognition Theory, Ph. D. Dissertation, University of Missouri-Columbia (1973)
3.24 Special Issue on Feature Extraction and Selection in Pettern Recognition, IEEE Trans. C-**20** (1971)
3.25 K. S. Fu, P. G. Min, T. J. Li: IEEE Trans. SSC-**6**, 33 (1970)
3.26 K. Fukunaga, W. L. Koontz: IEEE Trans. C-**19**, 311 (1970)
3.27 T. R. Vilmansen: IEEE Trans. C-**22**, 381 (1973)
3.28 J. T. Tou, R. P. Heydorn: *Computer and Information Sciences II* (Academic Press, New York 1967) pp. 57–58
3.29 W. G. Wee: IEEE Trans. IT-**15**, 47 (1970)
3.30 R. Gordon, G. T. Herman: Comm. ACM **14**, 759 (1971)
3.31 E. Hall, R. Kruger, S. J. Dwyer III, D. Hall, R. McLaren, G. S. Lodwick: IEEE Trans. C-**20**, 1032 (1971)
3.32 F. X. Roellinger, Jr., A. E. Kahveci, J. K. Chang, C. A. Harlow, S. J. Dwyer III, G. S. Lodwick: Computer Analysis of Chest Radiographs, U.S.-Japan Seminar on Picture and Scene Analysis, Kyoto, Japan (1973)

3.33 P. BRODATZ: *Textures* (Dover, New York 1966)
3.34 B. JULESZ: IRE Trans. IT-**8**, 84 (1962)
3.35 A. ROSENFELD, M. THURSTON: IEEE Trans. C-**20**, 562 (1971)
3.36 A. ROSENFELD, E. B. TROY: Proc. UMR-Mervin J. Kelly Communication Conference, Rolla, Missouri (1970) pp. 10.1.1–6
3.37 R. M. HARALICK, K. SHANMUGAM: Proc. Symposium on Computer Image Processing and Recognition, University of Missouri-Columbia, Columbia, Missouri (1970). pp. 12.2.1–10
3.38 R. S. MICHALSKI: Report **461**, Dept. of Computer Science, University of Illinois-Urbana (1971)
3.39 D. A. AUSHERMAN: Texture Discrimination within Digital Imagery, Ph. D. Dissertation, University of Missouri-Columbia, Columbia, Missouri (1972)
3.40 S. A. EISENBEIS, C. A. HARLOW, P. K. TSIANG: IEEE Conf. on Decision and Control, New Orleans (1972) pp. 611–612
3.41 M. K. HU: IRE Trans. IT-**8**, 179 (1962)
3.42 E. D. SACERDOTI: Technical Note **78**, SRI (1973)
3.43 C. A. HARLOW, S. E. HENDERSON, D. S. SCHRUNK: Proc. 3rd Conf. on Computer Applications in Radiology, University of Missouri-Columbia, Columbia, Missouri (1972) pp. 108–141
3.44 Y. YAKIMOVSKY: Scene Analysis Using a Semantic Base for Region Growing, Ph. D. Dissertation, Stanford University (1973)
3.45 Z. MANNA: J. ACM **16**, 244 (1969)
3.46 A. V. AHO, J. D. ULLMAN: J. Comp. System Sci. **6**, 125 (1972)
3.47 C. A. HARLOW, J. LEHR, R. PARKEY, L. GARROTTO, G. S. LODWICK: Radiology **97**, 269 (1970)
3.48 G. S. LODWICK: *The Bones and Joints* (Year Book Medical Publishers, Chicago 1971)
3.49 R. N. SUTTON, E. L. HALL: Technical Report IAL-TR 2-72, Image Analysis Laboratory, Departments of Radiology and Electrical Engineering, University of Missouri-Columbia (1971); also in IEEE Trans. C-**21**, 667 (1972)
3.50 R. P. KRUGER, W. B. THOMPSON, A. F. TURNER: IEEE Trans. SMC-**4**, 40 (1974)
3.51 D. L. HALL, G. S. LODWICK, R. L. KRUGER, S. J. DWYER III: Radiological Clinics of North America **9**, 533 (1971)
3.52 P. P. TSIANG, C. A. HARLOW: Technical Report IAL-TR 28–74, Image Analysis Laboratory, Department of Electrical Engineering, University of Missouri-Columbia (1974); also Proc. ACM Natl. Conf., San Diego (November 1974)
3.53 M. D. KELLY: *Machine Intelligence*, Vol. **6** (American Elsevier, New York 1971) pp. 397–411
3.54 D. BALLARD, J. SKLANSKY: Comp. Biomedical Res. **6**, 299 (1973)
3.55 J. W. PAUL, M. D. LEVINE, R. G. FRASER, C. A. LASZLO: IEEE Trans. BME-**19**, 444 (1974)
3.56 Y. P. CHIEN, K. S. FU: IEEE Trans. SMC-**4**, 145 (1974)
3.57 J. TORIWAKI, Y. SUENAGA, T. NEGORO, T. FUKUMURA: Computer Graphics and Image Processing **2**, 252 (1973)
3.58 E. L. HALL, R. P. KRUGER, A. F. TURNER: Optical Eng. **13**, 250 (1974)
3.59 R. P. KRUGER, W. B. THOMPSON, A. F. TURNER: IEEE Trans. SMC-**4**, 40 (1974)
3.60 C. K. CHOW, T. KANEKO: Comp. in Biology and Medicine **3**, 13 (1973)
3.61 T. KANEKO, P. MANCINI: IEEE Trans. BME-**20**, 413 (1973)
3.62 R. H. SELZER, D. H. BLANKENHORN, E. B. BECKENBACH, D. W. CRAWFORD, R. BRANDT, S. H. BROOKS: Proc. 4th Conf. on Computer Applications in Radiology, Las Vegas, Nev. (1975)

4. Image Processing in High Energy Physics*

R. L. McIlwain, Jr.

With 16 Figures

Image processing in high energy physics presents several features which are not common to other image-processing techniques: 1) There is relatively little emphasis on pattern recognition of complex geometrical shapes. The elements of the patterns can be accurately represented as arcs of circles, and the pattern is either simply connected or can be broken into simply-connected segments. 2) Even to recognize the pattern elements, it is necessary to determine the circle parameters to high precision. 3) The pattern recognition must ultimately occur in three dimensions, using data from stereo images. 4) The images contain both random noise which is removed early in the processing, and noise in the form of spurious elements, which often cannot be eliminated until the final stages.

Because of these unique problems the processing occurs in stages: first recognizing and joining short segments to form the circular arcs, then extending the arcs as far as possible, joining the arcs associated with the pattern at a common vertex (or vertices), and finally associating the elements in the stereo images to form the pattern in space. These stages are described in detail in this chapter.

The introduction describes the high energy physics problems to be solved and the different types of detectors currently employed. The development of early image processors is described to show the evolution of the image-processing techniques, followed by a description of modern processors. Finally there is a brief discussion of current problems and possible ways of solving them.

4.1 The Role of Image Processing in High Energy Physics

This section discusses the methods and techniques used to extract data about high-energy interactions of elementary particles using visual techniques. Most of the methods to be discussed have been developed to process bubble chamber film. The bubble chamber has the highest

* Work supported in part by U.S. Energy Research and Development Administration.

spatial resolution and the most noise of all visual detectors, except nuclear emulsions. Hence problems of analysis tend to be harder than for other detectors, and the programs are of more interest.

Before proceeding to the analysis techniques themselves, we will briefly examine the general problem of extracting data about the interactions of elementary particles from the record of the visual detector. Because of the very short range of the forces ($\sim 10^{-13}$ cm) it is possible to detect their effects only in collisions in which energy is exchanged between particles. A large fraction of our knowledge about these forces comes from studying correlations among measured quantities for the individual particles which participate. The quantities which we can measure are the three components of the momentum of each particle and occasionally its kinetic energy.

As an example consider a study of the ϱ-meson [4.1], which was strongly suggested by high-energy electron scattering experiments. This particle (or perhaps resonance) is very unstable and interacts strongly with its decay products. Its lifetime is estimated at about 10^{-22} s; so if it is produced inside the nucleus its decay time is comparable to the time required to traverse the nucleus. In no case can it ever leave the atom in which it is produced, and our knowledge of it can only come from a study of its decay products, which are two pions (or π-mesons). If we examine a reaction in which pions are produced, it may be possible to detect the ϱ. For example in the reaction

$$\pi^- + p \to \pi^+ + \pi^- + n \tag{4.1}$$

we may suspect that sometimes it proceeds by a two-stage process, namely

$$\pi^- + p \to \varrho^0 + n \tag{4.2}$$

followed by

$$\varrho^0 \to \pi^+ \pi^- . \tag{4.3}$$

To detect the ϱ^0 as a particle we must verify the following properties: a) It has a set of well-defined quantum numbers and b) it has a definite ground state (or state of lowest energy). The energy of the ground state is called the rest energy, the rest mass, or simply the mass of the particle, and can be determined by the following procedure. If we imagine that in (4.1) the pions are produced independently, the energy of each pion will be determined by chance except that the total energy of all particles must be conserved. From the energy and momentum of each pion (E^+

and $\boldsymbol{p}^{+}$) we can compute the energy and momentum of the center of mass of the two:

$$E = E_{+} + E_{-}, \tag{4.4}$$

$$\boldsymbol{p} = \boldsymbol{p}_{+} + \boldsymbol{p}_{-}. \tag{4.5}$$

If we use the relativistic energy-momentum relation

$$E^2 = c^2 p^2 + m^2 c^4, \tag{4.6}$$

then the effective mass of the two-pion system is

$$m = [E^2/c^4 - p^2/c^2]^{1/2}. \tag{4.7}$$

The probability of producing a pair of pions $(\boldsymbol{P}_{+}, E_{+})$ and $(\boldsymbol{p}_{-}, E_{-})$ is proportional to the density of states available, a quantity which can be computed. We can thus compute the probability that the pion system will have an effective mass $\underline{m}$. Such a "phase-space" distribution is shown in Fig. 4.1 together with a histogram of measured events. Notice that the phase space distribution has a smooth variation from zero mass up to a maximum which is determined by the energy available. The measured events show a large enhancement near a mass of 750 MeV/c^2 and this is interpreted as evidence that some structure with this mass exists for a finite time interval and that it decays into the observed pions. The peak shown in the data is wider than can be explained by the resolution of the measuring system. After correcting for resolution, the peak is still more than 100 MeV/c^2 wide. This mass uncertainty can be related to the lifetime of the state by using the Heisenberg uncertainty-principle relation

$$\Delta E \Delta t \gtrsim \hbar, \tag{4.8}$$

where $\hbar$ is Planck's constant $h/2\pi$, and we interpret ΔE as $(\Delta m)/c^2$ and Δt as the lifetime τ. Using the equality yields an estimate of the lifetime of 10^{-22} s.

To be really classified as a particle, this state must be shown to have definite quantum numbers. To establish all the quantum numbers requires extensive analysis of many reactions. As an example we consider the spin, or intrinsic angular momentum, of the state. The only other particles with spin are the proton and neutron each with spin $\frac{1}{2}$ (intrinsic angular momentum $= \hbar/2$); so the particle must have integer spin. If we transform to a coordinate system fixed in the center of

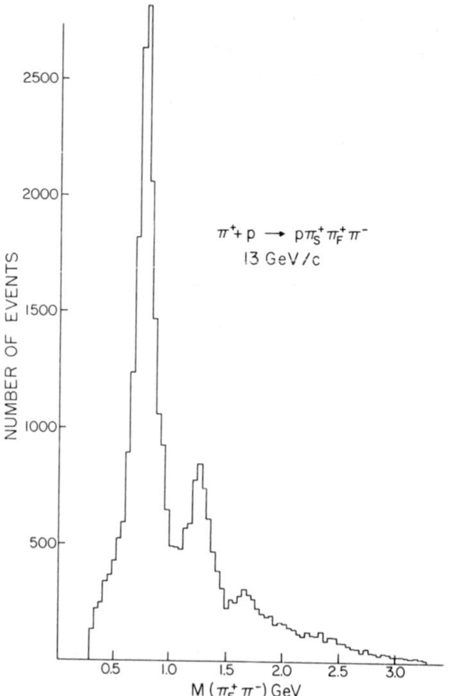

Fig. 4.1. The effective mass of the $\pi^+ - \pi^-$ system. The histogram shows the number of events in each mass interval. The large peak is the ϱ^0 meson and the second is the f^0 meson. The backgrounds are the events under the peaks

mass of the pion pair, the decay angular distribution depends only upon the spin. For spin-0, the distribution must be isotropic; for other spins it may be enhanced in certain directions. The observed angular distribution has a strong $\cos^2 \theta$ term, which is characteristic of spin-1 particles. Hence this quantum number was established.

Since most of our knowledge of elementary particles has come from such experiments, several features should be emphasized: a) The final conclusions are based upon statistical averages of quantities, and the precision depends very much on the total number of events in the sample. b) The quantities themselves must be computed for each event. There is no way to compute the (π, π) effective mass distribution from separate energy and momentum distributions of the π^+ and π^-. c) It is necessary to know in detail what is happening; to know which particle is the π^+, which the π^-, and to know that no undetected particle (other than the neutron) participated in the reaction. d) It is necessary to study all channels of the reaction to be sure observed effects are not due to kinematic coupling from another channel. For example, suppose there were a strong force between the π^- and the neutron. Since the neutron usually goes backward in the center of mass system, it would tend to pull the π^- with it and thus bias the angular distribution. e) The measured quantities enter the final result in a complicated way, and knowing the precision of the result requires knowing all the errors and their correlations, i.e. the error matrix. f) The resolution of the system depends on the measurement errors, and poor measurements can mean that some features will be smeared out and undetectable.

In the remainder of this section we will examine the general nature of visual detectors and give some examples, develop briefly some of the kinematic concepts, and outline the tasks required to go from the output record of the detector to a computer record which can be used by the experimenter. Section 4.2 will describe in some detail the hardware problems associated with the processing, and 4.3 will do the same for the software. Finally in 4.4 we examine a few problems which are just beginning to be attacked.

4.1.1 Visual Detectors

All visual detectors of charged particles use some means of amplification to raise the energy deposited by the particle to a level high enough to be recorded. It is an unfortunate fact that this amplification always degrades the signal, usually by losing some of the resolution inherent in the primitive process in the detector. The detection process always starts with the ionization of the material of the detector by the charged particle. This trail of ions is initially only one or two atomic diameters wide, i.e. a few nm (10^{-9} m). The highest-resolution detector, the nuclear emulsion, has a grain size of a few μm (10^{-6} m). The large loss in resolution is associated with the large amount of energy which must be supplied to enhance the signal. A typical high-energy charged particle loses about 2 MeV-cm^2/gm when passing through matter. If all this energy loss produces ions there will be 10^5–10^6 ions produced per gm/cm^2 and if these are all collected the total charge is about 10^{-13} coulombs. One gm/cm^2 represents a path length of 1 cm in water and about 20 cm in liquid hydrogen. This very small signal must either be amplified locally or collected locally and transmitted to an external amplifier, to retain the spatial information. The amplification must be performed quickly because the ions either recombine or drift away from their original positions.

The earliest visual detector was the Wilson cloud chamber. A container of gas is nearly saturated with vapor under a slightly elevated pressure. When an external detector system detects the passage of a charged particle, the chamber is expanded adiabatically and the gas becomes supersaturated, i.e. a non-equilibrium or metastable state. It can return to equilibrium only by condensing some of the vapor. The ions left by the charged particle act as centers upon which droplets can condense. Droplet growth can be explained satisfactorily by the electrostatic repulsion of ions on the surface. A photograph taken some time after the expansion will record these droplets and hence the path of the particle. The chamber is then recompressed and allowed to return to thermal equilibrium in the unsaturated state. It is thereafter ready for the next cycle.

There are several problems which limit its utility as a detector. The most serious is its long cycle time, required because of its slow return to equilibrium. The net heat lost during the cycle must be returned to the gas by thermal conduction. The recombination time for the ions is very long; so they must be removed by an electric field. Large chambers may require several minutes between expansions. Droplet formation and growth is also a slow process; so there must be considerable delay before the photograph is taken, and any bulk motion of the gas during the expansion will distort the tracks.

Some of these problems were reduced by the development of the diffusion cloud chamber in which a large thermal gradient establishes a layer of gas which is always sensitive. Even in the best designs this layer was always very thin; so large chambers were required, to achieve reasonable sensitive volumes. Its duty cycle remained low because of the need to sweep out the old ions before the next particle entered. Since it is always sensitive it must be used either with a pulsed source of particles or with a very small flux. Because of these problems, which are inherent in the device, cloud chambers no longer play a significant role in high energy physics.

The bubble chamber of [4.2] also employs a thermodynamic system in a non-equilibrium state. The working fluid is a superheated liquid in which bubbles of vapor form along the track of the particle. The operating temperature is about two-thirds of the critical-point temperature, and the quiescent pressure is considerably above the saturation vapor pressure. It is made sensitive by lowering the pressure suddenly so that the liquid becomes superheated. In a clean glass container this state is metastable and can be maintained until it is disturbed; observed times are consistent with the frequency of cosmic rays. The system returns to equilibrium by boiling, and, if left expanded, will repressurize to the saturation pressure. The bubbles which form along the track are initiated by the heat spikes produced by ion recombination [4.3] (Fig. 4.3). The growth of the bubbles depends only on the thermodynamic conditions, i.e. upon the degree of superheat. For every condition there is a critical size which a bubble must have if it is to grow; if it is under this size it will collapse. The size of the initial bubbles depends upon the amount of heat deposited in a small volume near the bubble. Recombination of individual ion pairs along the track does not provide a sufficiently high energy density to form bubbles of critical size. Rather the nucleation centers are formed by short δ-rays (atomic electrons which receive enough energy in the ionizing collision to produce additional ionization). Thus the bubble density depends upon the thermodynamic conditions, which determine the critical size and the density of δ-rays. Compared to cloud chambers, the bubble growth is very rapid; they become large enough to photograph in about a millisecond, and an expansion-recompression cycle can be completed in 10–20 ms. The return to thermal equilibrium is somewhat slower because the cycle is not reversible and some heat transfer must occur. Modern chambers are not clean glass containers, but are fabricated of stainless steel, glass, and other materials. During the expansion bubbles grow around gaskets, piston rings, etc. and these must be recompressed. These effects limit the cycle rate even for small chambers to about 20/s[1].

In the compressed state the chamber is not sensitive, and an initial bubble of any size will collapse very rapidly. It is not possible to expand the chamber after the initial bubbles are formed and have any remain. The bubble chamber works because the recombination time of the ions is very short compared with thermal diffusion times, and efforts to delay recombination to allow time for expansion after the particle passes have not been successful. Experimental chambers have been constructed which use ultrasonic waves to expand the fluid, with the hope of building a continuously sensitive chamber, but these have not proved practical. All operating chambers must be expanded at the time the particle enters.

Many different fluids have been used since GLASER's original experiments with diethyl ether. Two classes of chambers have emerged: cryogenic chambers which use hydrogen, deuterium, helium or neon as the liquid, and heavy liquid chambers which use hydrocarbons, fluorocarbons and other heavier liquids. The technologies are quite different since the former operate at temperatures far below ambient and the latter somewhat above. By far the most useful chamber has been the liquid hydrogen chamber, which can use hydrogen, deuterium, neon and mixtures of hydrogen and neon, by changing the operating conditions. Hydrogen is unique in being a pure proton target, and deuterium is the simplest neutron target available. The relatively low density of hydrogen and deuterium yields less multiple Coulomb scattering of the particles and hence better momentum measurements.

Hydrocarbons such as propane (C_3H_8) have a higher density of free protons (not bound in a nucleus) than liquid hydrogen, but the difficulty of interpreting events from the carbon has limited their usefulness for studying strong interactions. They have been very useful in studying the

[1] Rapid-cycling chambers are being developed which should operate at 60 expansions per second [4.4].

weak decays of mesons where the event rate is low and the large density is a great asset. The fluorocarbon chambers (usually Freon CF_3Br) contain relatively high Z (atomic number) atoms of fluorine and bromine and are good detectors of γ-rays. Other fluids useful for γ-ray detection are xenon, methyl iodide (CH_3I), tungsten hexafluoride (WF_4) and neon. Helium chambers have been used in special experiments requiring an α-particle target.

The detector with the highest spatial resolution is the photographic emulsion, which has been used for many years. Special emulsions are produced in plates less than a millimeter thick and usually assembled in stacks to obtain the desired thickness. After the exposure the plates are separated, developed, and mounted on glass for scanning and measuring. Grain size in some emulsions is less than 0.5 μm. Their high density results in many particles stopping because they lose all their energy by ionization, and the high resolution allows a precise measurement of track length and hence an excellent energy resolution. However, the short tracks do not produce sufficient curvature in magnetic fields normally available to make their use practical. The main problem with emulsions is the difficulty of scanning and measuring them. They are measured in three dimensions using precision microscopes with calibrated stages and focusing mechanisms. Although some progress has been made in automating the process, it is still slow [4.5]. As with other complex nuclear targets, event identification is difficult.

All of the detectors above have essentially no time discrimination and are very limited in their ability to handle many particles in a short time interval. The scintillation counter has the best time discrimination of all detectors in use today, but its space resolution is of the order of a centimeter at best. The scintillation chamber [4.6] was an attempt to combine good time and space resolutions. It has been limited by the very high cost of collecting and amplifying the light output. The device became feasible with the development of effective image intensifiers which used electronic amplification of an image on its photocathode to produce an enhanced optical image on a fluorescent screen. Early attempts employed a very fast lens to focus the light produced in a plastic scintillator onto a chain of several image intensifiers. Images of the scintillations were photographed, but the small depth of field of the lens precluded practical applications. Better results were obtained with scintillators fabricated in small fibers and assembled in bundles which could be directly coupled to the fiber-optic face plate of an image intensifier [4.7]. However, the cost was too great and the device has been abandoned.

The spark chamber [4.8] is a detector which combines moderate space resolution with a high duty cycle and fast cycling time. The physical basis of its operation is the electrical discharge in a gas, and it is related to the ionization chamber, the proportional counter and the Geiger-Müller counter. These counters all use an electric field to attract the ions produced by the passage of a charged particle to collector electrodes and measure the current produced. If the field is small, as in the ionization chamber, only the primary ions are collected and the current is very small. As the field increases the electrons from the primary ionization are accelerated enough to cause secondary ionization, and the current depends upon the applied fields as well as the primary ionization. This is the operating region of proportional counters. As the field is increased further, the current becomes nearly independent of the primary ionization, and this is the region of the Geiger counter. Significantly higher fields result in a spontaneous discharge which is not suitable for counters.

However, the discharge does depend on the presence of the primary ionization. If a pulsed electric field is applied immediately after the particle goes through, the field can be large for a short period of time and the discharge will be strongly influenced by the ions. Early spark chambers were constructed from parallel metal plates immersed in a gas. When the chamber is pulsed after a charged particle passes through perpendicular to the plates, the spark which develops follows the path of the particle. The path of the spark depends upon the geometry of the chamber; for small gaps (~ 1 cm or less) the sparks are nearly perpendicular to the plates, but for wide gaps (~ 10 cm or more) they tend to follow the ion trail of the track which may be inclined as much as 45° to the plates. The electric field required is about 10 kV/cm; so the cost of a chamber will increase rapidly with the gap size. The gas used in nearly all spark chambers is a mixture of neon and helium. Plates have been constructed of steel, aluminum, carbon, lead, and other materials. The spark develops very rapidly; so the chamber can be used in a high flux of particles. There is typically a dead time of a few milliseconds before the chamber can be pulsed again.

By suitably shaping the pulse applied to the chamber the spark avalanche can be frozen at different stages of its development. The avalanche starts with small short streamers which originate at the site of the primary ions. These then connect and form an ionized path between the plates which then conducts the full spark. If a very fast-rising pulse of about 50 ns duration is applied, the streamers only grow to a length of a few millimeters and can be measured to better than 1 mm. Streamer chambers [4.9] have a spatial resolution which begins to approach that of the bubble chamber, have about the same time resolution as spark chambers and are sensitive to particles in any direction.

The plates in narrow-gap chambers have been replaced by planes of very fine wires. The spark current flowing through these wires is used to record the spark position directly (in one dimension) rather than having it recorded on film. The recording can be achieved by running the wires through ferrite beads which are read out later. The most popular method today uses a magneto-strictive delay line. All the wires pass over a magneto-strictive rod, and the spark current produces a magnetic field which generates a sonic wave in the rod and which is detected by a transducer at the end. The delay in the signal is proportional to the distance of the wire with the spark from the transducer. Planes of wires can be placed at right angles to provide two coordinates of the particle and a third at 45° can resolve ambiguities from multiple sparks.

4.1.2 Kinematics and the Measuring Process

The output of the visual detectors is usually a photographic film which records the tracks of all particles involved, in two or more stereoscopic views. The measuring process reduces the tracks to a series of (x, y) coordinates in each view. From a knowledge of the geometry of the detector and its cameras, the measurements in the stereo views can be transformed into coordinates in space (x, y, z). In a sense this marks the end of the measuring process; however the early stages of analysis have a strong effect on the measuring since it is here that the quality of the measurements is determined. We will present a brief discussion of what happens to the measured points (x, y, z) before they are presented to the experimenter.

Each track in space is approximately a circular helix with its axis parallel to the magnetic field direction (the z-axis). If the axis of the helix passes through (x_c, y_c) as in Fig. 4.2, the helix parameters can be related to the space coordinates of the i-th point by

$$(x_i - x_c) = \varrho \cos \phi_i \,, \tag{4.9}$$

$$(y_i - y_c) = \varrho \cos \phi_i \,, \tag{4.10}$$

$$z_i = \omega \phi_i \,, \tag{4.11}$$

where ϱ is the radius of the cylinder of the helix, $\omega = \tan \lambda$ is the pitch and λ is the dip angle. The angle ϕ_i simply locates the point on the helix. The most stringent test of the measuring process is how well the set of points (x_i, y_i, z_i) fit the helix. Measuring systems produce 10–100 points to determine the helix parameters ϱ, λ, and ϕ_i. Thus there are nearly two degrees of freedom per measured point and the problem is well over-determined. The fit is usually made using the method of least squares, and the quantity f_{RMS}, which is the square root of the sum of the squares of the deviations divided by the number of points, is a measure of the quality of the meas-

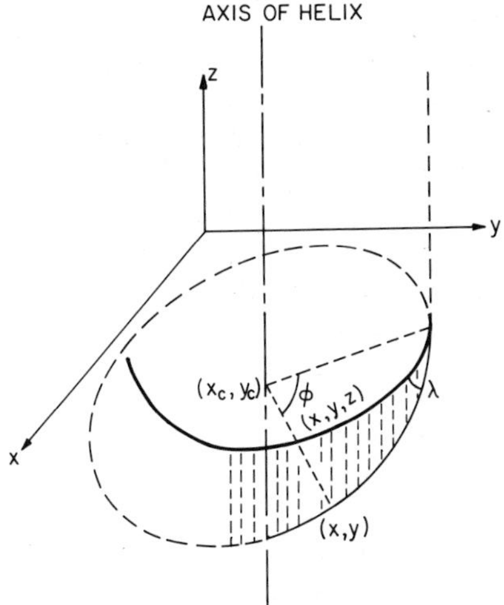

Fig. 4.2. The helical path of a typical track in a magnetic field

urement. The momentum components of the particle are determined by the magnetic field. The magnitude is

$$p=(C\varrho B)/\cos\lambda\,, \tag{4.12}$$

where C is a constant and B is the strength of the field. The components change as the particle changes direction, and the interesting values are the components at one or both ends of the tracks. Two corrections must be made to this simple model. First, the particle loses energy by ionization as it traverses matter. The fitted value of ϱ is assumed to correspond to the momentum at the center of the track and corrections are made at each end to account for the energy loss. These corrections depend upon the mass of the particle and separate corrections are made for each possible mass, taken from a table supplied by the experimenter. Second, the magnetic field is not uniform over the chamber and is not always exactly along the z-axis. This correction yields slightly different values of ϱ and λ at different points along the track. These corrections are made before f_{RMS} is computed. The output of this phase is a set of momentum components at each end of each track for each possible mass, together with the corresponding error matrices.

Having the parameters for each track, we next analyze the event itself to determine if it is consistent with the measured tracks, if it is complete, and what particles participate (mass and charge). The constraints which provide this information are conservation of energy and momen-

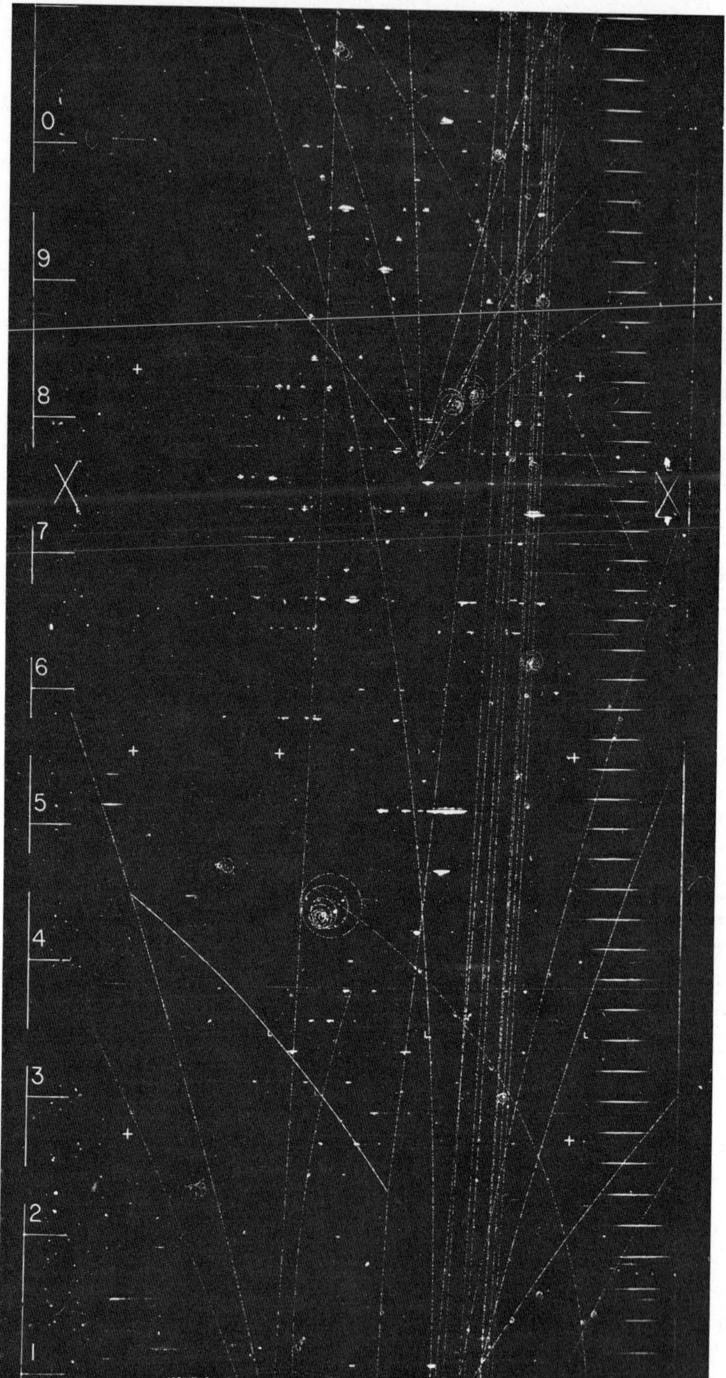

Fig. 4.3

tum. The energy of each track is found from the relativistic energy-momentum relation

$$E^2 = c^2 p^2 + m^2 c^4 , \tag{4.13}$$

where c is the velocity of light, p is the magnitude of the measured momentum, and m is the assumed mass of the particle. The total energy and each component of the momentum are conserved during the reaction, which provides four constraint equations. The initial state is usually well known; so if there are N particles measured in the final state, there will be $3N$ components of momentum measured, of which ($3N$-4) are independent. The hypotheses tested by the program are supplied by the experimenter and usually include all possible mass permutations for each track. Under the hypothesis that all secondary particles have been measured, the problem has four degrees of freedom ($f=4$) and the fit is a four-constraint (4C) fit. The fit to the hypothesis is made by least squares, in which the measured quantities are altered ("pulled") to obtain the best over-all solution to the constraint equations. The quality of the fit is expressed by the quantity $\chi^2 = \Sigma(\Delta p_i)^2/\varepsilon_i^2$, where the Δp_i are the changes in the p_i (pulls) and ε_i are the corresponding errors. There is usually only one hypothesis which has an acceptable χ^2 for a 4C fit. The equations can still be satisfied if up to four momentum components are missing or too poorly measured to be usable. The most frequent case is for one particle (e.g. a neutral) not to be measured at all. These 1C fits are usually much poorer than 4C fits and there are often ambiguities among different hypotheses. If no hypothesis has an acceptable fit, it is probably because more than one neutral participated in the reaction.

There are cases where there is no fit because of poor measurements, on mismatched tracks (the wrong track sets used from the different stereo views); these may yield unphysical results. Define the visible final-state energy and momentum by

$$\boldsymbol{P}_F = \Sigma_j \boldsymbol{P}_j \quad \text{and} \quad E_F = \Sigma_j E_j , \tag{4.14}$$

where the sum is over all measured particles, and let $\boldsymbol{P}_0$ and E_0 be the initial state quantities. Then the momentum and energy missing from the reaction are

$$\boldsymbol{P}_M = \boldsymbol{P}_0 - \boldsymbol{P}_F \quad \text{and} \quad E_M = E_0 - E_F . \tag{4.15}$$

Fig. 4.3. Example of a bubble chamber photograph. A 13.5 GeV/$c\,\pi^+$ beam is incident on the SLAC 2 m hydrogen chamber

If this were due to a single particle its mass would be

$$m_M = \pm [E_M^2/c^4 - P_M^2/c^2]^{1/2}, \tag{4.16}$$

where the positive sign is used for $E_M \geqq cP_M$. If m_M is small and negative or imaginary, it may be caused by normal measuring errors, but a large negative or imaginary value indicates gross mistakes.

4.1.3 Tasks in Visual Processing

The first task is to scan the film to find the interesting events. The events are usually separated into topological classes, and each class has its own set of scanning instructions. Frequently the film may be scanned for several classes at once, but scanning efficiency suffers if too many classes or classes which are too different are included. Problems in scanning are: achieving a high rate and good efficiency, and avoiding, or at least understanding, biases. In the absense of biases scanning efficiencies are easy to measure. If p is the probability of finding a particular event in one scan, $(1-p)$ is the probability of missing it, and is also the probability of missing it in a second independent scan. The probability of missing it in both scans is $(1-p)^2$ which is small unless the scanning efficiency is very low, and the list of events found in either one scan or the other is close to the number which exists. The two scans can be compared with the joint scan to obtain the two efficiencies. For events with low efficiency or where it is important to find them all, two complete scans may be required, but usually 10–30% are scanned twice to measure efficiency.

Scanning biases are much harder to detect and correct. Events tend to be missed when they blend with the background of beam tracks and stray tracks in the chamber. Biases arise when the form of the event itself causes it to blend in, rather than some particular confusion in one frame. Examination of the measurement results may reveal some biases by showing that certain regions of some variables are depleted. These may be geometrical biases in which, for example, certain regions of the chamber have few events. This is not a serious bias in the physics results, and corrections are easy. Biases in the physical variables, for example the scattering angle, are much harder. It may be hard to decide if the observed effect comes from a bias or from a physical cause. It is dangerous to try to correct data biased in this way because the correction nearly always involves a model which may not be correct or complete. If the bias is understood, the data from the biased region can be eliminated. The problem of measuring efficiencies and biases is one of the more compelling reasons for developing automatic scanning. Efficiency varies from operator to operator and even with time for the same operator. A machine would be more predictable and would only require a recalibration from time to time. Also a machine is a much simpler system than a human and could perhaps be modeled in the computer so that biases could be better understood and perhaps predicted. As we shall see, scanning is the hardest function to automate, and only a few systems have been successful.

The output of the scanning phase is a file which contains the list of frames scanned and data about the events found. From this file, a list is compiled of events to be measured. It goes, together with the film, to the measuring machine. Each event must be located and the information from the scan file must be verified. Then each track in each view must be measured together with fiducial marks which define the chamber coordinate system. Measurement is the slowest process in the chain of tasks unless it is automated. Considerable progress has been made here. A completely manual machine with a digital recorder measures 3–10 events per hour. The same machine under tight computer control can measure up to 50 events per hour, and an automatic machine can measure 150.

Once the events are measured they appear as digital records on tape or disk files, and all subsequent processing is by computer. The next step is to check the quality of the measurement. The event is checked to see if it agrees with the event-type code, has the correct number of tracks and fiducials measured, etc. Each track is checked for monotony and continuity. Monotony is a crude check of the measuring apparatus and requires that the points be in order in either x or y and that there not be any large gaps. Continuity tests that the points define a smooth curve, usually a circle or parabola. Since more points are measured than are required to fit the track, a small number may be discarded; else a failure requires re-measurement.

If the tracks must be matched in different views, it is done next. If the events are measured manually or automatically with checks by the operator at the end of an event, the tracks may have been matched manually. With automatic measuring, the tracks are not in order, and the program must decide which track in each view corresponds to a particular track in space. If the measurement is completely automatic, some views may contain spurious tracks, which are eliminated by this program. Next the tracks are reconstructed in space, using the measured points in several views and the geometry of the chamber and camera system. Tracks which are mis-matched or which are poorly measured give high values of f_{RMS} and are rejected and sent back for re-measurement. Events which are accepted are processed by the kinematic program and go to the physics summary file. The experimenter may decide that some events should be re-measured, but there are no automatic rejects after geometry.

Re-measurement time can seriously erode the measuring rate unless re-measurements are handled carefully. The most efficient procedure is to do all the processing, through geometry, on-line, and to do any required re-measurement there and then. The advantages are: a) the film is in place and need not be moved, b) the failure is usually on a single track or fiducial and only it need be re-measured, and c) the failure probably occurred because it was hard to measure, and repeated trials can be made without a great expenditure of time. This mode of operation requires at least time-sharing access to a sizeable computer and may not be possible. A carefully programmed small computer can make monotony and continuity checks which are good enough to catch most measuring errors. In most experiments it is not necessary to measure every event, provided no biases are introduced by the rejects.

4.2 Evolution of Film Measuring Devices

The development of machines to help analyze and measure film has occurred gradually since 1956 [4.10], but has approached full automation only within the last few years. It is not likely that the process will ever become completely automatic; the complexity of the film is so great that it is impractical to program every contingency. However, a few experiments [4.11, 12] have been completed in which the film was completely scanned and measured with human assistance only in confused regions. This mode of operation is called "automatic scanning" or "zero guidance" in the literature, since no human intervention is required to initiate the process.

Most film being measured today is measured in less automated modes which are called "minimum guidance" and "road guidance". In both modes the film is pre-scanned and indicative and topological information recorded. For the minimum guidance mode, the vertex position is measured to moderate precision with respect to a fiducial; while for road guidance several points are measured on each track. The pre-measurement process is slow, in many cases comparable to the measurement time on some manual machines. Minimum guidance systems are just beginning to reach reliable production, whereas road guidance has been used for many experiments at many laboratories.

All modes require human intervention either during the measuring process or afterward to correct mistakes made by the system or to resolve ambiguities. If the intervention occurs after the measurement it is called "off-line" or "fix-up". It has the advantage that expensive computing and measuring machinery is not tied up waiting for the human decision. However, it is effective only if all the pertinent information on the film has been extracted and stored in digital form. On-line intervention occurs right at the time the system has trouble or immediately after it thinks it has successfully completed an operation. Hence all the data are available for a definite decision, and the data can then be condensed to the form required for subsequent analysis. If problems arise from poor digitization, the fault can be corrected and a re-measurement performed immediately. With care, it is possible to obtain high acceptance rates with one pass over the film.

The achievement of this degree of automation is the result of careful attention to all phases of the measuring process, rather than to a single "break-through" in either hardware or software. In this section we will examine the evolution of the measuring hardware, concentrating on the main developments and illustrating their applications to specific machines. Many of the basic ideas were first applied in primitive form to manual machines; so the line between manual and automatic is not completely sharp. As used here, manual machines are guided by an operator who places a reticle on the point to be measured; automatic machines digitize either all of the film or selected portions under computer control, and semi-automatic machines automatically digitize a portion of the film selected by an operator.

4.2.1 Automation of Mechanical Functions

The earliest measuring machines were completely manually controlled. A precision stage used to move the film in two dimensions was activated by lead screws of a very fine pitch, typically 1 mm. A projection system formed a magnified image of the film on a screen viewed by an operator. Encoders mounted on the screws were read out under operator control and stored on punched cards or tape. Indicative information was encoded in switches and read out with the data. Because of the precision required, the machines were heavy and very expensive.

Tracks in bubble chambers are formed of strings of bubbes, each 0.1–0.3 mm in diameter. The bubble size depends on the operating conditions of the chamber when the pictures are taken. Smaller bubbles produce lower contrast and hence are difficult to scan. To save film cost, the image is demagnified as much as possible without losing significant resolution. Most chambers use a demagnification of 10X–20X, resulting in bubble images of about 13–30 μm diameter. It is desirable to measure the center of the bubble to within 5–10% of its diameter, i.e. to a few μm. The encoders on the lead screws usually have a least count of about 1 μm, somewhat smaller than the setting error of human operators.

Measuring to 0.1 of the diameter of a 0.2 mm bubble in a chamber 2 m long requires a precision of 10^{-5}. A machine which is stable against random vibrations of this magnitude must be very massive; the stage which carries the film has about the size and weight of the bed of a small milling machine. The fine screws which drive the stage are heavily loaded, and their maximum speed is limited to rather low values. The early encoders had mechanical contacts (brushes) which were also subject to excessive wear at high speeds. Since the encoder is coupled to the stage motion through the screw, wear of the screw threads means a deterioration of the measuring precision. Major improvements in the stage drive have been made by replacing the angle encoder with a linear incremental encoder of the Moiré fringe type directly on the stage, and by replacing the lead screws with recirculating ball screws, which are both precise and fast, or with hydraulic or pneumatic actuators. These improvements have increased the maximum stage speed by about a factor of 50, from 2–3 mm/s to 100–150 mm/s. The new encoders were made possible by improvements in optoelectronic devices, particularly phototransistors, which have large signal to noise ratios, and by reliable integrated circuits which can count at an undetectable error level.

The early machines had stages which were positioned by hand or with simple motor drives. These were followed quickly by servo drives which allowed the operator to easily move the stage to any point. The servo system was then used to implement an automatic track centering and track following circuit. From the operator's view this appeared as a lighted reticle of two parallel lines and an arrow (Fig. 4.4) which could be turned to any orientation. When a track was positioned between and parallel to the lines and the circuit activated, the stage would move so as to center the track between the lines. A foot pedal was used to control the speed of the stage in a direction parallel to the reticle while the circuit kept the track centered. As the track direction changed the operator had to adjust the orientation of the reticle to keep it parallel.

The mechanism, in its simplest form, consists of a slit which moves in a direction perpendicular to its length. A small portion of the projected image is focused on the plane of the slit and the light which passes throught the slit is collected by a photomultiplier. The photomultiplier output is differentiated to form the error signal for the servo system. The zero-crossing of the derivative is required to be at the center of the slit motion. If E [volts] is its displacement from the center, drive voltages $E \sin \alpha$ and $-E \cos \alpha$ are applied to the x and y servo amplifiers. If v is the track follow voltage (from the foot pedal) voltages $v \cos \alpha$ and $v \sin \alpha$ are added to the x and y amplifiers. The track follower on the MPII machine at LBL [4.13] can follow a track at several cm/s, recording points as it goes.

Since most of the measuring time is spent moving the stage and recording points, these improvements provided the major gain in efficiency over the simpler machines. However, other routine chores require con-

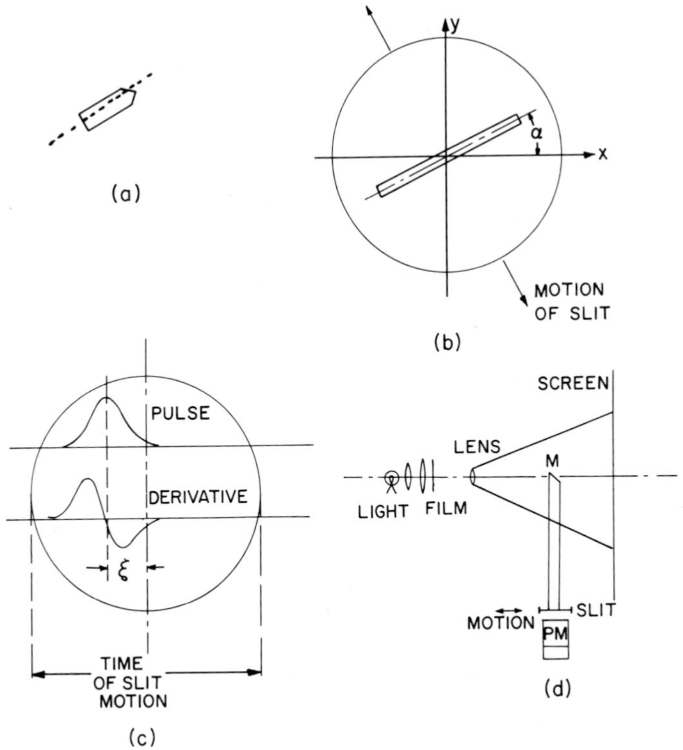

Fig. 4.4a–d. An electro-optical track centering system. (a) A track properly aligned in the reticle. (b) The geometry of the slit. The motion perpendicular to the slit may be linear or sinusoidal. (c) The track signal and its derivative with error signal ε. (d) The geometry of sampling the image

siderable time and can be improved by suitable automation. Moving the film is one of these, and is particularly time consuming for rare events with many uninteresting frames between them. Manual film advance requires much starting and stopping because the operator cannot read the frame numbers when the film is moving at high speed. Modern bubble-chamber films have frame markers located in specific positions on each frame and also have the frame number recorded in a binary code. Hence photoelectric detectors can be used to count frames, and even to read the numbers, as the film is moving. The problems of film handling are very similar to those of magnetic tape drives and similar approaches have been used on both. A typical drive consists of a supply reel with a torque motor which feeds film into a loop formed of a vacuum column or mechanical idler. A capstan or pinch roller pulls it from the supply loop through the film gate and into the take-up loop. Another torque motor pulls it from this loop and stores it on a take-up reel. These film drives can advance film at several m/s and stop at a predetermined spot on a predetermined frame.

Another time-consuming task is recording all the indicative information about the event. It is necessary to uniquely identify every event which is measured and to provide information to all the processing programs about the topology, the type of incident particle, etc. The first method of doing this was to record the data on scanning sheets at the scan tables; then the operator of the measuring machine encoded this information in switches which were recorded along with the data. Significant improvements in this procedure only occurred when the machines were put under computer control. With computer control all active events can be kept in a central file which is available at each stage of processing. The initial entry is made at the scanning table and it is up-dated at each stage and finally cleared when the output file is generated for the physicists. This procedure not only saves a great deal of time, but also avoids errors which occur every time data are transcribed.

The use of a computer also allows great time saving in the measuring process. The computer knows the relative positions of all the fiducials on the film. From the scan file it can learn the number of frames to be advanced, the approximate location of the vertex, and the number of secondary tracks. Thus it can advance the film and move the stage approximately to the position of the first fiducial. When the operator positions and records the first fiducial, it knows the others very well and can move to each in turn. It then moves to the approximate vertex position. The operator centers on the vertex and starts track-following on the first track. Track following can be done with the opto-electronic device described earlier or it can be done by computer. After the first few measured points, the program can extrapolate to the next point quite accurately and move the stage to it. At the end, the computer returns to the vertex and the next track is followed, etc. The Cobweb system [4.13] at Berkeley is a good example of computer controlled measuring machines. The measuring rate is increased by about a factor of five over completely manual control.

4.2.2 Automation of Precise Measurements

Further increases in performance and the development of automatic scanning and event recognition require a means of measuring points automatically. With current technology, the only practical way to proceed is to scan the film serially and to present the results sequentially to a computer. A spot of light is programmed to traverse the film along a specified path and the opacity of the film is measured with a detector. The information on a bubble chamber film is binary, i.e. at a particular point there is or is not a bubble. However, the film itself is grey, usually with different levels of background on different areas of the film. The detector must extract the binary information from local variations in opacity.

The functions required of a flying spot digitizer (FSD) are: a) generating an intense spot of light of the required shape, b) forming an image of the spot on the film, c) deflecting the image so that it follows the required path, and d) detecting the light which passes through the film and extracting the binary information. The different machines differ primarily in the way they implement the functions a) and c). There are two broad categories: systems which use conventional light sources and deflect the beams by mechanical motion, and systems which use a cathode-ray tube (CRT) for generation and deflection of the light. Function b) is usually performed by a commercial lens, which may be expensive but does not present any unusual design problems.

Illumination is always a problem because of the small spot size and the large scan speed. Bubble images are typically 20–30 μm in diameter, with edges which are not sharply defined. Good signal to noise ratios require a spot size of 15 μm or less. Scan speeds vary from a few meters per second for some CRT devices to several hundred meters per second for some mechanical scanners. Thus the time required for the spot to traverse a bubble is of the order of a few μs or less. For mechanical scanners the light intensity is limited by heating effects in the source and scanning speed is limited only by light intensity or by the mechanical strength of the elements. For CRT devices the intensity may be limited by heating (burning the phosphor) or by saturation of the phosphor; i.e., above a certain level, an increase in electron beam current does not result in an increase in brightness. Scan speed may also be limited by the phosphor decay time. If the speed is too large, the spot can move a bubble diameter before the light from the previous position has died away, resulting in a smeared spot. Different deflection schemes will be presented by describing the machines in which they originated.

The first machine constructed to scan bubble chamber film with a spot of light (the HPD) was built by Hough and Powell [4.14] at CERN. A diagramatic sketch of the device is shown in Fig. 4.5. The entire frame is scanned by a combination of mechanical stage motion and a flying spot of light. A disk containing radial slits rotates at constant speed. The intersection of a radial slit with a fixed vertical slit generates a spot of light in the shape of a parallelogram which traverses the film in a straight line at approximately constant speed. During the scan the stage carrying the film moves at constant speed in a direction perpendicular to the scan

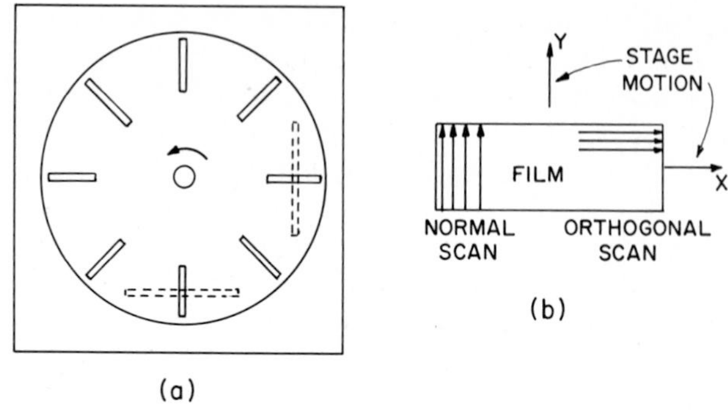

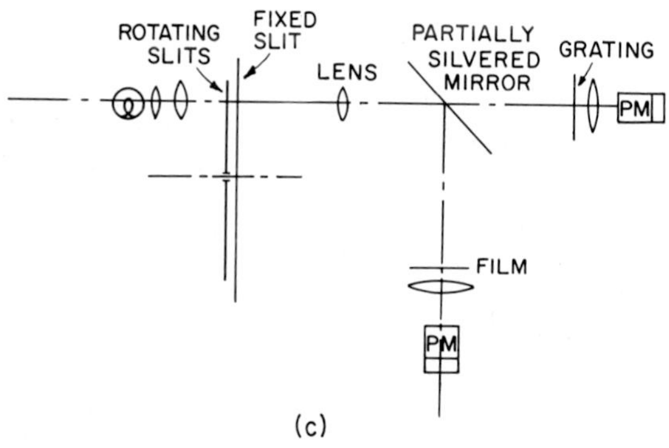

Fig. 4.5a–c. The HPD optical system. (a) The arrangement of radial and fixed slits. (b) The scan patterns for normal and orthogonal scans. The stage motion is perpendicular to the scan lines. (c) The optical system

line (the X-direction). Typically the disk contains eight slits and rotates at 60 rev/s to produce 480 scan lines/s, each separated by about 50 μm. This scanning rate is 25 m/s for the spot and 25 mm/s in the direction of the stage. A $50 \times 150\ \mathrm{mm}^2$ image requires about six seconds to scan; three views require about 20 s; so the maximum rate is 180 triads per hour.

All raster scan systems have a problem with tracks which make too small an angle with the scan line, since only a few scan lines will cross the track and signals from several bubbles may coalesce and not be detected. Tracks which make an angle of less than 45° with the scan line are measured on a second scan in the orthogonal direction. A fixed horizontal slit is used instead of the vertical one and the stage is moved in the orthogonal direction (Y-direction).

The coordinates of the spot on the film are derived from two devices. A Moiré fringe encoder on the stage measures its position (X for the normal scan and Y for the orthogonal scan). Part of the light from the spot is diverted before it reaches the film and focused onto a precision grating. A detector and counter count the number of grating lines it crosses (the W-coordinate). The least count for both coordinates is of the order of 1 μm. The W-coordinate may have a different least count than X and Y. Because the stage is always in motion, the scan lines are slightly tilted with respect to the stage, and the coordinates are not quite orthogonal.

A mechanical scanner which has become very popular recently is the Spiral Reader. In Europe it is called the L.S.D. (Lecteur a Spirale Digitizée). Originally proposed by McCormick [4.15] at Berkeley, it lay dormant for several years until it was put under computer control and brought into production in 1966, when it measured 400000 events. It was inspired by the simple observation that near the vertex of an event all the tracks are nearly radial. A plane polar coordinate system replaces the usual (x, y) system, with the origin at the vertex. Figure 4.6 shows the appearance of an event in the usual (x, y) system and transformed to the (r, ϕ) polar coordinates. Notice that the region near the origin $(r \approx 0)$ is very uncluttered in the polar system and very confused in the rectangular system. Also crossing tracks even near the vertex do not reach the origin. This is a great advantage in the early stages of track following.

The machine must be classed as a semi-automatic machine since the operator must place the origin of the polar coordinates (a reticle) on the vertex. The film is mounted on a movable stage with x and y encoders, and an optical projection is provided for the operator. The spiral scanner is composed of a rotating periscope which collects a portion of the image and transmits it to a detector (Fig. 4.6). The ϕ-scan and r-scan are separate motions. The path of the light is from the objective lens L to a long, narrow plane mirror A mounted on the inner surface of a cone, to a small mirror B which reflects it into a light pipe to the photomultiplier PM. To generate the ϕ-scan this assembly rotates about the common axis of the cone and lightpipe, which is also the optic axis. The small mirror B which forms the aperture is in the shape of a narrow slit. When a track segment parallel to the slit sweeps across it, the PM signal is strongly modulated. The PM and light pipe are driven linearly along the optic axis to produce the r-scan. When the mirror B is near the apex of the cone, r is near zero; as B moves up the optic axis, r increases. The result is that the slit B sweeps outward in a spiral from the vertex, sampling tracks as it goes. The positions of the points are computed from x and y, the coordinates of the vertex, and r and ϕ, the coordinates of the track with respect to the vertex. The (r, ϕ) coordinates are measured on a projected

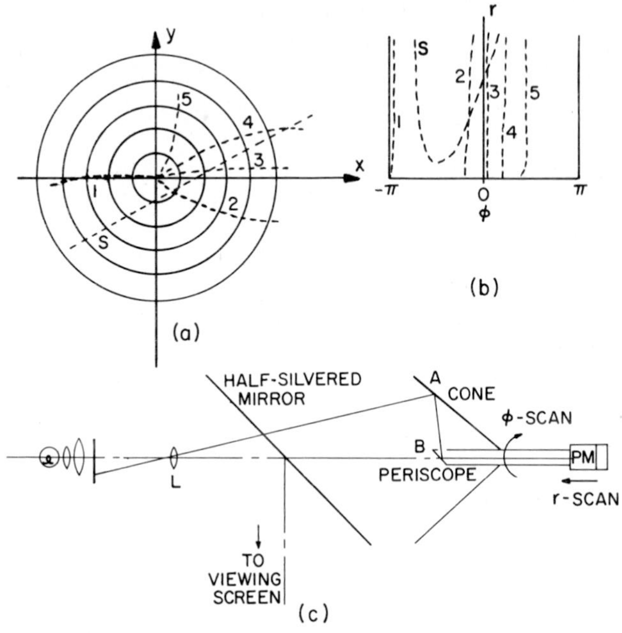

Fig. 4.6a–c. The spiral reader scan. (a) A four-prong event with crossing track in the (x, y) system. (b) The same event transformed to the (r, ϕ) system. (c) The optical path and scan motions

image rather than on the film; hence there must be corrections for optical aberrations. For good optical components, the effects are small and only the final average points (not each digitization) must be corrected.

There are limits to the quality of its results; so it cannot be used for all measurements. It is severely limited in following tracks with large curvature since they soon make too large an angle with the slit to be digitized effectively. The excursion in the r-direction is limited by the size of the rotating equipment required, by the fact that even high momentum tracks do not stay parallel to the slit, and by poor signal to noise ratios at reasonable scanning speeds.

Another mechanical scanning device, also from Berkeley, is the Scanning and Measuring Projector (SMP) [4.16]. It was designed to eliminate the massive film stage and its expensive control components, so that it could be constructed at modest cost. Initially it was used only for measuring because it was felt that it was too valuable to be wasted as a scanner. However, for event topologies which occur frequently, the scanning takes no more time than finding the pre-scanned events. It was designed from the principle, borrowed from surveyors, of making moderately precise

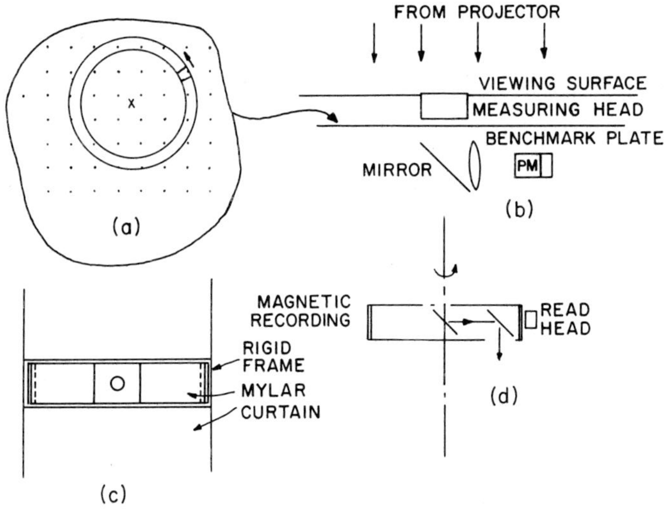

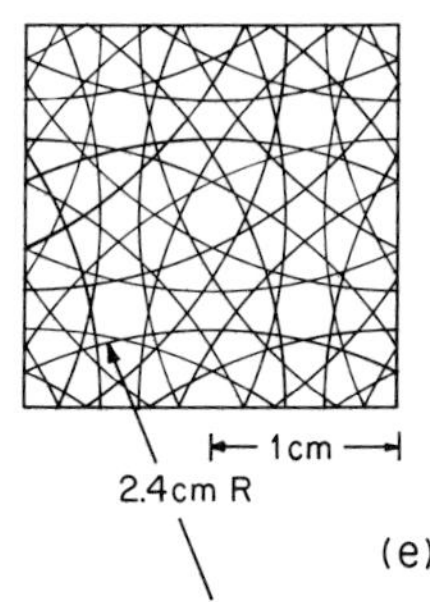

Fig. 4.7a–e. The SMP measuring system. (a) The locus of the periscope aperture over the benchmark plate. (b) Mechanical arrangement of the measuring head, benchmark plate and detector. (c) The viewing surface. The measuring head is free to move on the surface. (d) The optical path through the periscope in the measuring head. (e) Locus of all points which can be digitized by an SMP

measurements from precisely known points or bench-marks. Measuring to 1 μm on a 10 cm frame requires a precision of 10^{-5}, whereas the same point measured with respect to a bench mark 1 mm away requires a precision of 10^{-3}, and finding the bench marks requires a precision of 10^{-2}. Problems of vibration were reduced by keeping the film stage fixed and making all measurements on the image plane.

The points are automatically digitized by a small, light-weight scanning head which is positioned by the operator on the projected image. The position of the scanning head is digitized by simple linear digitizers to within $\frac{1}{4}$ of the benchmark spacing. The benchmarks (see Fig. 4.7) are transparent holes 0.1 mm in diameter in an opaque glass plate, evenly spaced on a 1×1 cm^2 grid. The plate is located in the image plane of the projector and covered by a set of opaque curtains which carry the scanning head. This latter is a rotating periscope with a $\frac{1}{2} \times \frac{1}{2}$ cm^2 aperture, which intercepts a portion of the image and sweeps it in a circle of radius R. When the image of the aperture sweeps over a benchmark, light passes through and is collected on a photomultiplier. Tracks which cross a benchmark modulate the light and are detected. A magnetic recording which rotates with the periscope contains the angle θ of the periscope (or equivalently $\cos\theta$ and $\sin\theta$). The points are computed from the position of the periscope center (x, y) and the periscope angle θ and known radius R to a precision equivalent to a few μm on the film. Notice that any point digitized must lie on a circle of radius R centered at the benchmark which digitizes it. The locus of all such points for $R=2.40$ cm is shown in Fig. 4.7e.

Lasers have not been used extensively because fast, precise deflection schemes have not been developed. The intensity and quality of the spot from even a low-power laser would enhance the signal to noise ratio attainable and increase the possible scanning speed. Frisch *et al.* [4.17] at Cambridge have built a mechanically deflected laser scanner called Sweepnik. The laser spot is astigmatically focused into a line which is caused to rotate at constant speed about an axis outside itself by a dove prism to produce a sunburst scan pattern. An encoder attached to the prism gives the angle at any time. Small mirrors driven by loudspeaker voice coils deflect the pattern under computer control. The mirror positions are obtained from interferometers which use the beam from the back of the laser. Track following uses the same techniques developed for manual machines and is initiated by the operator. With computer control, some automation has been possible in the measurement of fiducials and in track searches. The slow scanning speed of about 0.1 m/s limits its ability to be automated effectively.

A programmed CRT scanner (PEPR) was proposed by Pless [4.18] shortly after the beginning of the HPD development. It was controlled by a small computer and a sophisticated analog and digital controller.

Comceptually the CRT scanner is very simple. The electron beam of the cathode ray tube is focused into a small spot on a fluorescent screen. A scanning pattern is generated by deflecting the beam, which also moves the spot. An image of the spot can be rapidly moved to any part of the film; so it can be programmed to produce nearly any scanning pattern. However, extreme care is required to produce a tube which is adequate for accurate scanning of bubble chamber film. The problems fall into the two categories of light intensity and uniformity, and of stability and linearity of the deflection. High intensity and small, uniform spot size are required for a good signal to noise ratio at high scanning speeds. Deflection linearity is not essential since the device is under computer control and nonlinearities can be corrected, but stability is required to reduce the need for frequent calibration.

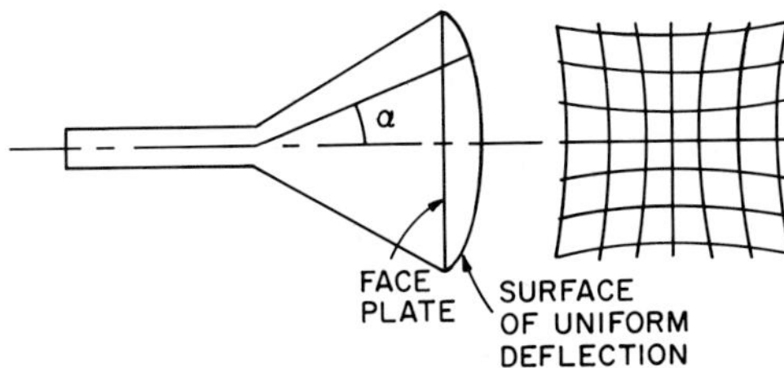

Fig. 4.8. Deflection geometry of a CRT and the resulting pincushion disortion

Nearly all precision CRT scanners use electromagnetic focusing and deflection, which has several advantages over electrostatic methods. The critical elements are outside the tube and can be adjusted in orientation and position, and the currents in these elements can be adjusted without disturbing the electrostatic field distribution inside the tube. The usual magnetic elements are: a focus coil which is used to minimize the overall spot size, two astigmatism correction coils which change the focus slightly to compensate for astigmatism in the tube structure, and two deflection coils which deflect the beam in two orthogonal directions. A small, low-inductance focus coil is often used for dynamic corrections to the focus, and some tubes benefit from a "halo coil", which is an auxillary focusing coil located around the electron gun. This latter coil improves the electron optics of the electron gun and helps reduce a halo effect which arises from electrons scattering from the edges of apertures. The PEPR system also has a highly stigmatic lens (a diquadrapole) which deforms the nearly circular spot into a short line of variable orientation and length.

A problem with all CRT's is that all focusing and deflection forces work by changing the angle of the electron beam, as shown in Fig. 4.8. The result is that the natural surface for such a device is a sphere, because with ideal focus and deflection fields, the surfaces of best focus and of uniform deflection sensitivity are spheres. Since chamber film is flat, it is necessary to transform the spherical electron field to a flat optical field. This can be done in electron optics by having a flat screen on the CRT and correcting the focus and deflection currents, or in the light optics by using a spherical screen and a special lens [4.19]. A good method might be to use a fiber-optic face plate with a spherical phosphor surface inside the tube and a flat polished surface outside; however, fiber optics do not have the required resolution. With a flat face-plate the distortion, called pincushion distortion, is shown in Fig. 4.8. The distortion in the deflection is not as serious as in the focus, because a simple transformation on the measured points will correct for any deflection errors, but a poorly focused spot produces poor digitizations. Both effects are corrected by reducing the currents in the coils at small radius and increasing them at large. Analog multipliers can be used to compute the radius from the deflection currents and make approximate corrections, assuming ideal cylindrical symmetry. The final corrections must be made by computer to compensate for aberrations in the tube and coils. The astigmatism coils have only a small effect which can be computed. The parameters in these correction formulae are computed from measurements on a known grid.

Several different schemes are employed to determine the spot position on the CRT face. The simplest is the approach used in POLLY [4.20] at Argonne, in which the system is assumed to be stable and transformation formulae connect the coordinates with the deflection coil currents. The parameters of the transformation are found by measuring the calibration grid. In practice the system is found to be stable for a period of days, but the calibration is run routinely during every shift. PEPR uses separate lenses and beam splitters to focus the spot onto precision gratings oriented along x and y. A stable oscillator locked to the grating frequency as the spot moves is used to interpolate between rulings. At Saclay [4.21] part of the light is split from the main beam after the lens and focused on a grid. The deflection is programmed so that grid crossings are detected first in one coordinate and then the other. The final position within the grid square is found from interpolation formulae, which need be only linear or quadratic rather than fifth order as is usually required.

All of the systems generate the scan pattern by first deflecting the spot to the region of interest and then starting a programmed sweep. The deflection is achieved by setting the currents in the deflection coils with a digital to analog converter (DAC) which can have a least count of from 2 μm to 40 μm, depending upon the system.

The main difference among the different CRT systems is in the method of forming the scan element and sweeping it over the region of interest. PEPR, as mentioned earlier, uses a highly stigmatic magnetic lens to form a line about 20 μm wide and up to 2 mm in length to detect track elements parallel to itself. The element is moved (swept) up to 2 mm in either x or y by a linear ramp current added to the main deflection current. The sweep speed can be selected as either 200 m/s or 20 m/s, and the scan element can be rotated in 1° intervals over the full circle. The area of interest is scanned at different orientations and hit coordinates are recorded with respect to the programmed starting position. Fig. 4.9 shows scan patterns for several systems.

POLLY took a somewhat different approach, which was inspired by a directional histogram algorithm used by some HPD programs (see Subsect. 4.3.1 below). In this algorithm a picture element, called a slice (Fig. 4.9c), is generated and a count is made of the number of digitizings within it. A large count indicates a line segment parallel to the slice. In the HPD program the digitizings are selected from an ordered array in computer memory; the idea with POLLY is to generate the slice as a

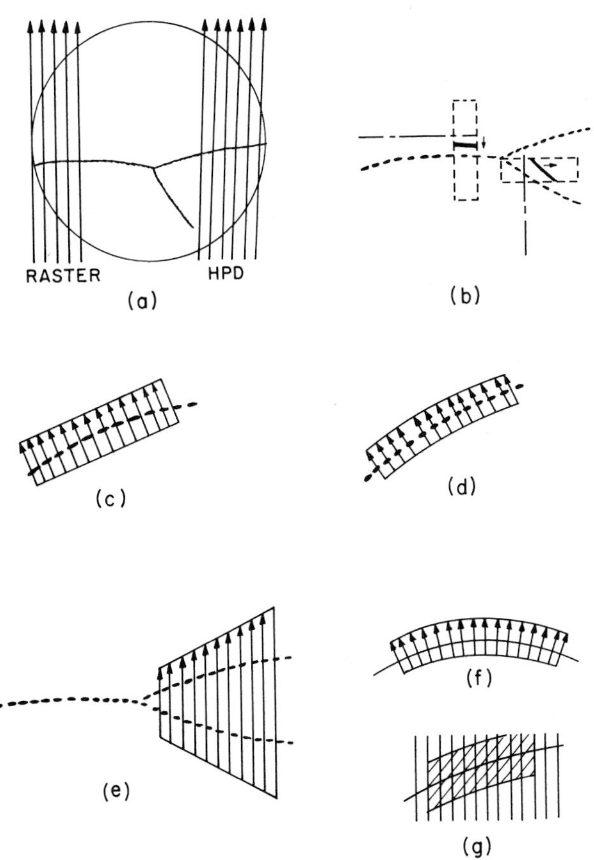

Fig. 4.9a–g. Some sample scan patterns. (a) A simple raster scan with step displacement between scan lines and the HPD scan with continuous displacement. (b) Two examples of the PEPR scan. (c) The POLLY slice scan. (d) A curved slice scan with parallel scan lines. (e) A wedge scan to look for secondary tracks from a known vertex. (f) A curved slice scan with radial scan lines. (g) A slice generated from raster-scan data

scan pattern on the film and to digitize only the points within it. The CRT is moved to the starting point by programmed deflection currents and is then swept in a series of short sweeps perpendicular to the slice direction (Fig. 4.9d). If the slice scan is oriented parallel to a track and centered on it, all the track hits fall at the centers of the scan lines. The position of the hit is read from a counter which is synchronized with the scan. Only three states are detected for each line: no hits, one hit, or multiple hits, and only lines with one hit are used. The average coordinates of the hits on these lines form the center coordinates of the detected line element. Notice that noise within the slice scan will be included in the average points.

The Illinois DOLLY [4.22] uses the same scanning scheme, but forms the scanning pattern digitally and deflects the spot with a digital to current converter. Two adders are used, the first to move the spot to the starting position for a sweep and the second to advance the spot during the sweep. A small diode memory contains a fixed program for generating the scan pattern, using parameters which are stored by the computer in specified registers. This allows the length, orientation, resolution and speed of the sweep to be easily varied. The program which moves the spot to the start of each scan line follows a parabolic or circular path, again using variable parameters. It is possible to use long slice scans, even when following highly curved tracks. The parabola can also contain a second-order correction to the pincushion distortion. The digitizings are stored in a first-in, first-out stack which has room for four hits per line. The hit coordinates are: the distance from the edge of the parabolic path for the first, and the scan line number for the second. If the slice scan is parallel to the track, these coordinates are the tangent and normal respectively, which is very convenient for forming histograms and average values.

At Purdue [4.23] the digital control has been extended considerably by using a digital processor to control the formation of the scan pattern, the buffering of points, and the pre-processing of hits. Both programs and data are stored in a fast semi-conductor memory by a small control computer. All components of the CRT and track detection systems are accessible to it and are addressed just as the memory is. An airhmetic unit and a number of counters under hardware and program control complete the system. The only hard-wired component is the scan line generator which uses parameters furnished by the program. These include the length, orientation, resolution and speed of the scan line. Any of these parameters can be changed between scan lines; so that except for the restriction of having straight scan lines, the scan pattern is completely flexible. The hits are buffered under program control to any part of memory. The most important pre-processing functions include forming histograms and average values. Usually the hits are histogrammed in bins along the slice (Fig. 4.10a), and the processor returns the number of hits as well as the average coordinates (in the tangent, normal coordinate system) of the hits in each bin.

The success of the spiral scan in the Spiral Reader has inspired a similar innovation in a CRT device at Duke [4.24] called RIPPLE. A circular scan of prescribed radius is generated by forming deflection currents proportional to sin ωt and cos ωt, where ω is the scan frequency, and the center of the circle is set by DAC's. Gates are provided to transfer

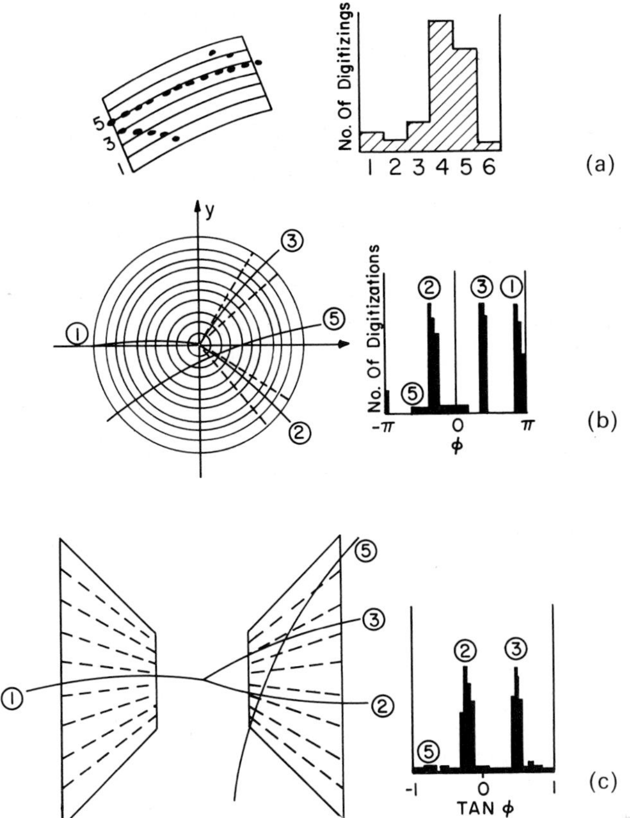

Fig. 4.10a–c. Examples of scan patterns and the resulting histograms. (a) A slice scan with noise. (b) The RIPPLE angular histogram. The Spiral Reader scan is similar. (c) A wedge with a crossing track

hits to the computer if they fall within either of two specified angular wedges (see Fig. 4.10b). The machine also forms angular histograms of all hits into 64 bins. Tracks which cross the center of the circle will form peaks in bins which are 180° apart; this is the case in following a track. If the center is near a vertex, all tracks connected to it will form pulses. Tracks far from the center populate many bins and contribute only to a background. As it approaches the center the pulse will sharpen. In track following the gate is narrowed to accept data only near the track direction. Extra pulses in the histogram warn of another aproaching track and the second gate can be opened to follow it in. The combination of an automatic histogram and polar coordinates forms a powerful tool for handling track intersections of all types. Its main disadvantage is that most of the scanned area is not used, and the scanning time is long for the amount of relevant data collected.

Light collection and track detection problems are similar in all of the FSD systems, although they may be more troublesome in some than in others. Either scanning speed or the signal-to-noise ratio or both are limited by the light available. The main objective lens must have large aperture and a fairly short focal length. This requires a thick lens with large elements; those in current use have an aperture of $f/3$ to $f/4$ for an image at infinity. Used in 1:1 to 2:1 demagnification the effective aperture is reduced to about $f/6$. To reduce aberrations the lens elements must be designed to the spectrum of the light source. All of the optical surfaces are coated to reduce reflections, but at best only about 25% of the light is transmitted.

Track detection is influenced by the quality of the spot and the size and shape of the bubble. All FSD systems scan at constant velocity, except for transients at the start and end of the sweep. A bubble on the film is transformed into a pulse in time at the output of the photomultiplier. The task of the track detection unit is to measure the center of the pulse and to provide auxiliary information which may aid in discriminating between bubbles and other marks which may be on the film. The pulse is approximately Gaussian; so it can be specified by its height and width. Several methods have been used to find the center of the pulse. For symmetric pulses they all give the same result, whereas for asymmetric pulses the meaning of "center" is ambiguous, and different methods give different answers. The peak of the pulse can be found by differentiation and detecting the zero in the derivative, or the centroid of the pulse can be found by integration techniques. The most common method used currently is simply to set a discriminator on the voltage comparator above the noise level and to measure the turn-on time and the width of the pulse. The last method is not good for asymmetrical pulses because it gives a pulse position which depends upon pulse height. Fortunately there is no evidence that such effects are significant.

Complications arise from noise which comes from several sources. Phosphor noise is not very serious with modern cathode ray tubes and can be largely eliminated by subtracting a signal derived from a reference photomultiplier which looks at the spot before it passes through the film. Since both signals have the phosphor modulation on them, it will be nearly cancelled. The photomultiplier noise is not very large if there is sufficient light. It also has a very broad frequency spectrum; so it can be greatly reduced by filtering. By far, most of the noise comes from the film itself, arising from imperfections in the lighting and the bubble chamber construction and operating conditions. Spurious, uninteresting tracks are not considered noise in this context. If the bubble size is uniform and the scanning speed is constant, then the signal pulse will have a well-defined shape in time, and its frequency spectrum can be determined. Tuned filters can be constructed whose transfer characteristic is very similar to the pulse spectrum, and noise outside this band is greatly reduced. As a practical matter, one usually keeps the dc component of the signal; so the filter reduces only the high-frequency noise. Much of the remaining noise can be eliminated by digital filtering techniques. For example, digitizations can be rejected if the hits are wider or narrower than certain limits, obtained from the known bubble size. This filtering may be done in hardware or programmed in the computer.

The only other large component of noise is a slow variation of film opacity over the frame. This has too low a frequency to cause spurious digitizations, but rather tends to push the digitizer system outside its operating range. Automatic gain control circuits are used to keep the signal within bounds. A number of schemes have been devised to determine the average opacity of the film in the neighborhood of the spot and to set the gain of the system. CRT systems have an easier problem than do raster scan systems since the area scanned in one operation is very small. The background opacity can be sampled using sampling or averaging techniques on the current scan line, by saving the value found in the previous scan line, or by making a dummy sweep with the spot defocused so as not to detect the bubbles. In raster scan systems, such as the HPD, the background varies during the scan line and only the first option is available.

4.2.3 Data Selection and Storage

The quantity of data which must be collected and stored from a measuring machine varies over a very wide range, depending upon the type of scan, the amount of pre-processing, and availability of feedback from subsequent analysis programs or from an operator. The total information content of the film varies greatly with chamber size, but estimates can be made. The highest estimate assumes a cell size comparable to the resolution of the machine and yields $\sim 10^9$ bits. In a bubble chamber film, most of the cells are empty; so this is not an efficient method of storing data. A raster scan of the HPD type inputs about 36 bits per hit. Assuming 40 hits per scan line and 2000 scan lines per frame leads to an estimate of 10^6 bits. This is close to the minimum requirement for digitizing all the significant data on the film. Because of the high cost of large memories, a great deal of effort has gone into reducing the quantity of data at the earliest possible time.

A process which greatly reduces the amount of data is the preselection of the region of film to be measured. In the full guidance mode, rough measurements are made manually of one fiducial, the vertex and two points on each track. The control computer constructs circular roads wide enough to contain all the track points and uses these to gate the desired digitizings from the measuring machine. All the data outside the roads are disgarded. SMP's use nearly the same procedure except that the roads are made in real time as the operator moves the measuring head to keep the track in the aperture.

The other method of reducing the amount of data to be stored is to process it as it comes in. This requires a great deal of computing power, especially if the data are from a fast scanner such as the HPD.

The CRT systems use the film itself as a random access memory. Since the scan pattern is variable, its area, and hence the number of spurious digitizings, can be reduced as more information is gained about the track. The scan patterns used almost never result in as many as 100 digitizings stored at one time, and these are quickly reduced by a factor of 10–20 into average points. The total amount of data within the computer at any time need not exceed a few thousand bits.

4.3 Programming for Automatic Machines

It is evident from the descriptions of the processes and machines involved that there is very little conventional pattern recognition required to analyze high-energy visual data. The tracks are very simple curves, well approximated by parabolas or circles, and the fiducials are simple crosses. The events are simple open topologies and except for neutral decays are simply connected. The problems are elimination of noise, making precise measurements of the points, and making a stereoscopic correlation of track images in different stereo views. These problems would be trivial except for the large quantity of both data and noise contained in the film. The success of the operating systems depend upon careful procedures at each stage. Some of these procedures are presented in this section.

4.3.1 Track Detection Procedures

Track detection proceeds in several stages: a) Correlations are made among raw digitizings to establish points belonging to the same track element, b) track elements are linked to determine the slope and curvature of the track, c) the track is extrapolated using these parameters, and a search is made for additional points, d) the parameters are refined and narrower limits placed on acceptable digitizings, and e) the tight criteria are applied to the entire track to choose the points to be averaged.

Steps a) and b) are the track initialization process, c) and d) are the steps in track following and e) is the precision measurement of the track.

Track following is easier than track initialization and was developed earlier, and some insight can be gained by examining the early electromechanical track follower (Subsect. 4.2.1). From the operator's viewpoint it appears as a reticle in the form of a boat which can be translated and rotated (Fig. 4.4). As long as the track is within the boat and roughly parallel to its beam, it will be followed. In the absence of a track signal the boat moves (with respect to the film) directly forward. When a track is parallel to the boat but displaced from the center an error signal displaces it laterally to bring it to center. The detector has a slit which moves across the width of the boat, and sums all the hits within it to get its signal. The position of the slit when the signal appears provides the error signal. If the operator does not keep the boat aligned with the track, the signal will disappear and the track will be lost. If the boat contains two tracks which are nearly parallel the servo will follow midway between them, and if they later separate it will follow one or the other at random. The signal from the slit is equivalent to the pulse in a histogram. Its problems are that it can handle only a single pulse and it cannot change its direction.

There have been attempts to combine the track finding and following tasks into a single global track recognition scheme [4.25]. Since the tracks are very nearly circular, the problem is conceptually very simple. The points are transformed to a coordinate system (r, ϕ, for example) which displays their circular nature, and then a search is made for points belonging to the same circle. The problem is that the parameters of the transformation must be determined from the points themselves in some iterative process. Every new iteration requires that all the points be transformed again; so the process is too slow with current processors. The development of fast parallel processors may make this procedure feasible in the future, but currently it is not competitive with the find and follow methods.

Track finding and following both make correlations among digitizings to detect the track elements, and the main problem is to decide which correlations are appropriate at different stages. If the correlation is too simple the system becomes unstable in the presence of noise and tracks are lost; if it is too complex, too much time is required to make the correlations. An obvious correlation to make is that points belonging to a track have a small spread in the coordinate normal to its direction. In general the direction is not known; so the correlation is hard to detect. There are cases, for example in finding beam tracks, where the direction is known *a priori* and the method is useful. In the HPD scan the points are ordered in W, the distance along the scan line, and then in N, the scan line number. Beam tracks near the entrance have about the same W-value for successive values of N.

A very simple correlation, which is independent of track parameters, is that of proximity. If n digitizings can be linked each to nearest neighbors in such a way that the distance between any two is not too large, it is probable that they belong to the same track. The method becomes unstable in the presence of noise. Electron spirals and δ-rays can lead it astray, and when two tracks cross it will follow one or the other at random. It is also unstable against changes in chamber operating conditions and drifts in the digitizer circuits. A maximum limit must be set on the separation of points to be linked; and if the density of points changes, the limit will be wrong.

One of the earlier successful filtering programs [4.26] used a combination of global and proximity correlations to filter SMP data. The disadvantages were recognized, but the program was found to be economical in computer storage and processing time. The SMP data enter roughly ordered along the track. Global parameters are determined by averaging clusters of points along the track, near the beginning, middle and end. These averages are used to define a parabola near the path of the track. Next all points are transformed to a curvilinear coordinate system with axes tangent (ξ) and normal (η) to the parabola and ordered in ξ. A search is made for "stepping-stone" points distributed at regular intervals along the track. A prospective stepping-stone is examined to see if it has an η value which is not too large and which is close to the value of its neighbors. If it satisfies these criteria, it is accepted and the next candidate farther down the track is examined. If it fails, other points in the neighborhood are tested. When stepping-stones have been found all along the track, limits are established in η for accepting digitizings and averages are formed. The average points are very sensitive to the crude average points used to generate the parabola; so the beginning, middle and end of the track must be in clear regions. Often stepping-stones are generated in the middle of noise, and the η acceptance-values become too large. The system worked because there was immediate feedback to the operator who soon learned to be especially careful in certain parts of the track and to avoid very confused regions.

The fault in this program was not in using global and proximity correlations, but in using them too soon. On well established tracks these methods will work and will produce good average points. However, at the very important stage when nothing is known about the track, more powerful correlations are required. The most important characteristic of a track at this stage is that a short segment of it is well approximated by a straight line. A powerful procedure for displaying this characteristic is the histogram. A small area oriented along the known or assumed direction of the track is divided into longitudinal strips, called bins. The digitizings within the area are examined and a count is made of the number which fall within each bin (see Fig. 4.10). A sharp pulse indicates a track parallel to the histogram direction, whereas a uniform population is indicative of an oblique track. The sensitivity of the method depends upon the length of the area and the width of the bins. Wide bins are less sensitive to the orientation of the track and to multiple tracks close together, and they are more sensitive to noise. It is advantageous to start track searches in regions of minimum confusion so that the initial histograms will not be too badly distorted. Narrow bins are very sensitive to track orientation and to crossing tracks and are relatively insensitive to noise. Final histograms have bin widths comparable to a bubble diameter and contain very little noise.

The method of constructing a histogram depends very much on the manner in which data are collected. In programmed machines such as POLLY the data have one coordinate (y) approximately along the track direction and the other (x) perpendicular to it. The histogram is then formed simply by dividing x into appropriate intervals and counting the points in each. In an HPD, which has all the digitizings in memory, a more complicated procedure is required. The area of interest, called a slice, is generated by imposing limits on the W coordinate (which may be analytic functions) and the scan line numbers. The nature of the raster scan requires that W be the coordinate which is binned. If the limits are $W_{\mathrm{MIN}}(n)$ and $W_{\mathrm{MAX}}(n)$ for the n-th scan line, all points used must satisfy $W_{\mathrm{MIN}} \leqq W \leqq W_{\mathrm{MAX}}$. The bin number is found from (W-W_{MIN}). Any shape slice can be programmed to optimize search strategies for different tasks (Fig. 4.10).

The Spiral Reader digitizes data in a convenient form to find tracks radiating from a vertex. In the (r, ϕ) coordinate system with origin at the vertex all tracks have nearly the same ϕ for not too large r (Fig. 4.10b). A histogram with bins in ϕ will have peaks at each track angle. RIPPLE has the same general scan pattern and accumulates the histogram in hardware as the scan progresses. The line scan-element of PEPR produces the same information as a short histogram; because points will not digitize efficiently if they are not aligned with the scan element, a hit is equivalent to a histogram pulse.

Several other devices have means of accumulating histograms in hardware. The Coccinelle at the College de France [4.27] uses a known scan speed and a series of delay lines to form histograms of tracks at different angles within a region about $\pm 45°$ from either the horizontal or vertical direction. The method relies on the fact that if a straight line is raster-scanned it will appear on each scan line at a time determined by the scan speed and the angle of the track. At zero degrees the delay

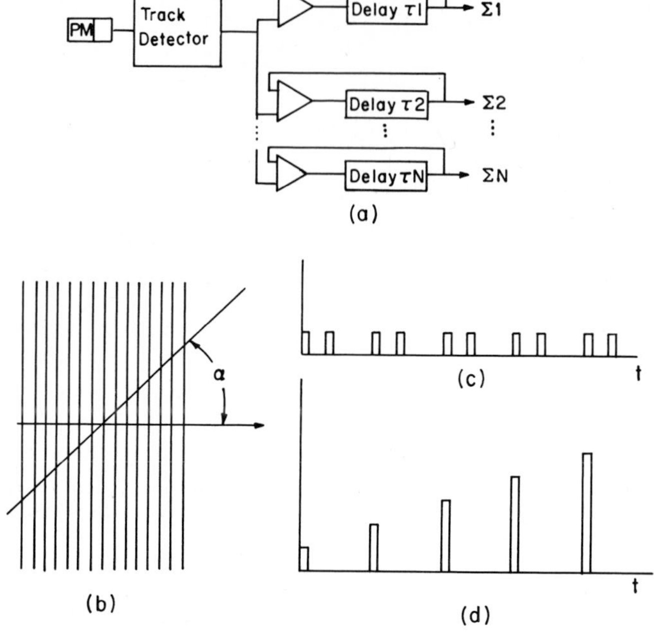

Fig. 4.11a–d. The Coccinelle electronic delay system which generates a tangent histogram. (a) Block diagram of the system. (b) The scan geometry with a track at angle α. (c) The signal from a channel with incorrect delay. (d) The signal from the correct channel

between the pulses on successive scan lines is just the period of the scan lines. The corresponding delay for a track at an angle α is

$$T = L/V + (l/V)\tan\alpha\,, \tag{4.17}$$

where L/V is the scan period, l is the space between scan lines, and V is the scan speed (see Fig. 4.11). Different angles are represented by different amounts of delay, and 32 channels produce an angular bin size of about 3°. The histogram pulse is formed by adding the delayed pulses to the incoming pulse. A track with hits on n scan lines will produce one pulse of amplitude nV_0 at the output of the delay line corresponding to its angle, but will produce n pulses of amplitude V_0 in channels corresponding to other angles. The histogram hardware is used with a CRT scanner which generates slice scans either along x or y.

A completely digital system to histogram in two dimensions is being developed at the Max Planck Institute in Munich [4.28] to work with HPD data. In BRUSH, a map of the digitizings is transferred to a special

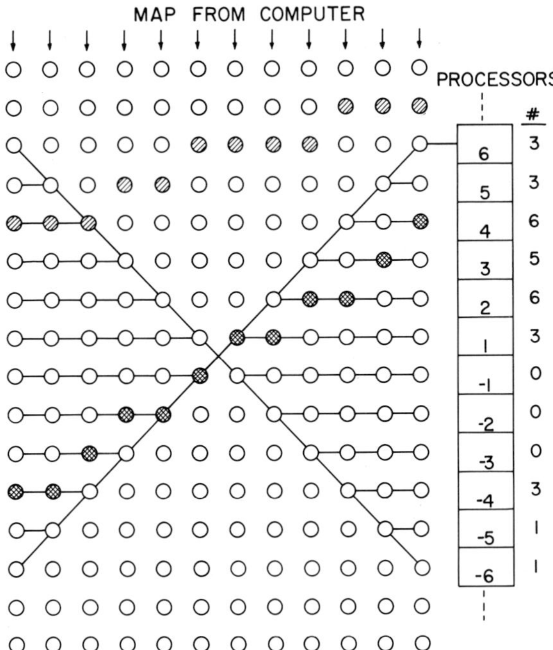

Fig. 4.12. The Brush shift register system which generates tangent histograms. Each processor examines a different angle

two-dimensional shift register. As the points are shifted through the register, special processors examine the bits in all directions to detect peaks. Figure 4.12 shows the shift register with a single track. The dotted lines between cells indicate the bits which are examined by the processors; a separate processor looks at each line of bits. The shift register is loaded from a computer which has already generated the map from the HPD data. At each clock pulse the map advances and all processors look for a pulse. To do this the processor sums the bits in all the cells connected to it and subtracts a threshold count. Because the track may be spread over two rows of cells, it then adds the sum it has saved from the previous clock pulse. Next the processor compares its sum with that of its two neighboring processors and sets a flag if it has the largest count of the three. The computer examines the flags and decides which digitizings are involved in the pulse and what their angle and center position are. With integrated circuit logic the shift register can be advanced at 1 MHz; so the system can easily handle HPD data in real time.

The next step in the track-find task is to link the track elements to form larger segments. Tables of data called track banks are kept in memory, which contain all the known parameters of all track candidates which have been found. The track element is first compared with all track banks to see if it is an extension of one or more of them. If so, the corresponding track parameters are updated; if not, a new track bank must be started for it. It is possible to assign an element to more than one track bank because the precision of the banks changes as more data accumulate, and in the early stages an element may be consistent with several of them. The extrapolation from a single element must be very broad and almost any element in close proximity and pointing in the same general direction will be linked. After two elements the direction is better defined, and after three, a curvature estimate can be made.

Once the track parameters have been defined, we enter the process of track following, although the distinction between finding and following is rather fuzzy. The track may be considered initialized or found, once enough data are accumulated to fit a second-order curve. A prediction of the expected position and angle of the next element is made, with uncertainties which depend upon the data already accumulated. These uncertainties form a road about the predicted position and angle, and as more data are accumulated, the road becomes narrower. The width of the road determines the maximum safe distance that the track can be extrapolated. Because the tracks are not exactly circular it is better to use the most recent track parameters to extrapolate. Once the track has passed a critical length S_0, only the most recent length $(=S_0)$ is used to compute the parameters.

As the road gets narrower, the histograms are not essential. The digitizings can be tested against the extrapolation directly because the road is strong enough to override most of the noise encountered. In systems for which histograms are cumbersome to construct it has the virtue of simplicity and speed. However, the directional content of the road data is not used as the data come in, and in confused regions with much noise the road may stray from the track. The problem is particularly bad in the case of tracks crossing at a small angle. Scanners always have a dead time after a track is detected during which they are not sensitive, and for any crossing track there will always be a time when only the wrong track is digitized. These points will be in the road but will come from a track with different parameters; hence it will distort the extrapolation. For small angle crossings the road may switch completely to the spurious track. If a histogram is used, the crossing track can be detected as it approaches, and evasive action can be taken to avoid getting lost.

The final stage of track-following is to clean up any false linkages and to reduce the data to a small number of average points for the geometry program. Because the track is known better at the end of the measurement, many false linkages can be eliminated by using the track's parameters and following it back to its origin. The construction of average points is done after the track parameters are known as well as possible for the entire track. The points can be formed during following or on a separate pass. Precision is required in track following only to avoid losing the track, and noise which does not distort the parameters too much has little effect. However, any noise included in the points to be averaged will contribute directly to the track measurement error. The amount of additional filtering required depends upon the final road width. If the road can be narrowed to a bubble diameter the noise contribution will be comparable to the scatter in the track bubbles and hence is not too important. Proximity tests can be used on short segments of tracks by averaging the points, computing the scatter in the direction normal to the track, discarding the points outside some tolerance and averaging again. The histogram is a better technique because the bins can be made much narrower than the road. They can usually be chosen so that the normal scatter of the points occupies one or two neighboring bins; then any noise present is statistically insignificant. Using a histogram technique, Oxford [4.29] has reported a reduction of the r.m.s. scatter of their points by a factor of two compared with just using a narrow road.

For curved tracks there is another effect which can yield poor data. If a segment of curved track has its digitizings averaged, the average will lie inside the arc of the track. The effect will be to generate a track with a larger curvature than the actual one. This error is small except for highly curved tracks, and it can be reduced by forming averages of the distances of the points from the edge of the curved road and of their distances along the road. Equivalently, the points can be averaged and then the average can be displaced slightly toward the center of the road.

4.3.2 Finding Vertices and Secondary Tracks

Problems of associating secondary tracks with a common interaction point, called a vertex, occur mostly in minimum guidance or zero guidance modes of operation. In road guidance all the tracks are specified on the pre-measurement record. The only problem is to project the tracks backward to verify that they originate at a common point. Track intersection is always done by extrapolation because there is only a small probability of having a bubble exactly at the vertex and the machine is not likely to digitize it even if it exists. The best values of angle and curvature are determined at the ends of each track near the vertex and they are extrapolated (Fig. 4.13). The vertex will lie near the center of the polygon formed by the tracks, and the best position is found by assigning weights to the tracks, according to how well they are measured.

In the minimum guidance mode [4.30] only an approximate vertex position and the number of secondary tracks are known, and it is necessary to devise a search pattern to find the tracks leaving the vertex. The problem is simplest for the Spiral Reader and suggests the approach used in other systems. Fig. 4.6 shows a typical event in the (r,ϕ) coordinate system. Notice that only a few tracks penetrate to the region near $r=0$, and these have a high probability of belonging to the event. A histogram in ϕ for small values of r will contain pulses for most of the track candidates. Histograms for several slices at different radii are required to find all the candidates. All candidates are then followed to find the real tracks, and these are extrapolated back to the vertex.

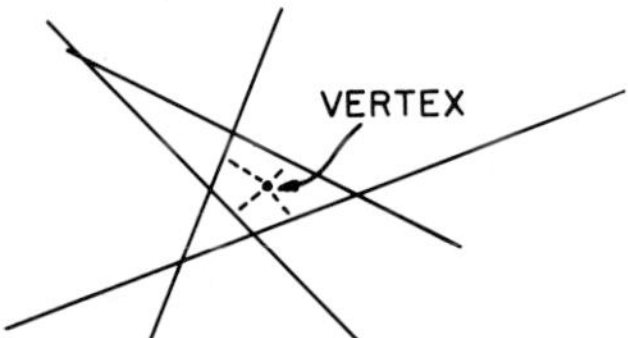

Fig. 4.13. Track extrapolation to determine the vertex

A wedge-shaped area is used to search for tracks in HPD data. Figure 4.10 shows the forward and backward wedges used with the normal scan. A similar pair from the orthogonal scan completes the search pattern. The wedges are divided into segments as shown, and the points in each segment are gated separately from the data. In each segment the quantity $(y_i - y_v)/(x_i - x_v)$ is histogrammed. Thus all the points with the same tangent will fall in a single bin or pair of bins. All pulses above a threshold are considered as track candidates.

A slightly different search method is used with POLLY because of the way it records hits. Since only one good hit is recorded per scan line, a very large number of narrow slice scans would be required to cover the region around a vertex. To avoid this, a special scan pattern, called an octagon search, is constructed, as shown in Fig. 4.14. Notice that the scan lines are approximately radial from the vertex; so a secondary track will produce either a very wide hit or multiple hits on a scan line which traverses it. Both of these conditions are flagged. The flagged scan lines from a series of several octagons around the vertex are used to find track candidates. All flagged lines at approximately the same angle are considered as belonging to the same track. When the angles of all candidates have been found, the usual track following routine is entered. POLLY-like devices which record more than one hit per scan line can use the more conventional slice scan to search for tracks.

The problem of finding vertices without any guidance is a difficult one because of the large increase in the quantity of data which must be

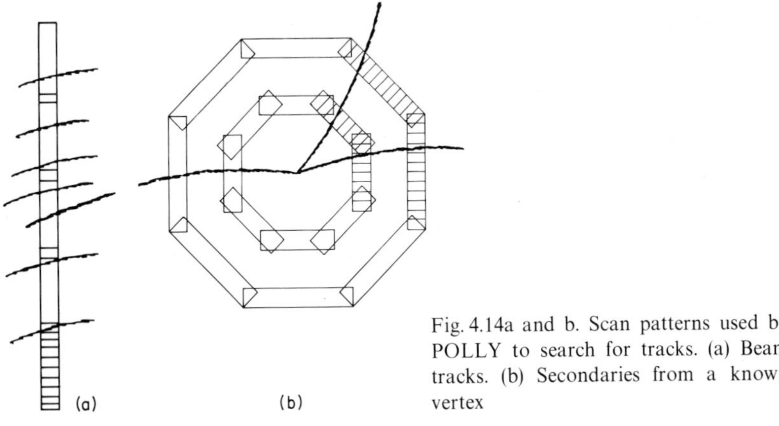

Fig. 4.14a and b. Scan patterns used by POLLY to search for tracks. (a) Beam tracks. (b) Secondaries from a known vertex

processed. The first successful experiment to process a substantial fraction of the film automatically was performed at Argonne on POLLY II. The system was not designed to find all vertices but only beam track interactions. Neutral particles which decay in the chamber are not found. First a beam-track search is initiated along the entrance window using overlapping slice scans with scan lines along the expected beam direction (Fig. 4.14). All track candidates are followed until they leave the fiducial region of the chamber or until they end and become vertex candidates. After all beam tracks have been followed, the vertex candidates are examined for possible secondaries using the octagon search pattern described above. If no secondaries are found it may be because the interaction produced all neutrals, or the track being followed may have entered a wall of the chamber.

The only program which is completely automatic in the sense of processing the entire film is the DAPR system [4.12] at Berkeley. The system extracts all of the information from the film, condenses it into tracks, track segments, fiducials, etc. and writes a record which can be processed by other programs to construct events and analyze them. The input data come from an HPD, although any complete scan of the film would be suitable. At least one PEPR group, at Heidelberg [4.31], is using the same general method. There are two levels of processing. The first level runs the HPD and generates track segments in real time, which are stored on a disk file. The second level is a multi-phase program which links segments into tracks, tracks into vertices, and finally matches tracks and vertices in three dimensions.

The first-level program controls the HPD and produces as many sweeps as required to cover the film with both normal and orthogonal scans. A sweep is composed of a specified number of scan lines of a specified length, and each sweep is processed independently. The input data rate of about 15000 words/s limits the complexity of the track-following routines. Tracks are started by proximity tests of the digitizings and then continued with extrapolation techniques. Since all the data on the film which can be reduced to track segments are retained, it is not necessary to follow the track completely in real time. The orbits of beam tracks are known, and they are followed with a special routine which uses the beam parameters. Fiducials are measured as short track segments. The digitizings which have been associated with track segments are averaged to form master points and saved on a disk.

When sufficient data have been accumulated in the first level, the second-level programs are fetched to complete the processing. These programs run in a time-sharing mode with the first-level program. Each program makes a complete pass over the data, updates it and saves it for the following program. The first program links track segments into tracks, recognizes the fiducial marks, and eliminates segments outside the fiducial volume. Certain known marks in the chambers, such as unused fiducials, are eliminated from the data. Tracks from the normal and orthogonal sweeps are merged and edited to form a single track list for the view. In the case of duplicate tracks the data are taken from the sweep which has the better angle with respect to the track. The output is a track list referred to known fiducials.

The next program uses the track lists to try to find all the vertices. It works first in a single view and finds vertex candidates by checking for proximity of track end-points. Several tracks within a certian proximity of each other define the candidates as the center of the polygon formed from their extrapolation (Fig. 4.13). A somewhat looser criterion is used to search for additional tracks which may be associated with this candidate. The result is a vertex candidate list for each of the stereo views. The candidates are next examined for compatibility in all stereo views. This eliminates false vertices caused by spurious tracks and crossing tracks, which will not have a good stereo reconstruction. All tracks are extrapolated and tested for proximity to the surviving vertices in order to search for tracks with long gaps near the vertex. Next a track-match program matches tracks from the different stereo views and eliminates tracks which appear to come from a vertex in one view but are not associated with it in space. The data from this program are passed to the experimenter to select events of interest and to complete the processing.

4.3.3 Track Acceptance and Track Match

The number and quality of tests of tracks in a single view varies with a number of factors. In fully automatic machines these tests will have been included either explicitly or implicitly in the procedures which extract the track information from the digitizings. For manual or semi-automatic machines with the geometry program operating in real time the tests are not so important because the geometry errors can be corrected quickly. The most extensive testing occurs in manual or semi-automatic machines for which the data are processed some time after the measurement. Quality tests on the tracks usually involve some form of continuity test which require that the points vary smoothly along the track. The test may include a requirement that the track follow some second-order curve, but the criteria cannot be too close because many effects cause departures of the track from either a circle or a parabola. If a track-match program is not used, the operator must associate tracks, and some tests of consistency can be made. Tests at this stage can reject poorly measured tracks and operator mistakes, but they cannot assure that only good measurements will pass.

The track-match program is used before the spatial reconstruction program to associate tracks in the different stereo views so that they will reconstruct correctly. An experienced operator can often tell at a glance which tracks are associated in different views, but he uses many of the fine details of the track to decide – details which are not available in the final measured points. The match program relies heavily on correlations among the views and the stereo properties of the optical system. Any two tracks will reconstruct in the sense that the projection cones of their optical rays from two views will intersect to form a space curve. Only if the space curves approximates the real track will the third view be compatible; thus the redundancy of the third view is very important. In addition, the reconstructed track is required to lie within the chamber volume and to be an approximately circular helix. In principle, the reconstruction program itself could match the tracks; the track-views would be permuted until a good χ^2 was obtained for a helix fit. However, this procedure is much too slow; so the track-match program must be a fast program which decides which track-views will reconstruct without doing the complete reconstruction.

Before discussing the program, it is necessary to examine the geometry of image formation, at least in a simplified form. We will assume simple thin lenses with no distortion, i.e. pin-hole optics, and neglect changes in the refractive index along the paths of the rays. Figure 4.15b is a view in the plane defined by two of the optic axes. Notice that any horizontal plane has the same appearance in all views. The top fiducial plane is a convenient two-dimensional coordinate system which is common to all views. A point $P(x, y, z)$ projects on this plane as the point $P_1(x_1, y_1)$ for view 1 and $P_2(x_2, y_2)$ for view 2. Any fiducial marks on this plane have the same coordinates in all views. Figure 4.15a is a view of this fiducial plane for a chamber with three cameras. Notice that the projections P_1, P_2, P_3

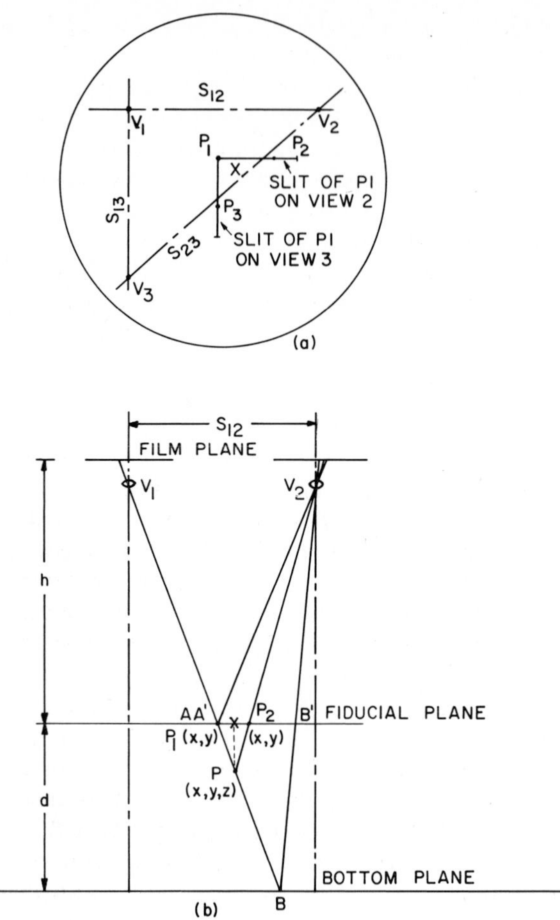

Fig. 4.15a and b. Geometry to reconstruct a three-dimensional point from stereo photographs. (a) Top view of the fiducial plane. (b) Section through cameras *1* and *2*

of any point $P(x, y, z)$ form a triangle similar to the optic axis triangle. Any point $P(x, y, z)$ which projects as $P_1(x_1, y_1)$ on view 1 must lie on the line AB in the chamber. The line AB projects as the line $A'B'$ in view 2; hence the point P must project somewhere on this line. The line $A'B'$ is called the slit of the point P_1. There is a similar slit in view 3. The lengths of the slits are the same for all points $P_1(x_1, y_1)$ and depend only on the geometry of the chamber; the length is

$$L_{ij}=S_{ij}d/(h+d)\,, \tag{4.18}$$

where S_{ij} is the separation of cameras i and j; d is the chamber depth and h is the height of the plane of the camera lenses above the fiducial plane.

This simple geometry allows several powerful tests upon possible matching tracks [4.32, 33]. First consider tracks in pairs of views, choosing as the primary view the one which has the best stereo axis with the other views (view 1 in Fig. 4.15). A track in the primary view is chosen and compared with all tracks in a secondary view, chosen such that the track makes the largest angle with the stereo axis. The two-view tests are: a) The sign of the curvature must be the same in both views. This is true for any geometry in which the optic axes are parallel, but is not true for some of the new large chambers. The test may not be applicable for short high-momentum tracks where measuring errors may be larger than the curvature. b) If the optic axis is translated to one end of the track, then the other end must be on the same side of the axis in both views. c) The ratio of the lengths of the tracks in the two views has a maximum (or minimum) value determined by the chamber geometry. d) The track in the secondary view must pass through the slits from all points in the primary view. Usually the test is applied only to the mid-point and the two end-points. e) The tracks must be consistent with the same space helix. The corresponding point in the secondary view (the intersection of the track with the slit) is constructed for the first point in the primary view. If the tracks in the two views are from the same helix, they will diverge linearly with increasing arc length. If the end-point separations are measured, the mid-point separation can be predicted. f) By constructing corresponding points at both ends of the track, the dip (angle of the track with respect to a horizontal plane) can be estimated. The chamber depth limits the length of tracks with large dips. g) The reconstructed end points must be inside the chamber. h) If there is a point where the track is tangent to the stereo axis, this is a corresponding point in both views (a space point which can be identified in both views) and must be the same distance from the stereo axis in each.

This list is a compilation of tests used by several groups and not all are used by each one. A clear failure of one of the tests removes the pair of tracks from further consideration. All pairs which pass these tests comprise a list of two-view matches, but the pairs are not unique, i.e. the same track-view may appear in several pairs. Next each pair is compared with all tracks in the third view (the view not belonging to the pair), and tests a)–c) are applied. With two views matched the space curve can be computed and projected onto the third view. Any point in one view generates a line (the slit) in the second and third views. The intersection of the track and the slit is an approximate corresponding point in the second view, which in turn has its slit in the third view. The intersection of the two slits in the third view is the projection predicted by the first two views. Two additional tests come from this construction: i) The space angles of the track at the vertex are computed from the two-view match, projected onto the third view, and compared with the angle of the track in this view. j) Three points on the track are projected onto the third view and define a circle. The radius of this circle is compared with the radius of the track circle. Tracks which pass these tests are placed in a list of three-view matches, but again they may not be unique. In practice there are very few ambiguities in this list, and these arise because two or more space tracks have very similar space angles and curvatures. Usually they are beam tracks or high-momentum forward secondaries.

The ambiguities are resolved in several ways. If a track appears in a three-view match, all of its two-view matches are discarded, and all unique three-view matches are retained in the list of final matches. The three-view ambiguities are resolved on the basis of total track length agreement among the views. If the match program is run after measuring, the same criterion is applied to the two-view ambiguities. If the program is run in real time, several things can be done to try to get unique three-view matches. The two-view match allows the projection in the third view to be made accurately; then it is possible to search for the track some distance from the vertex. The track may be new or it may be one which was rejected in some test because of poor measurement. Or the track may be obscured in one view by some feature of the chamber; in these cases the two-view match is accepted. A very small fraction of tracks may remain ambiguous and these are either examined visually by the experimenter or discarded. In many experiments it is helpful to match beam tracks in a separate program, because the parameters of the beam can often be computed from measured magnetic fields much better than they can be measured in the chamber. This increased knowledge allows tighter criteria to be used in the tests and reduces the number of ambiguities.

A system called SATR, being developed at Wisconsin by Thompson [4.34], pushes the match process forward into the measuring procedure and follows tracks in three dimensions. This has the advantages that track confusion is less; that parts of a track obscured in one view can be followed in the other two; there are no spurious two-dimensional tracks left over; and the rather lengthy match program is not needed. However, the construction of points in space from the stereo projections is a lengthy procedure which is feasible only with parallel processors.

The three-dimensional track recognition starts with a two-dimensional scan of each view, searching for track elements in a conventional way. An element found in two or more views defines a small volume in the chamber. All digitizings in each view which could possibly fall within this volume are sent to a three-dimensional track recognition unit. The first module in the unit discards points which are outside the volume

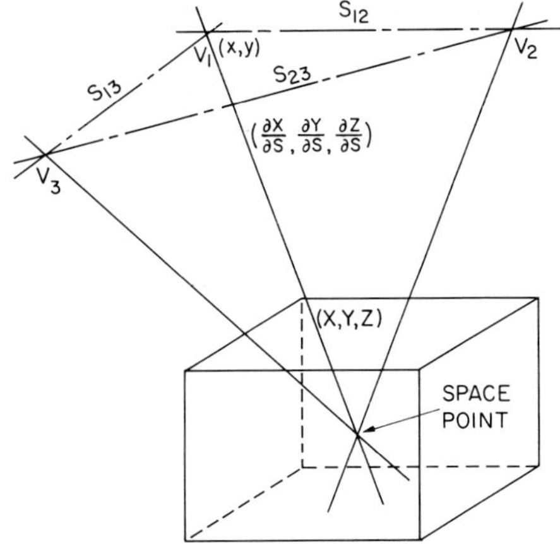

Fig. 4.16. SATR geometry to reconstruct a space point. The cube is a three-dimensional register and the three rays are found by interpolation

element. Next each point which remains is transformed into light rays. Standard rays have been computed for grid points within each view and stored in the unit. The rays from each point are found by interpolation between standard rays. Figure 4.16 shows the transformation of a point (x, y) in one view into a ray $((x, y, z, \partial x/\partial s, \partial y/\partial s, \partial z/\partial s)$ in space. Each ray is used to generate a series of addresses to access cells in a three-dimensional register which represents the volume element. Each cell contains a flag bit A and three counter bits BCD, which will contain the number of rays which passed through that cell. The cell counts are used to find the center and direction of the track element. The track is started by using the center and direction of the element to predict the next volume element to be processed. With more data a second-order curve is used to extrapolate, and when the track is established, a helix is used. The processors employ a combination of fast analog and digital circuits to construct the rays and read out the histograms.

4.4 Future Prospects and Problems

Current research in high-energy physics presents severe challenges to all detection techniques, but especially to the visual techniques. The three major problems all arise from the very high energy produced by the newer accelerators. The first problem is simply the large momentum of both the primary and secondary particles which have much stiffer trajectories. The curvature or amount of bending per unit of path length varies as the field strength and inversely as the momentum. A large increase in momentum causes a much straighter track. The second problem is that all the secondaries lie in a very small cone about the forward direction. Because of the nature of the relativistic transformation most of the increase in energy goes into the motion of the center of mass of the system, which greatly exceeds the motion of the secondary particles about the center of mass. As a result, even a particle which goes backward in the center of mass coordinate system will be swept forward by the center of mass motion and will emerge in the forward direction in the laboratory. Finally there is a large increase in the production of neutral particles, which can only be detected by charged particles produced in secondary reactions. Roughly one-third of all secondaries produced are neutral. As mentioned earlier one neutral can be found from the energy and momentum conservation equations, but more than one leaves the event undetermined. At low energy only a small fraction of events contain more than one neutral, and the problem is not serious.

These problems can be attacked either within the framework of conventional techniques or by the development of new methods. The next subsection will examine the possibilities available in current visual detectors, where the prognosis is rather gloomy. In the final subsection we will look at possible new detection methods and hybrid systems using different types of visual and non-visual detectors.

4.4.1 Improvements in Visual Detectors

The problem of small curvature can be approached in several ways: a) For a given momentum the curvature is proportional to the magnetic field strength, and the ten-fold increase in momentum could be matched by a ten-fold increase in field to retain the present curvature. This approach is limited by strong economic and structural considerations. The field strength required is far above the saturation fields of all ferro-magnetic materials; so an air-core magnet is required. Bubble chamber magnets of conventional construction require 3–6 megawatts of power, and increasing this by an order of magnitude is not feasible. Superconducting coils are being used on the new chambers, but current technology does not allow a substantial increase in field strength throughout the large volume required for a bubble chamber. The magnetic forces increase with the field strength and put severe strength requirements on the structural materials used. Present technology probably limits attainable fields to not more than twice the fields in current use. b) The accuracy of momentum measurements depends upon the integral of the field over the path of the particle; so a longer path could compensate for smaller curvature. The largest chambers today are four or five meters in their longest dimensions, and a significant increase is not likely for both economic and technical reasons. Photography, expansion, heat shielding, etc. are all more efficient with a chamber which is approximately spherical. Also the very important neutrino experiments require as large a mass of material as possible; so future chambers are likely to be close to spherical. The cost of constructing and operating these chambers is approximately proportional to the volume, and a ten-fold increase in length requires a thousand-fold increase in cost. There is a fundamental limit on size, which is being approached by current chambers; this is the interaction length of nuclear particles, which is about 4.5 m in hydrogen. A secondary particle of this length has about a 30% chance of interacting, and the momentum measurement can only extend to the interaction point. c) Finally, the momentum accuracy could be increased by increasing the measurement accuracy of the points. Some improvement can perhaps be made by photographing smaller bubbles and scanning with a smaller spot of light, but substantial improvement will be very difficult. Overall precision now is about 10^{-5}, which puts severe requirements on all parts of the measuring system. At great expense an overall improvement of a factor of ten is perhaps possible by a combination of these improvements; so a 200 GeV particle could be measured with the same precision that a 20 GeV particle is measured today.

The small angle cone (or jet) of secondary particles presents problems both for the track-following and track-match programs. Track following requires hardware changes to separate tracks as clearly as possible and software changes to follow through regions where two tracks may be superimposed. Track match problems arise because most of the tests in Subsection 4.3.3 become very weak. Track curvature is hard to measure, and the stereo tests will leave many ambiguities since the tracks in the forward jet are all very similar. There has been some success in matching tracks visually by using a microscope to match the spatial distribution of bubbles, but techniques for doing this automatically have not been developed.

The detection of neutral particles becomes very important at high energy. Except for a few experiments especially designed to study the production of neutrals, this problem has been ignored at low energy. The production of more than one neutral is rare enough that such events can be discarded and small corrections made to the final result. At several hundred GeV where the average number of particles produced is 15–20 and where one-third of them are neutral, the problem cannot be ignored. Experimental and theoretical techniques are being developed to study interactions when only a few secondary particles are measured. Such "inclusive reactions" yield important data about high-energy phenomena, but there is still a need to measure as much of an interaction as possible.

The most copiously produced neutrals are π^0 mesons, followed by K^0 mesons, neutrons, and neutral hyperons. Neutrons can be detected with very low efficiency in visual detectors; hyperons have a large probability of decaying and hence being detected in the detector. K^0's can be detected with moderate efficiency from their decay, and this efficiency can only be enhanced with external detectors. Most important is the detection of the π^0 which decays in about 10^{-16} s into two photons; thus the problem is one of detecting γ-rays at an energy where the only significant interaction with matter is the production of electron-positron pairs by conversion in the electromagnetic field of a nucleus. The radiation length or mean path length for conversion varies from 11 m in hydrogen to about 3 mm in uranium metal. Early detection attempts used plates of a heavy metal inside the chamber; a few cm of lead will convert nearly all the γ-rays. However, the data on the γ-ray energy and hence π^0 momentum is poor. To be efficient the plate must be several radiation lengths thick, and in this thickness the initial γ-ray produces several secondary γ-rays and many pairs, i.e. an electromagnetic shower develops. The emergence of this shower from the plate provides poor information on even the direction of the γ-ray. The use of several thin plates helps to concentrate the shower, but only at the expense of requiring more space in the chamber. Neon is a relatively dense liquid which can be used as a working fluid in a bubble chamber; it can be used alone or mixed with hydrogen. It has a radiation length of 25 cm; so it makes an efficient γ-ray detector. Since neon is a complex nucleus the interpretation of the primary interaction is not easy, and the large density increases the multiple Coulomb scattering contribution to the measuring error of all tracks. Bubble chambers have been operated with a neon or neon-hydrogen mixture to detect γ-rays, but containing a separate volume of pure hydrogen as the target and low-density region for measuring secondary tracks. Both volumes are sensitive and visible to all cameras. This may prove to be a very effective way to detect neutrals.

The general problem of particle identification is probably not soluble in a visual detector by itself. Two methods are used at low energy: a) The event is fitted with constraints with the particle masses chosen for all possible particles. A single good fit to the event implies that the masses of that fit are the real masses. b) An estimate of ionization can distinguish among various possible masses if the momentum is known. This estimate is significant only for momenta less than about 1.5 GeV/c. The secondaries from high energy collisions all have the same ionization.

4.4.2 The Use of Auxiliary Detectors

The best hope for solving the problems of high energy collisions probably lies in combining visual and non-visual detectors. The narrow jet of particles allows the use of modest-size detectors to detect all the high momentum secondaries; those of low momentum can be measured in the chamber. The average density of the detectors can be small, and long path-lengths can be used to get very precise momentum measurements. Gas Čerenkov counters can measure the velocities of the secondaries to moderately high momentum, and transition radiation counters become effective at very high momentum. These aid greatly in identifying secondary particles. Neutral π^0's can be detected in counters which completely absorb the electromagnetic showers and produce a pulse proportional to their energy. In addition, an electronic trigger derived from the counter signals can be used to flash the lights only for interesting events.

Visual detectors play a central role in hybrid detection schemes. Firstly, they provide a detailed view of the vertex, which is not available to the counter system alone. The total number of charged particles is easily found, and this is important in studying many reactions. The counter results may be distorted by secondary reactions and also by reactions occuring in the walls of containers, etc.; these are easily detected in the visual detector. Finally, the measurement of the momentum vector is greatly improved by using the visual detector measurement as well as the counter data. The extrapolation to the vertex in a bubble chamber is accurate to a small fraction of a millimeter, and the angles of emergence of a particle can be measured very precisely. The vertex measurement determines the starting point of the orbit very precisely and improves the measurement of the magnitude of the momentum, and the angle measurement provides an accurate decomposition of the momentum vector into its components.

Other possible components of the hybrid system will be described very briefly. The momentum measurements are usually made by a series of wire spark chambers, proportional wire chambers, or drift chambers and large bending magnets. Wire spark chambers have a spatial resolution of a few millimeters, whereas the other two have resolutions of about 1 mm or slightly less. The fields of the magnets are carefully measured so that the momenta of the particles can be computed from their trajectories. The chambers are placed before, after and between the magnets to determine the trajectories. Since the path lengths may be many meters, the orbit can be measured with good precision.

Scintillation counters are used primarily to provide time resolution and to define the trigger which activates the spark chambers and flashes the lamps. Although 15 to 25 tracks may traverse the bubble chamber during the time it is sensitive (a few milliseconds), they are spread out in time. Since the scintillation counters have a time resolution of a few nanoseconds (10^{-9} s), the particles are clearly separated. The scintillation counters supply tags which tell which particles were associated with a particular event.

Čerenkov counters are able to measure the velocity of a particle which traverses them. When a charged particle travels faster than the local velocity of light in a medium, a "bow-wave" of electromagnetic radiation is generated. The angle of this wave is determined by the ratio of the velocity to the local velocity of light. As the particle momentum increases its velocity approaches the velocity of light in vacuum and becomes nearly independent of its mass. This limits the usefulness of Čerenkov counters. A relatively new counter detects the radiation emitted when a charged particle crosses the boundary between two media. This radiation depends upon the relativistic parameter γ ($=[1-v^2/c^2]^{-1/2}$) and hence does not saturate at large momenta. Both of these devices can be used with a momentum measurement to obtain the mass of a particle which passes through them. They also have very good time resolution and can be used to tag particles.

Neutral detectors make use of a high-Z dense material to convert γ-rays into showers, or of hydrogenous material to provide a large cross section for neutron induced reactions to detect neutrons. In both cases the charged reaction products are detected. If the neutral need only be detected the detector can be simply a plate of material placed before a scintillation counter. A small thickness of lead makes an efficient γ-ray detector, but neutron detectors must be thick to have even a modest detection efficiency. Detectors which measure the energy of the neutrals must be large enough and massive enough to completely contain all the particles produced in secondary interactions. Scintillators are usually sandwiched within the material to detect a substantial fraction of the charged secondaries.

Large scale use of hybrid systems is just beginning, and many problems must yet be solved to make them effective. One of the first problems is to "hook up" or connect the tracks in the chamber to the trajectories measured by the external detectors. In most cases the hook-up procedure also resolves track match ambiguities since the tracks must not only satisfy the stereo tests but must also project to the external trajectories. The output of the external detectors is usually easier to analyze than the visual data; so in practice the external trajectories are projected back into the chamber to place rather tight limits on the space angles of the tracks. Mass detection schemes are still rather primitive and usually can detect only one particle at a time, which is very limiting for the large number of particles produced at high energy.

4.5 Guide to the Literature

Developments in visual techniques in high-energy physics tend to be published in the form of internal reports of individual laboratories or in proceedings of conferences on instrumentation. Journal publications usually occur much later and tend to be descriptions of complete systems. The best sources for reports of current developments are the conference proceedings.

The most recent general conference on instrumentation was the *1973 International Conference on Instrumentation for High Energy Physics* held at Frascati, Italy, on May 8–12, 1973. Sessions related to visual techniques were:

1) Bubble chambers. Invited review papers were presented on large bubble chambers and on the use of track-sensitive targets and rapid-cycling chambers. Contributed papers described work on ultrasonic chambers, rapid-cycling chambers, track-sensitive targets and the use of counter-derived triggers for operating the flash tubes.

2) Electrically-pulsed chambers. Review papers were given on streamer chambers and flash-tube hodoscope chambers. Contributed papers included these topics as well as reports on spark chambers operated at cryogenic temperatures and possible use of laser techniques to enhance the light output of streamer chambers.

3) DC operated chambers. These are not visual devices but are important because they both compete with and enhance the usefulness of visual detectors. These chambers are multi-wire proportional chambers and drift chambers.

4) Data acquisition. This included papers on collecting data from large counter systems as well as visual detectors. Papers describe the current status of several operating measuring systems as well as a few new ones.

The *Oxford Conference on Computer Scanning* held at the University of Oxford, UK., on 2–5 April, 1974, contains contributed papers from most laboratories which are using CRT devices to scan film from visual detectors. Recent developments include an emphasis on programmable controllers and better displays for the operator, a new type of deflection of the CRT beam, and the application of CRT scanning to problems outside of high energy physics.

Earlier conferences include the *International Conferences on Data Handling Systems in High Energy Physics* held at Cavendish Laboratory, Cambridge, on March 23–25, 1970 (CERN 70–21); the *Informal Colloquium on* POLLY *and* POLLY-*Like Devices* at Argonne National Laboratory on January 11–12, 1972 (ANL-7934); the *European Spiral Reader Symposium* at Stockholm on May 30–June 1, 1972 (CERN 72–16); and the *International Conference on Advanced Data Processing for Bubble and Spark Chambers* at Argonne National Laboratory on October 28–30, 1968 (ANL-7515).

A good review of both software and hardware before 1967 is contained in the two-volume *Bubble and Spark Chambers* edited by R. P. Shutt. The articles contain extensive bibliographies.

References

4.1 D. H. Miller, L. Gutay, P. B. Johnson, F. J. Loeffler, R. L. McIlwain, R. J. Sprafka, R. B. Willmann: Phys. Rev. **153**, 1423 (1967)

4.2 D. A. Glaser: Phys. Rev. **91**, 762 (1953)

4.3 F. Seitz: Phys. Fluids **1**, 2 (1958)

4.4 R. D. Watt: "1973 International Conference on Instrumentation for High Energy Physics"; ed. by S. Stipcich; Frascati (1973) pp. 44–49

4.5 H. Brettel, C. Carathanassis, L. Kollmeder: "International Conference on Data Handling Systems in High Energy Physics"; vol. II, Cambridge (1970); CERN Rept. 70–21, p. 851

4.6 E. K. Zavoisky, M. M. Butslov, A. G. Plakhov, G. E. Smolkin: J. Nucl. Energy **4**, 340 (1957)

4.7 G. F. Reynolds, R. Giacconi, D. Scarl: Rev. Sci. Instr. **30**, 497 (1959)

4.8 J. W. Cronin: Spark chambers. In: R. P. Shutt (Ed.), *Bubble and Spark Chambers* (Academic Press, New York, London 1967), vol. I, pp. 316–406

4.9 D. C. Fries, M. Davier, J. Derado, F. F. Liu, R. F. Mozley, A. C. Odian, J. Park, J. Park, W. P. Swanson, F. Villa, D. Yount: Nucl. Instr. Meth. **107**, 141 (1973)

4.10 M. Alston, J. V. Franck, L. T. Kerth: Conventional and semiautomatic data processing and interpretation. In: R. P. Shutt (Ed.): *Bubble and Spark Chambers* (Academic Press, New York, London 1967) vol. II, pp. 52–139

4.11 W. W. M. Allison, F. Beck, G. Charlton, D. Hodges, J. A. Loken, B. Musgrave, H. B. Phillips, R. Royston, R. A. Sunde: "International Conference on Data Handling Systems in High Energy Physics"; Vol. 1, Cambridge (1970); CERN Rept. 70–21, p. 325

4.12 H. S. White, B. Britton, J. Franz, W. S. Gee, D. Hall, N. Jontulovich, F. Windorski: "Advanced Data Processing for Bubble and Spark Chambers"; Argonne (1968); Rept. ANL-7515, p. 275

4.13 H. C. Albrecht, E. P. Binnall, R. W. Birge, M. H. Myers, P. W. Weber: "Advanced Data Processing for Bubble and Spark Chambers"; Argonne (1968); Rept. ANL-7515, p. 173

4.14 P. V. C. Hough, B. W. Powell: Nuovo Cimento **18**, 1148 (1960)

4.15 E. Rossa: Rev. Phys. Appl. **4**, 329 (1969)

4.16 P. G. Davey, R. I. Hulsizer, W. E. Humphrey, J. H. Munson, R. R. Ross, A. J. Schwemin: Rev. Sci. Instr. **35**, 1134 (1964)

4.17 O. R. Frisch, D. J. M. Davies, A. S. B. Street: "International Conference on Data Handling Systems in High Energy Physics", vol. I; Cambridge (1970); CERN Rept. 70–21, p. 285

4.18 L. Pless, L. Rosenson, P. Bastien, B. Wadsworth, T. Watts, R. Yamamoto, M. H. Alston, A. H. Rosenfeld, F. T. Solmitz, H. D. Taft: A precision encoding and pattern recognition system. In: Smordinsky, Y. A. (Ed.): *XII International Conference on High Energy Physics* (Moscow: Atomizdat, 1966) pp. 409–413

4.19 M. E. Sekeley: "Proceedings of a Conference on Programming for Flying-Spot Devices"; New York: Columbia University, (1965), p. 62

4.20 W. W. M. Allison, F. Beck, D. Hodges, J. G. Loken, B. Musgrave, H. B. Phillips, R. Royston, R. A. Sunde: Nucl. Instr. Meth. **84**, 129 (1970)

4.21 M. Bourdinaud, F. Cadiet, E. Coulareau, A. Dillet, M. Goldwasser, M. Guyonvarb, J. F. Le Fur, C. Lewin, G. l'Huissier, J. C. Michau, R. Miche, J. Mullie, B. Pichard, J. F. Renardy, G. Riols, G. Seite, J. C. Sellier: "Oxford Conference on Computer Scanning"; vol. I, Oxford (1974) p. 321

4.22 R. W. Downing, J. Simaitis, D. Zander: "Informal Colloquium on POLLY and POLLY-like Devices"; Argonne (1972); Rept. ANL-7934, p. 4

4.23 R. L. McIlwain: "Oxford Conference on Computer Scanning"; vol. I, Oxford (1974), p. 123

4.24 L. R. Fortney: "International Conference on Data Handling Systems in High Energy Physics"; vol. I, Cambridge (1970); CERN Rept. 70–21, p. 215

4.25 M. J. Bazin, J. W. Benoit: IEEE Trans. NS-12, 219 (1965)

4.26 J. H. Munson: "The Scanning and Measuring Projector"; Techn. Rept. UCRL-11154, Lawrence Berkeley Laboratory (1963)

4.27 B. Equer, C. Guignard, G. Reboul, A. Volti: "Advanced Data Processing for Bubble and Spark Chambers"; Argonne (1968); Rept. ANL-7515, p. 6

4.28 H. Billing, A. Rüdiger, R. Schilling: "Advanced Data Processing for Bubble and Spark Chambers"; Argonne (1968); Rept. ANL-7515, p. 48

4.29 P. G. Davey, J. F. Harris, B. W. Wawes, P. R. Lamb, J. G. Loken, U. West: "Oxford Conference on Computer Scanning"; vol. I, Oxford (1974) p. 149

4.30 M. Ferran, J. Gerard, A. Head, G. Moorhead, A. Sambles, J. Searle, J. Burren: "Programming for Flying-Spot Devices"; Munich (1967) p. 346

4.31 K. Brokate, R. Erbe, E. Keppel, D. Kropp, E. Grimm, H. Schneider, J. Wells: Nucl. Instr. Meth. **96**, 157 (1971)

4.32 D. Hall: "Advanced Data Processing for Bubble and Spark Chambers"; Argonne (1968); Rept. ANL-7515, p. 299

4.33 J. M. Gerard, W. Krischer, D. O. Williams: "Advanced Data Processing for Bubble and Spark Chambers"; Argonne (1968); Rept. ANL-7515, p. 431

4.34 M. A. Thompson: "Advanced Data Processing for Bubble and Spark Chambers"; Argonne (1968); Rept. ANL-7515, p. 81

5. Digital Picture Analysis in Cytology

K. PRESTON, JR.

With 37 Figures

5.1 Introduction

In the field of diagnostic cytology, i.e., in the study of the constituent parts and functions of organic human cells, an army of from 20000 to 40000 pathologists, hematologists, and geneticists (both professional and technologists) perform several hundred million visual examinations in the United States each year. These examinations are of cells of the human body fluids and tissues and are a vital part of programs for health screening, diagnosis, and treatment. The massive effort expended in cytologic examinations and the associated effort expended in radiologic examinations (the examination of X-ray films) represent the major pictorial pattern recognition work in the United States and, for that matter, in the world. This effort far exceeds that expended in other picture processing applications, such as the visual examination of photographs taken by civilian earth resources satellites and military reconnaisance systems. The cash flow associated with these examinations in the United States is a few billions of dollars annually, i.e., about 5% of the annual national health budget.

Due to the minute size of the objects being visually studied (cells are of the order of 1 to 100 μm in diameter), all cytologic examinations are performed using the microscope. Because of the necessity of examining cells individually in a single focal plane, the cell-containing body fluid or tissue is spread as a thin film or as a thin slice of tissue on a glass substrate (the microscope slide). In order to simplify the visual examination by adding significant chromatic labels to the cells and their constituent parts the film or slice is stained with various organic chemicals. Thus the way in which the specimen is mechanically and cytochemically processed prior to the visual examination is of great importance (see Section 5.3).

The worker doing the examination mentally catalogues the shape, size, texture, and color of each cell and its constituent parts (i.e., nucleus, cytoplasm, etc.) with the purpose of recognizing its type, age, and pathologic state. The average worker catalogues thousands of cells per day and, during or after each visual examination, keys alphanumeric

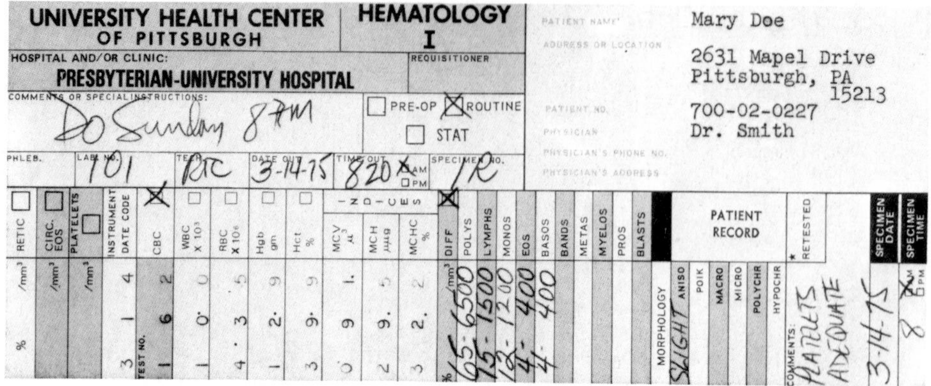

UNIVERSITY HEALTH CENTER OF PITTSBURGH | HEMATOLOGY I

HOSPITAL AND/OR CLINIC: PRESBYTERIAN-UNIVERSITY HOSPITAL | REQUISITIONER

COMMENTS OR SPECIAL INSTRUCTIONS: Do Sunday 8AM

PRE-OP | ROUTINE [X] | STAT

PHLEB. | LAB NO. 701 | TECH. RTC | DATE OUT 3-14-75 | TIME OUT 820 AM [X] / PM | SPECIMEN NO. 1R

PATIENT NAME: Mary Doe

ADDRESS OR LOCATION: 2631 Mapel Drive Pittsburgh, PA 15213

PATIENT NO.: 700-02-0227

PHYSICIAN: Dr. Smith

PHYSICIAN'S PHONE NO.

PHYSICIAN'S ADDRESS

RETIC	CIRC. EOS	PLATELETS	INSTRUMENT DATE CODE	CBC [X]	WBC $\times 10^3$	RBC $\times 10^6$	Hgb gm	Hct %	MCV μ^3	MCH $\mu\mu g$	MCHC %	DIFF [X]	POLYS	LYMPHS	MONOS	EOS	BASOS	BANDS	METAS	MYELOS	PROS	BLASTS	PATIENT RECORD	RETESTED	SPECIMEN DATE	SPECIMEN TIME
% /mm³	/mm³	/mm³	3 1 4	TEST NO. 1 6 2	1 0. 0	4 . 3 5	1 2. 9	3 9. 9	0 9 1.	2 9. 5	3 2. 2	% /mm³	65 6500	15 1500	12 1200	4 400	4 400						MORPHOLOGY: ANISO SLIGHT; POIK; MACRO; MICRO; POLYCHR; HYPOCHR. COMMENTS: PLATELETS ADEQUATE		3-14-75	8 AM [X] PM

Fig. 5.1. Typical patient record prepared in the clinical laboratory when a blood cell analysis is requested. The machine printout is produced by the Coulter Model S which measures blood cell size and volume. The manual entries are performed during the visual examination of the blood smear indicating, in this case; 65% Neutrophils (Polys), 15% Lymphocytes (Lymphs), 12% Monocytes (Monos), 4% Eosinophils (Eos), and 4% Basophils (Basos). Comments on morphology and the adequacy of platelets have, in this case, also been entered by the technologist

data into simple mechanical counters which are then manually transcribed onto the patient's record along with pertinent written comments. In a few large laboratories this information is keyed directly into a central computer. A typical patient record is shown in Fig. 5.1.

This time-consuming work is tedious and often poorly performed. In fact the National Communicable Disease Center has found that 40% of laboratories tested nationwide do unsatisfactory work as reported by SENCER [5.1]. Yet, unfortunately, the outcome of the analysis carried out is considered by the physician as an important indicator of the patient's state of health. Any type of mechanization or automation, whereby some sort of computer-controlled microscope scanner could be used to replace or aid the eye-brain system of the examiner in "picture analysis", would be a boon.

It is the purpose of this chapter to review the state of the art in the automation of digital picture analysis in cytology as of the mid-1970's. This is particularly appropriate since, after literally three decades of research, automation of certain aspects of the cytologic examination is now becoming a reality. Several apparently successful automatic television microscopes have been announced and are becoming commercially available. Although somewhat cumbersome and very expensive, these systems approach the capabilities of the human examiner.

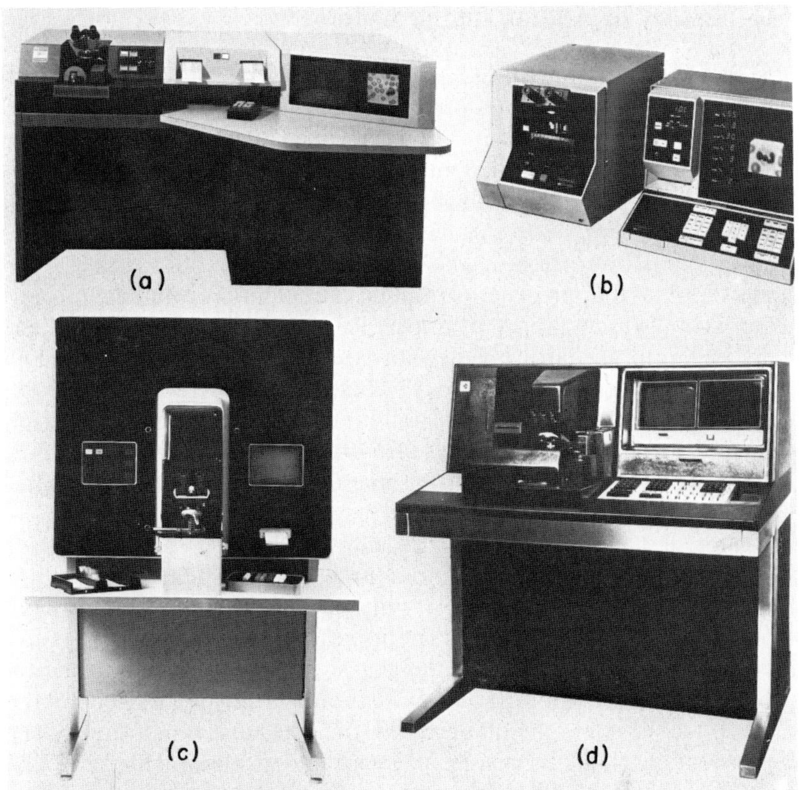

Fig. 5.2a—d. Current commercial digital picture analysis systems for use in aiding the technologist in the automatic analysis of cells. The instruments are produced by (a) Perkin-Elmer, (b) Corning Electronics, (c) SmithKline Inc., and (d) Coulter Electronics

All of them use some form of digital picture analysis to extract measurements from the images of cells generated by the automatic television microscope (see Fig. 5.2).

As one would expect, the story of automated microscopy is still unfolding. We have yet to reach the "height of the party". Much work is yet to be done in the automation of sample preparation, in computer-directed scanners, in digital cytologic picture processing, and in pattern recognition for the purpose of cytologic analysis. The need for automation becomes more acute daily as demands for more detailed cytologic examinations increase while the pool of highly skilled manpower required for this complex work diminishes. The few automatic machines which are advertised are still borderline in cost-effectiveness and have hardly been applied and tested. Hence the reason for this chapter.

5.2 History of Automation in Cytology

The recent eruption of successes in the application of both analog and digital electronics to image processing in cytology, and the publicity associated with commercial entries into the field, may lead the uninitiated reader to feel that automated cytology is a new and only recently developing field. This is far from the case, as can be seen by the extensive reference list to this chapter, and it is important for both the present and past researchers who desire to succeed in this field to survey, understand, and appreciate its gradual and painful evolution. As early as the late 1940's and certainly in the 1950's, preliminary, yet important, work began to appear in the development of the television microscope for use in quantitative metrology. These early television microscopes employed electronic circuits to analyze video signals using relatively elementary pulse-height and pulse-length counters. At this time—more than 20 years ago—objects on the microscope slide could be counted and sized at full video rates.

Later, in the 1960's, transistorized circuitry led to the same revolution in automatic microscopy as it did in general-purpose computation. Sophisticated, high-speed, special-purpose computers using thousands of components could function sufficiently reliably to make possible almost any degree of sophistication (at a price). At the same time the programmable general-purpose computer (especially the minicomputer) made possible, through software, the rapid testing, modification, and sophistication of digital picture processing algorithms. This led to the development of several highly complex research systems in the 1960's whose performance, at least in some areas of cytology, came close to equaling that of the human technologist.

In the 1970's a new era began with the announcement of several commercial computer-controlled scanners which automated the visual examination of human white blood cells. Assuming that these systems will be successful, it is expected that at least two more eras will follow: 1) an era in which the initial commercial instruments, which are presently slow and expensive, are further developed, modified, and refined to the point of cost effectiveness and 2) a new era—in the 1980's—which will see a breakthrough in speed and a dramatic reduction in cost per cell analyzed with, simultaneously, an extension of capabilities into the field of genetics (the analysis of chromosomes) and into the analysis of cervical smears (for cancer detection).

The late 1970's and the 1980's will also see for the first time the development of large, machine-generated data banks in cytology including both multi-color video images and, more importantly, quantitative morphologic and colorimetric data on the patient's cells and cell constituents. This information will be gathered on tens of millions of patients, i.e., on billions of cells. This breakthrough will inevitably lead to certain extraordinary and unanticipated discoveries in the health arena, as have other mass population studies which have been carried out under the general umbrella of what is called "multiphasic health testing".

For the uninitiated reader who wishes to rapidly up-date himself in cytology automation (without reading all of the one thousand or more articles published to date), several valuable surveys should be addressed. In the 1950's, PARPART [5.2] and FLORY [5.3] surveyed the medical applications of television microscopy and BERKLEY [5.4] treated the state of the art in object counting and sizing by means of the television microscope. In the 1960's several important compendia were published, edited by MONTGOMERY [5.5], RAMSEY [5.6], and TOLLES [5.7], which spanned the entire field. These documents cover cytology automation in the analysis of chromosomes, bacterial colonies, and white blood cells and for monitoring in bacterial warfare. At the same time certain individual authors reviewed the field, such as BUSHOR [5.8] and SHACKEL [5.9], who specifically treated medical applications of the television microscope, and EAVES [5.10] and NEURATH [5.11], who wrote general surveys of image processing in medicine. LEDLEY and RUDDLE [5.12] and INGRAM and PRESTON [5.13] discussed progress in cytology automation, specifically in connection with chromosome analysis and white blood cells, respectively.

In the 1970's PRESTON and RUTOVITZ [5.14] and MELLECKER [5.15] published still further surveys as part of conferences on automatic cytology sponsored by the Engineering Foundation (New York City). Then MAYALL [5.16] edited a collection of papers from the 1973 Engineering Foundation Research Conference on Automatic Cytology. At the same time, books on the subject appeared edited by EVANS [5.17] and by WIED et al. [5.18]. The reader familiar with this literature is well on his way to acquiring a firm foundation in the subject.

The next few pages expand on the above remarks by providing the reader with a summary of progress, decade by decade.

5.2.1 Developments in the 1950's

The early history of automatic picture processing in cytology is directly allied with the development of the television microscope. Early work centered on the use of cathode ray tube (CRT) flying-spot scanners which imaged the face of the CRT onto the microscope slide. The flying spot scanned objects on the microscope slide, and boundary and shape descriptors were then extracted from the video signal using custom-designed electronics. In some cases Nipkow-disc scanners were utilized which caused the image plane of the microscope to be scanned with a mechanically generated raster, as in the primitive television systems which existed prior to the availability of television imaging tubes (such as the orthicon and the vidicon). This early work in television microscopy as related to cytology was contemporaneous with similar work related to automatic specimen analysis in metallography. It also occurred at the same time that flying-spot scanners were being used for automatic character recognition.

Scanning Microscopes

Parpart [5.2] was one of the first to use television microscopy in cytology. He studied movements of red blood cells. At the same time Young and Roberts [5.19] developed a flying-spot scanner for use in particle size analysis as further reported by Walton [5.20], Hawksley et al. [5.21], and Causley and Young [5.22]. Also Barer [5.23] used television techniques to study living cells.

During the same period, the quantitative photometric aspects of television microscopy were treated by Ramberg [5.24]. Simple gating circuitry was devised for selecting a single line of the total television raster which crossed a particular portion of the cell being examined (such as nucleus and/or cytoplasm). Measurements made in this manner could be employed for determining the optical transmission coefficient (or extinction coefficient) of the cell and its constituent parts. This was of particular interest using ultraviolet wavelengths because these wavelengths are specifically absorbed by the major nuclear proteins DNA and RNA (deoxyribonucleic acid and ribonucleic acid). Work in this area is described in many publications such as Pollister and Moses [5.25], Moses [5.26], Skeggs [5.27], Williams [5.28], and Montgomery and Bonner [5.29].

Picture Processing—Stereology

In the 1950's it was generally assumed that the microscopic objects being scanned appeared on the microscope slide at high contrast, i.e., with well-defined boundaries. Thus, as the television scanner interrogated these objects, the video signal generated was expected to appear as a series of pulses. The length of each pulse would be directly related to the extent of the object (left boundary to right boundary) being scanned. If the object were small, an analysis of pulse lengths taken on consecutive scans would show a multiplicity of short pulses, whereas if the object were large, there would be a multiplicity of long pulses. In a sense, each object was being "sectioned" by the scanner.

Real-time circuits for pulse length analysis were designed and the results of the measurements taken and their study were called "stereology" as discussed in greater detail in Subsection 5.5.2. By adding pulse height analysis, descriptive measurements could be generated of the optical transmission of the objects being scanned at the wavelength (or wavelengths) selected. Although pulse length and pulse height measurements were relatively crude, the special-purpose electronic circuitry required was simple and, therefore, extremely fast and could operate at full video frame rates. This led to the development in the 1950's of very high speed but rather trivial picture analysis systems for cytology. Much of this effort at that time was concentrated in Great Britain. The status of this work is well described by COATES [5.30] and TAYLOR [5.31].

Finally, BERKLEY [5.4] describes a significant additional degree of sophistication called "intercept association". By holding the pulses generated during one television scan in a recirculating "delay line" memory and recalling them in synchronism with the next scanning line, it is possible to determine whether a particle which was intercepted by one scanning line is also intercepted by the next scanning line. If intercepts take place in synchronism, they are "associated" with, i.e., belong to, the same particle. If not, the particles must be separated. Using intercept association, particle counting circuitry operating at video rates was developed and applied to bacterial colony counting as described by MANSBERG [5.32]. These circuits were "fooled" by particles of complex shape. A particle having protrusions orthogonal to the direction of television scanning would be counted as several particles. However, in colony counting, where the colonies were regular in shape, such circuitry was adequate and could be used for rapid automatic analysis. By accumulating data on the average object transmission associated with the intercepts corresponding to each particle, it was possible to measure both particle area and transmission as a function of the optical wavelength. This led to an important development called the "Cytoanalyzer", which is described in the next paragraph.

The Cytoanalyzer

The Cytoanalyzer development commenced with the publication by MELLORS et al. [5.33] of the fact that cancer cells in cervical smears could be distinguished from normal cells by measuring nuclear size and transmission using the Papanicolaou stain. Characteristically, cancerous cells were claimed to have larger and more heavily stained nuclei than normal cells. This discovery led TOLLES [5.34] of Airborne Instruments Laboratory to develop, under the sponsorship of the American Cancer

Society, a fully automated television microscope for screening cervical smears. The system incorporated microscope slide feed, automatic focus, and had special-purpose circuits for performing a two-dimensional histogram of nuclear size versus nuclear transmission. This ambitious program as detailed by TOLLES and BOSTROM [5.35] was the precursor of all automatic image analyzers in cytology. The rise and fall of the Cytoanalyzer (as with the rise and fall of the Roman Empire) is a worthwhile case study in what often goes wrong in ambitious digital picture processing programs in this complex field. It was not until ten years after the initiation of this program that the reasons for its collapse and failure were clearly evident (see below).

5.2.2 Developments in the 1960's

The 1960's saw the continuation of the use of the flying-spot microscope (especially in the ultraviolet) as described by BARER and WARDLEY [5.36]. Also see the proceedings of a conference of the New York Academy of Science called "Scanning Techniques in Biology and Medicine" edited by MONTGOMERY [5.5]. In addition to this, ZWORYKIN and HATKE [5.37] had developed the "color-translating microscope" which formed three simultaneous images of the microscope slide at three ultraviolet wavelengths. The transmission of cells in the field of view at these wavelengths was then displayed using the three primary colors of a color-television monitor. This development of "pseudo-color" long antedated similar developments in infrared and earth satellite reconnaissance displays which are often referred to as "new developments" in the recent literature.

However, the most significant trends of the 1960's commenced with the introduction into automation in cytology of two new interests, namely, automation in hematology and in genetics. Also, in the 1960's, another major event occurred, which was the introduction of the first commercial system for digital picture analysis in television microscopy. This system was the Quantimet apparatus of Image Analyzing Computers Corp., a subsidiary of Metals Research Ltd. in Great Britain. But first let us treat the last years of the Cytoanalyzer.

The Demise of the Cytoanalyzer

By the early 1960's it was clear that the Cytoanalyzer program was in trouble due to two major difficulties. First, the assumption made earlier that cancer cells in cervical smears could be completely identified by their nuclear size and contrast was being disproved. Second, although the Cytoanalyzer was able to measure nuclear size and contrast at high speeds (several square centimeters of microscope slide area per minute), it was soon found that artifacts were a major problem in creating false data for scanning cervical smear preparations. The simplistic approach which assumed that every object on the microscope slide was a well-defined cell nucleus was soon disproved when it was found that clumps of blood cells, strands of tissue and mucus, and overlapping epithelial cells were causing a multiplicity of false positives.

At first it was hoped that, despite a high false positive rate, it would be possible to use the Cytoanalyzer to screen cervical smears at rates faster than those accomplished by the human technician and, thus, improve overall throughput in the clinic. Extensive work on improvements to the system are reported by PRUITT et al. [5.38], by BOSTROM et al. [5.39], and by DIACUMAKOS et al. [5.40–42]. Despite mechanically sieving cervical smear material and application of same to microscope slides by aerosol spray, and replacing the Papanicolaou stain by a new cytochemical (Weigert Hematoxylin-Tartrazine), it soon became impossible to obtain tolerable false positive rates without an intolerable number of false negatives. At this point, the entire Cytoanalyzer program ended.

Other efforts in cervical smear analysis were attempted later in the 1960's, such as those by ROSENBERG et al. [5.43] and SPRIGGS et al. [5.44], but they also were unsuccessful. It was unfortunate for all involved that the most difficult problem had been attacked first and that failure had given the topic of automatic cytology a bad name. Ten years of hard work in hematology and genetics would be required to overcome this initial reversal in cytology automation.

Systems for Hematology Automation

Under Atomic Energy Commission sponsorship, a program was commenced in 1960 to automate the reading of smears of human blood on microscope slides. This development was called the CELLSCAN program. By 1961, a special-purpose computer and television microscope had been constructed as reported by IZZO and COLES [5.45]. Its initial use was in research on the feasibility of screening blood smears for a rare type of blood cell whose occurrence appeared to be significant as an indicator of low levels of radiation damage to humans. Then, in 1964–1966, CELLSCAN was revised and applied to general human white blood cell recognition as discussed by INGRAM et al. [5.46]. Automation of blood smear analysis turned out to be far easier than automation of the analysis of cervical smears, and the CELLSCAN program showed immediate signs of success. This success, and the growing interest in the automation of white blood cell picture analysis, was due to work by PREWITT and MENDELSOHN [5.47], CHENG and LEDLEY [5.48], and YOUNG [5.49] in this field as reported in the mid and late 1960's. More details are given in Section 5.6.

Systems for Automatic Chromosome Analysis

In the field of genetics, discoveries in the late 1950's had led to new methods of sample preparation which permitted, for the first time, the detailed observation and analysis of human chromosomes (called "karyotyping"). This spurred the same team of scientists at Airborne Instruments Laboratory who had developed the Cytoanalyzer to design a new image logging microscope. This system, called "CYDAC", was installed at the University of Pennsylvania. Its features are described by NADEL [5.50]. It recorded the full microscope field on magnetic tape in digital format. The use of this system in genetics is described by MENDELSOHN et al. [5.51–55] who devised computer algorithms for automatic micro-photometry for the analysis of the DNA content of individual chromosomes.

Other major genetics programs using automated television microscopes with associated on-line general-purpose image processors were simultaneously developed by the University of Pittsburgh, as described by WALD and PRESTON [5.56], and by the Medical Research Council in Great Britain, as described by RUTOVITZ [5.57, 58]. At the same time, the National Biomedical Research Foundation (Washington, D.C.) and the New England Medical Center (Boston) also established major programs in chromosome picture processing using photomicrograph scanners rather than television microscopes as input devices to general purpose computers. These programs are described by LEDLEY et al. [5.59] and NEURATH et al. [5.60, 61], respectively. Other work in this field in the mid-1960's is reported by GILBERT [5.62] and GALLUS et al. [5.63].

General-Purpose Digital Picture Analyzers for Cytology

The first general-purpose system for the digital analysis of cell images was brought into being at the University of Chicago by WIED et al. [5.64]. It was called "TICAS". Also, the National Institutes of Health in a cooperative program with the Perkin-Elmer Corp. developed the Spectre II system for general-purpose cell image analysis as described by STEIN et al. [5.65]. Finally, the first commercially available system also appeared in the 1960's. This was the Quantimet System of Image Analyzing Computers, Ltd. Its characteristics and applications are discussed by BEADLE [5.66] and FISHER [5.67]. Other commercial systems for general-purpose automated microscopy were then announced and marketed by Zeiss, E. Leitz, Inc., Wild Heerbrugg, Ltd., and Bausch and Lomb. See publications by ZIMMER [5.68], MULLER [5.69], WEIBEL et al. [5.70], and MORTON [5.71], respectively.

Other Applications

Other activities in the 1960's also stimulated the development of automatic picture processing microscopes. The Department of Defense commenced the construction of scanning microscopes for use in the detection of bacterial warfare attacks as reported by NELSON et al. [5.72]. MANSBERG and SEGARRA [5.73] developed equipment for the enumeration of neurons in brain tissue slices. Work in bacterial colony counting continued as reported by SALOMON and MUSGRAVE [5.74]. A scanning system for high-speed grain counting in what are called "autoradiographs" was developed by NORGREN [5.75]. SOFFEN [5.76] investigated the design of an automatic television microscope for use in investigating the presence of life on Mars. OBERST et al. [5.77] and MAWDESLEY-THOMAS and HEALEY [5.78–80] commenced an investigation by scanning microscopy of the structure of goblet cells in sections of lung tissue. These and other programs indicate the depth of involvement of workers in cytologic picture analysis at the end of the sixth decade of this century.

5.2.3 Recent Developments

In the late 1960's the CELLSCAN program of the Perkin-Elmer Corp. was totally revised under DoD auspices and became the first entirely automated research system for the analysis of blood smears on microscope slides. The new system was called "CELLSCAN/GLOPR" and included automatic search, focus, and cell classification modes as reported by INGRAM and PRESTON [5.13]. It was first demonstrated to the Public Health Service in the summer of 1969 and announced publicly at the 1970 Engineering Foundation Research Conference on Multiphasic Health Testing (Davos, Switzerland) as a "landmark which points the way" according to NEURATH [5.11].

Simultaneous work on white blood cell analysis by MILLER [5.81] and BACUS [5.82] led to an outpouring of funds in automatic cytologic picture processing in hematology by several industries who saw the money-making possibilities of commercial instrumentation. MILLER established the Geometric Data Corp. (now a subsidiary of Smith, Kline, and French, Inc.) and BACUS became allied with Corning Electronics, Inc. (a subsidiary of Corning Glass Works). In 1972, both Geometric Data and Corning announced commercial blood smear analyzers, viz., the Hematrak and the LARC systems, respectively. Unfortunately, data on digital picture analyzing algorithms and on specific test results on these systems are cloaked in industrial secrecy, although some information is available in publications by COTTER [5.83] and CARTER and COTTER [5.84] as regards the LARC. MILLER [5.85] also described Hematrak in a presentation at the 1972 Engineering Foundation Conference on Automatic Cytology. MEGLA discussed the LARC system at the same conference and prepared papers for open publication; see MEGLA [5.86, 87]. Other commercial systems were announced in late 1974 by Coulter Electronics, Inc., and by Honeywell, and in 1976 by Perkin-Elmer. The Honeywell system performs only the search and focus operations. Other systems yet to be commercialized have been patented by SMITHLINE [5.88] and BRAIN [5.89].

Research still continues in the general field as is reported in *Research Efforts* and *Other Applications*, below.

Commercial Systems

First let us compare the commercial systems which have appeared on the market in the 1970's. The LARC employs a *general*-purpose minicomputer to perform picture analysis. The Hematrak employs *special*-purpose digital logic to carry out its picture analyzing algorithms. Both of these systems are interactive and require constant attendance by a hematological technician. The technician must perform manual tasks such as loading and unloading each microscope slide and be prepared to observe (on a television screen) and classify those cells which the machine scores as unusual. Neither machine is particularly fast and, with the help of the technologist, carries out one examination (100 cells total) every few minutes. Both machines are expensive (in the neighborhood of $ 100000 per system) and it is not clear from an economic standpoint whether they are cost effective.

The Coulter Electronics system removes the necessity of manual assistance from the operator in handling the microscope slides, as it employs a magazine loader. The Honeywell system performs no cell classification at all, and simply acts as a high-speed cell locator which presents a cell each 0.25 s to the technologist (on a color television screen) for visual classification.

Clinical trials of these instruments are in the early phases and it will be one or two years before complete evaluations are available. Further details on some published cell classification success rates are given in Section 5.6.

Research Efforts

The major research programs in digital picture processing in cytology which commenced during the 1960's (see *Systems for Hematology Automation*) continued on into the 1970's. One major change was the transfer of the CYDAC program from the Department of Radiology, University of Pennsylvania, to the Lawrence Livermore Laboratory (California), where a new facility has been established by MENDELSOHN which is described by MAYALL [5.90].

At the New England Medical Center the work led by NEURATH in chromosomes has expanded into blood cell analysis; see NEURATH and BRENNER [5.91]. An allied program in the analysis of blood smears was carried out in conjunction with CERN (Switzerland); see GELSEMA and POWELL [5.92] and GELSEMA et al. [5.93]. Other joint efforts with NEURATH in the field of chromosome analysis have been carried out in conjunction with Gallus (Italy), as described by GALLUS and NEURATH [5.94] and NEURATH et al. [5.95]. Another cooperative research effort between NEURATH and the group of RUTOVITZ in Great Britain is described by GREEN and NEURATH [5.96]. Finally, NEURATH is producing an automatic microscope for blood smear processing with the Massachusetts Institute of Technology; see DEW et al. [5.97]. Also at the Massachusetts Institute of Technology work in digital picture processing as regards cytology continues, as has been reported by YOUNG [5.98, 99].

The program at the National Institutes of Health under Lipkin continues and is documented by LEMKIN et al. [5.100] and by CARMAN et al. [5.101]. The TICAS program at the University of Chicago has expanded widely; see WIED et al. [5.18]. It is involved in the first major effort in generating a data base of cytological images and image measurements as described by BIBBO et al. [5.102].

The program at the National Biomedical Research Foundation became allied with Georgetown University. It has evolved from a photomicrograph scanning system to one employing an automatic microscope-computer system called MACDAC, as described by GOLAB [5.103]. A major new program has commenced at the Jet Propulsion Laboratory (Pasadena), as described by CASTLEMAN and WALL [5.104], applied to the automatic analysis of chromosomes. Other efforts in chromosome analysis continue in France; see LE GO [5.105]. The CELLSCAN/GLOPR system applications have been extended into chromosome analysis and cervical smear analysis, as described by PRESTON [5.106].

Other Applications

Although the major efforts in digital picture analysis in cytology still are chromosome analysis, blood smear analysis, and cervical smear analysis, certain other efforts are worthy of mention. WANN et al. [5.107, 108] reports work in both grain counting in autoradiography and in the analysis of neurons. ZAJICEK et al. [5.109] reports on the analysis of cells extracted from breast carcinomas by means of needle biopsy. VOSS [5.110] has worked on methods for counting and sizing overlapping red cells; AUS et al. [5.111] has done an analysis of HeLa cells; EDEN et al. [5.112] has studied the feasibility of detecting malarial parasites in blood; TER MEULEN et al. [5.113] and AUS et al. [5.114] have worked on problems associated with virology; while CHRISTOV et al. [5.115] and MOORE [5.116] have performed automatic analyses of cells in thyroid tissue. Clearly the field is active and is expanding into many interesting areas.

5.2.4 A Major Omission

In all of the above discussions there has been one major omission. There has been a lengthy and successful effort during the 1960's and 1970's in the development of what are called "flow systems" for cytological analysis. These systems have not been discussed because digital image processing is not incorporated. However, because of the extensive work on this approach and because it is considered by many to be competitive with digital picture processing it must be included in this chapter.

The general mechanization of a flow system (including cell sorting) is shown in Fig. 5.3. Cells in liquid suspension are delivered one at a time to a sensing aperture where either optical or electronic detection (or both) is used for cell classification. If cell sorting is employed, the individual cell is encapsulated in a droplet carrying an electric charge. This droplet moves at high velocity and may be deflected electrostatically to a collecting port.

The earliest reports of flow systems are by CROSLAND-TAYLOR [5.117] and by COULTER [5.118]. These early systems employed electronic sensing where either the resistance or capacitance of the flow stream through a microaperture was monitored. Changes in either the resistance or capacitance of the contents of the aperture produced pulses of electrical current whose length and height are related to the volume of the cell passing through the aperture. This process makes possible high-speed cell counting in a fluid stream. It is capable of handling red and white blood cells at several thousand per minute.

The next major advance in flow systems for cell analysis was made by KAMENTSKY et al. [5.119], who introduced optical sensing (in the ultraviolet) at the flow aperture. By measuring absorption at two wavelengths, electronically generated scatter diagrams showed clusters of points characteristic of the specimen (see Fig. 5.4).

At the same time that Kamentsky was applying this technique to general cell analysis, ANSLEY and ORNSTEIN [5.120, 121] were working on a variety of enzyme stains which would be specific to the most commonly occurring types of white blood cells. The success of their

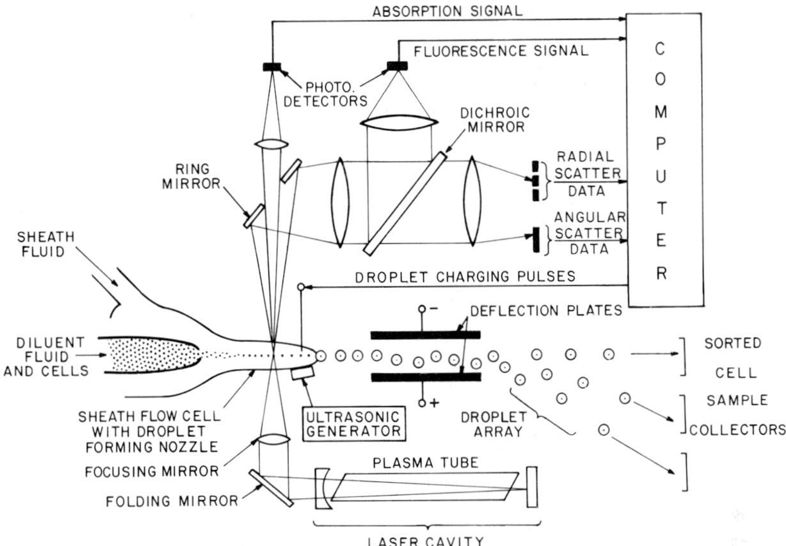

Fig. 5.3. General schematic diagram of a sheath flow system for delivering cells one at a time to a region illuminated by a laser. Absorption, fluorescence, and both radial and angular scatter are measured and a computer processes these measurements to provide a droplet charging pulse to the electrostatic cell-sorting portion of the system

research led to the development of a multi-channel, multi-aperture flow analyzer introduced by the Technicon Corp. in 1970 and called the Hemalog D. The operation of Hemalog D is covered by the ANSLEY and ORNSTEIN patent [5.122]. Each flow channel uses a biochemical treatment of the flow stream designed to tag a particular type of blood cell. The reader interested in the evaluation of this instrument as a white blood cell analyzer is referred to papers by SAUNDERS et al. [5.123], MANSBERG [5.124], PIERRE and O'SULLIVAN [5.125], MANSBERG et al. [5.126], and SAUNDERS and SCOTT [5.127]. Still other articles describing this system have been prepared by SAUNDERS [5.128–130].

The final facet of flow system development was to add electronic cell sorting by borrowing from technology developed for "ink-spray" computer printers as described by SWEET [5.131]. These devices break the flow stream into droplets that are electronically charged. Because of their velocity, these droplets may be deflected so as to arrive for collection purposes at well-separated points in space. Work in this area was initially carried out by FULWYLER et al. [5.132, 133] and further developed as described in papers by VAN DILLA and FULWYLER [5.134], VAN DILLA et al. [5.135], and STEINKAMP et al. [5.136]. A computer

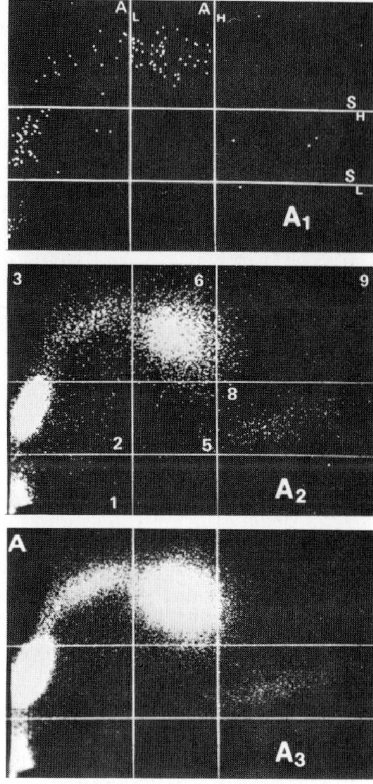

Fig. 5.4. Two-dimensional scatter diagrams (formed on a display CRT) showing (on the abscissa) transmission and (on the ordinate) narrow angle light scatter as plotted using a flow system. Plots are shown (top to bottom) for about 100 cells, for about 10000 cells, and for about 100000 cells. These recordings were made in 0.2 s, 20 s, and 200 s, respectively. The numbered regions correspond to: 1) platelets, red cell "ghosts" and some electronic noise, 2) small and medium lymphocytes (unstained), 3) large lymphocytes, basophils, blast cells (if present) and unstained and weakly stained monocytes, 6) neutrophils, and 8) eosinophils. (Courtesy L. ORNSTEIN, Mount Sinai School of Medicine, City University of New York)

controlled flow system with cell sorting has recently been built in Germany by ARNDT-JOVIN [5.137].

Besides Technicon, which introduced Hemalog D for clinical use specific to hematology, other companies have been formed to exploit the commercialization of this field. KAMENTSKY established Bio/Physics Systems Inc. to produce flow instrumentation for research purposes. Some applications are described in KAMENTSKY and MELAMED [5.138–141] and in KAMENTSKY [5.142]. FULWYLER has been funded by COULTER to establish Particle Technology Inc. to build cell sorting flow systems. Nuclear Research Associates has attempted to combine flow techniques with image processing in cancer cell screening as described by FINKEL et al. [5.143]. Other users of flow systems are SALZMAN [5.144] for volume analysis and LEIF [5.145], LEIF and THOMAS [5.146], and LADINSKY et al. [5.147] for cancer cell screening for cells gathered in cervical smears. Recent modifications of the Leif system are given in THOMAS et al. [5.148]. Theoretical studies of such systems have been carried out by GROVER et al. [5.149] and GOLIBERSUCH [5.150]. Other papers by DITTRICH and GOHDE [5.151] and HOLM and CRAM [5.152] discuss flow system using optical fluorescence detection, as do YATAGANAS and CLARKSON [5.153].

The optical sensing in flow systems was based on light absorption in earlier versions, but now measurement of light scattering has been added. Scattering includes the measurement of the redirection of incident light as a function of angle. The general theory of light scattering is discussed in a book by KERKER [5.154], and GRAVATT [5.155] has surveyed general applications. Since light scattering theory relates to the general field of optical Fourier transform techniques, books by GOODMAN [5.156] and PRESTON [5.157] are also pertinent. A discussion of this method of optical sensing and of digital processing in the Fourier domain is given in Subsection 5.5.7.

5.3 Sample Preparation

Before moving on to the description of scanners and digital picture processors for cytology it is important to treat the problem of sample preparation at least as related to the major cytologic examinations, namely, examinations of 1) cervical smears, 2) blood smears, and 3) chromosome preparations. As every project in automatic cytologic picture processing has demonstrated, sample preparation for machine analysis is the key to success. Any author would be delinquent if he failed to discuss this subject. Unfortunately, the treatment here must be brief as this chapter is devoted to digital techniques in picture analysis.

Recall that in *The Demise of the Cytoanalyzer*, the failure of the Cytoanalyzer was attributed to a large extent to its inability to distinguish the cells to be analyzed from artifacts in the cervical smear (such as mucus, blood cells, clumped cells, etc.) and that the true significance and magnitude of this problem were not recognized until the program was some six years past its point of initiation (assuming that this point occurred with the publication of MELLORS et al. [5.33]). The final year of the Cytoanalyzer program saw intense efforts in both mechanical and chemical preparation of the sample. Once again, refer to DIACUMAKOS et al. [5.42]. Clearly these efforts should have been made at the onset of the program.

The fact that the National Cancer Institute has just recently committed more than one million dollars in research grants and contracts to further study sample preparations of cervical smears for automatic screening indicates that this is still considered one of the major roadblocks to success in this field. Philosophically we can say that the electrical engineer, who has been the pioneer in applying digital picture processing in biomedicine, but who is usually untrained in biochemistry and biomechanics, consistently has placed the cart before the horse in automating cytology.

5.3.1 Sample Preparation for Automation in Hematology

Sample preparation in hematology is far simpler than that required in the preparation of cervical smears or chromosome samples. In hematology blood is drawn from a vein or subcutaneously by skin puncture and is applied directly to the microscope slide without biochemical treatment. Traditionally a few drops of blood are placed at one end of the microscope slide and are smeared into a thin layer using a second microscope slide as a smearing tool. The result is a film of blood which is thick at one end and thin at the other, as shown on the right hand side of Fig. 5.5. After smearing, the blood film dries rapidly. When dry, it is stained with a mixture of dyes called "Wright's stain", then rinsed and fixed. The traditional method of blood smear preparation yields a smear in which at the thick end cells are too clumped to be satisfactory for visual analysis, and at the thin end are too widely separated and disrupted. However, since only a few square millimeters of blood film are observed by the technologist in performing a 100-cell blood analysis, there is usually some region of the smear (between the thick end and the thin end) which is satisfactory. This region is located by the technologist when the microscope slide is loaded on the microscope stage. The reader wishing to become familiar with the details of this process should obtain the standard classical texts, pamphlets, and manuals by DIGGS et al. [5.158], WINTROBE [5.159], HARPER [5.160], MIALE [5.161], and the U.S. Army [5.162].

Little effort was applied to automatic blood smear preparation until the mid-1960's, when PRESTON and NORGREN [5.163] used a "spinner" to produce extremely uniform films of blood on microscope slides, as shown in the left-hand view in Fig. 5.5. Spinning is a technique borrowed from the semiconductor industry where it is used to spread photoresist films on the surface of silicon wafers. In spinning, the microscope slide is accelerated quickly to a high rotational velocity (5000 rpm) about an axis orthogonal to the plane of the slide. Before spinning, a few drops of blood are placed near the intersection of the spin axis with the plane of the slide, and spinning is carried out for approximately one second. Modifications of spinning blood have been patented by STAUNTON [5.164] which concentrate on producing uniform air flow in the vicinity of the rapidly rotating slide. The results of spinning on the morphological

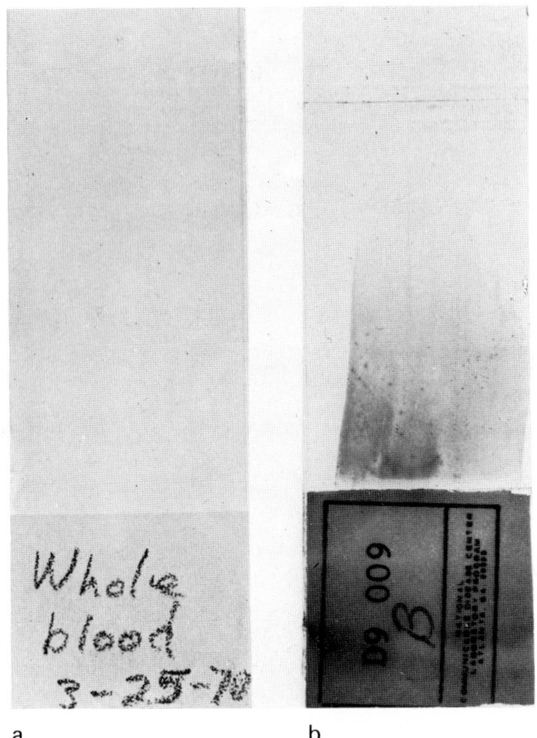

Fig. 5.5a and b. Typical smears of human blood on microscope slides produced using both (a) mechanical spinning and (b) manual smearing

properties of both red and white cells have been extensively studied by INGRAM and MINTER [5.165], by ROGERS [5.166, 167], and by BACUS [5.168]. Commercial spinners are available from both the Corning Electronics Corp. and the Perkin-Elmer Corp.

The Geometric Data Corp. has commercialized a blood smearing device also. Here the traditional smearing technique has been automated. This course of action has been followed because, in certain patients, spinning has been found to distort red cell morphology in ways that traditional smearing appears to avoid.

Photographs of the three commercially available blood film preparation units are shown in Fig. 5.6 along with one device currently used for automatically staining the blood film (or smear) on the microscope slide. A photomicrograph showing cells on the microscope slide is given in Fig. 5.7. The status, therefore, of automatic sample preparation in hematology is that it has now reached automation on a

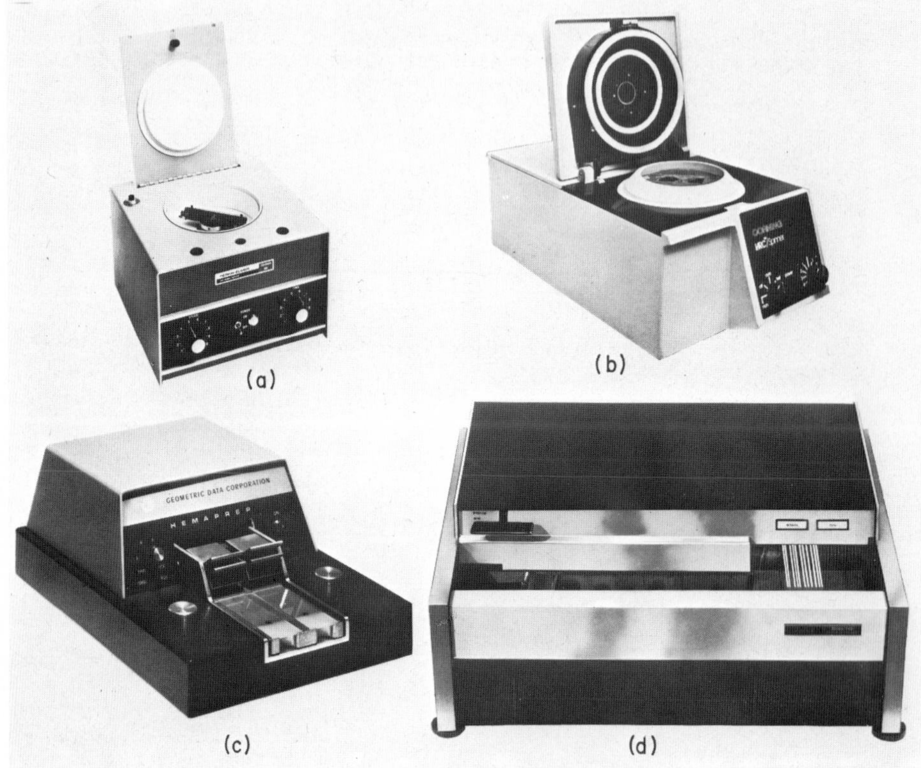

Fig. 5.6a—d. Present commercial units for the preparation and staining of human blood smears: (a) Perkin-Elmer spinner, (b) Corning spinner, (c) SmithKline (Geometric Data Corp.) smearer, and (d) Ames HEMA-TEK® stainer

commercial basis. Whether spinning or smearing is to be preferred will not be determined until thorough clinical trials are completed.

5.3.2 Sample Preparation in Chromosome Analysis

In the field of chromosome analysis little has been done to specifically tailor the preparation of samples to machine reading. Such an effort is included in a recent contract between the National Institute of Child Health and Human Development and the Jet Propulsion Laboratory and is being carried out by the City of Hope laboratories. At present manual sample preparation is time consuming but not particularly laborious. It requires the culturing over several days of blood extracted from the patient using biochemicals which cause the cultured cells to be

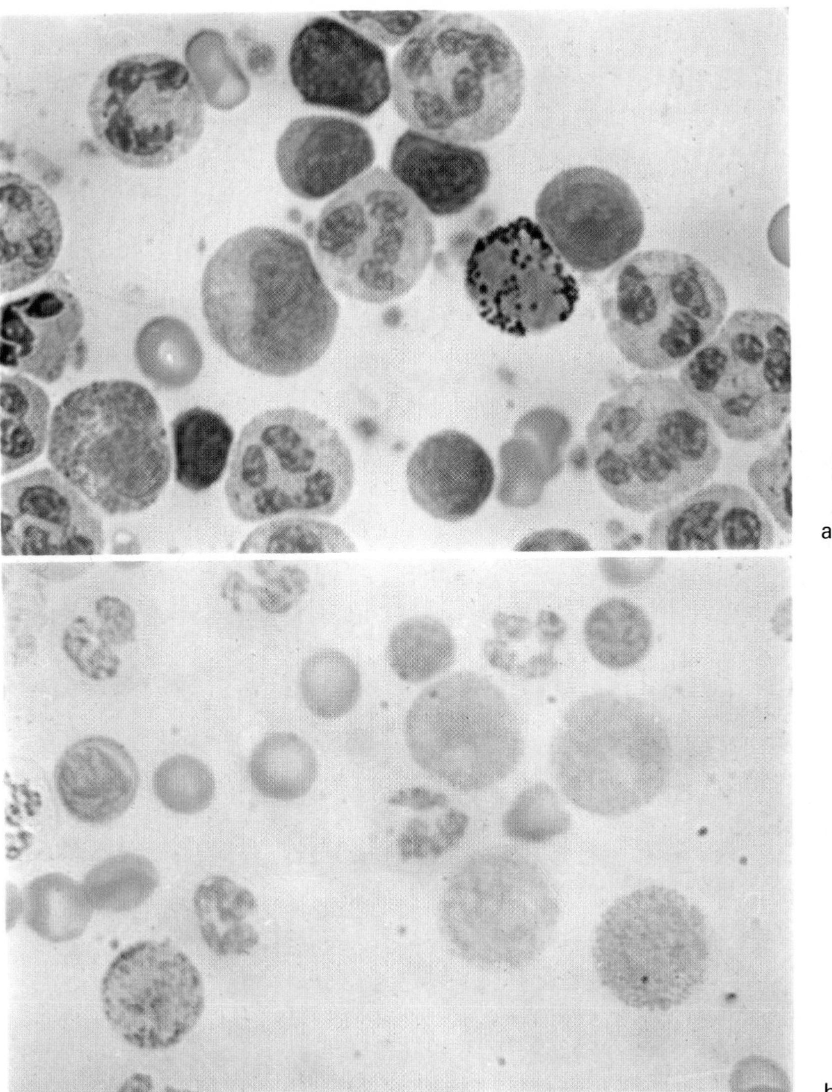

Fig. 5.7a and b. Stained buffy coat smears of human blood showing (a) the standard Wright-Giemsa stain and (b) the trichrome stain developed by ORNSTEIN [5.200]

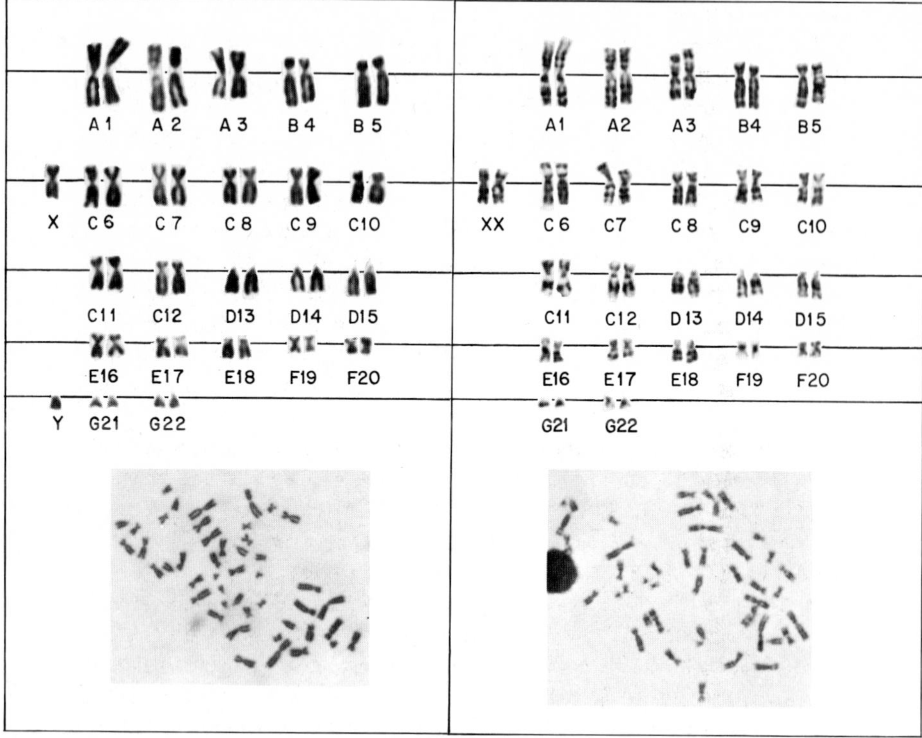

Fig. 5.8

Fig. 5.9

Fig. 5.8. Photomicrograph of a chromosome spread and its karyogram using conventional staining and classification techniques. (Courtesy N. WALD, University of Pittsburgh School of Medicine)

Fig. 5.9. Photomicrograph of a chromosome spread and its karyogram prepared using a high pH Giemsa stain which elucidates banding patterns. (Courtesy N. WALD, University of Pittsburgh School of Medicine)

arrested at the metaphase stage of cell development. Once the cultured blood specimen is ready, it is treated with a hypotonic solution which enlarges the metaphase cells so that, when they are spread on a microscope slide, the minute chromosomes (a few micrometers in length) are well separated. This entire technique for sample preparation is a relatively new one which was discovered and refined by HUNGERFORD et al. [5.169].

Once the smear of metaphase cells is produced, machine analysis consists of image processing routines which 1) isolate each chromosome, 2) locate the arm junction (the centromere), and 3) calculate the centromeric index, i.e., the ratio of the two short chromosome arms to the two long ones (either on a length or area basis).

Chromosome analysis has suddenly undergone an upheaval due to the discovery and perfection of a new biochemical treatment devised by CASPERSSON et al. [5.170–172]. The visual analysis of chromosomes using the Caspersson stain is radically altered because of the appearance of "banding patterns". A comparison between ordinary cytochemical staining and the Caspersson method of preparation is shown in Figs. 5.8 and 5.9.

The new method of CASPERSSON has been found to be more specific in the identification of chromosomes than the earlier method of staining. Additional discussions of this new development are provided by BREG et al. [5.173, 174]. Alternate staining methods for the elucidation of banding are compared by SUMNER and EVANS [5.175] and a discussion of microphotometry of such material is presented by LUNDSTEEN and PHILIP [5.176]. Whether the geneticist will demand that machine analysis be done on banded chromosome preparations or on the older, non-banded chromosome images is still in question. The cost-effectiveness of automation has yet to be established. Finally, work on automatic sample preparation is still far from being completed.

This situation exists despite the fact that most geneticists would insist that every newborn in the United States should be analyzed for chromosomal abnormalities. To accomplish this task would require approximately 3 million analyses per year. At present, although about 600 laboratories are equipped for cytogenetic analysis, less than 100000 such analyses are carried out in the United States annually.

5.3.3 Automatic Sample Preparation for Cervical Smears

As mentioned above, the first major effort to devise methods of preparation of cervical smear material for machine analysis was carried out in conjunction with the Cytoanalyzer program in the early 1960's. Methods for mechanically sieving cervical smear material for the purpose of removing debris and cellular clumps are described by DIACUMAKOS [5.41], as well as methods for applying the sieved material to the microscope slide by aerosol spray techniques. Other efforts in preparing smears for the Cytoanalyzer are discussed by PRUITT [5.38].

The original development of biochemical stains to assist in cancer cell detection in cervical smears is due to PAPANICOLAOU [5.177], and it was his formula that was initially used for the Cytoanalyzer. However, in the final stages of the Cytoanalyzer program, the Papanicolaou stain was discarded in preference to the Weigert Hematoxylin-Tartrazine stain described in DIACUMAKOS [5.40]. This is hardly the end of the story, and the reader interested in pursuing the automatic analysis of cervical smears should review the survey paper on the irrigation smear by DAVIS [5.178] as well as papers by PARRY et al. [5.179] on the dispersal of cell clumps using ultrasound and by LEIF [5.145] on the separation of cells by buoyant density.

The new push to succeed where the Cytoanalyzer failed has led to recent efforts by TUCKER and GRESNAM [5.180] on the preparation of cervical material for automatic analysis, and by BAHR et al. [5.181], who evaluate the Papanicolaou stain for use in automatic pattern recognition. One should also read the survey by WHEELESS and ONDERDONK [5.182]. Certain other papers by WHEELESS and PATTEN [5.183] and by ADAMS [5.184] treat the use of the biochemical acridine orange for analyzing epithelial cells in cervical smears in a flow system environment using fluorescence.

5.3.4 Other Sample Preparation Techniques Relating to Image Analysis in Cytology

Before ending this section on sample preparation, let us point out that there is a wealth of information on the enhancement of biological information in the cell image by means of biochemical dyes (or "stains"). With the push to automation in cytology in recent years we have seen considerable effort to obtain cell-type or cell-constituent specific stains and (in relation to the determination of amount of the genetic material in chromosomes) stains specific to the nucleic acids.

Serious investigation in the use of such stains for elucidating the nucleic acid content of cells (both RNA and DNA) dates back to work of FEULGEN and ROSSENBECH [5.185]. Work on DNA and RNA staining was performed by POLLISTER and RIS [5.186], by DISTEFANO [5.187], and by DEITCH [5.188], who extended the work to protein specific stains (naphthol yellow S). More recent work is by DEITCH [5.189] using methylene blue, by DECAPOA et al. [5.190] as regards genetics, and, as regards fluorescence, by BOHM and SPRENGER [5.191]. Work in the 1970's using the fluorescence of stains is reported by TRUJILLO and VAN DILLA [5.192], by LATT [5.193], and by GOLDEN and WEST [5.194], as regards fading rates in fluorescence. Consecutive staining has been studied by DAVIDSON [5.195]. Staining of a viable cell population has also been undertaken in the study of cell kinetics reported by KIEFER et al. [5.196] and by GRAY [5.197].

Recent papers which are specific to the use and influence of stains on pattern recognition have been prepared by DECAMPOS et al. [5.198], GILL and JOTZ [5.199], and by ORNSTEIN and ANSLEY [5.200]. SANDRITTER et al. [5.201] and STEINKAMP and CRISSMAN [5.202] have reported on staining for pattern recognition and image analysis relating to cancer cells in both breast carcinoma and cervical smears.

Many other papers are specific to particular areas. BREMESKOW et al. [5.203] report on DNA in L-cells, and HENRY et al. [5.204] on stains specific to hemoglobin. DAVIS and ORNSTEIN [5.205] and SWEETMAN et al. [5.206] report on enzymatic stains specific to various types of leucocytes. Other reports on stains specific to leukocyte recognition are given by GILBERT and ORNSTEIN [5.207] as regards basophils, and by ATAMER and GRONER [5.208] as regards perioxidase chemistry. Extensive reports on the dye acridine orange are provided by ADAMS and KAMENTSKY [5.209], MELAMED et al. [5.210], and STEINKAMP et al. [5.211].

From this extensive list of papers it is obvious that the biochemistry of sample preparation is an active and complex field. The reader should be aware of the existence of an excellent review volume on stains entitled *Biological Stains*, and of the journal *Stain Technology,* which provides a vehicle for new reports. The *Journal of Histochemistry and Cytochemistry, Histochemica Acta, Histochemie,* and the *Histochemistry Journal* also serve as references. Finally, since the human eye is still frequently the primary evaluator of color imagery (although being supplemented by machine analysis), the reader should review topics in human color perception as discussed by such authors as SHEPPARD [5.212].

5.4 Scanners and Scanning Mechanisms

Once the cytological sample is mechanically prepared and stained with cytochemical dyes, the information which it contains is transferred to the computer for the purpose of digital picture processing by means of the television scanner. Such scanners transform the colorimetric information present in the cell (and its background) into a digital number array. In current practice it is almost universal to use standard non-coherent microscope illuminators for the purpose of interrogating (illuminating) the specimen and causing the transfer from the object to the image plane. In some cases, such as reported by KAUFMAN and NESTER [5.213] and by DAVIDOVITS and EGGER [5.214], coherent scanning systems are utilized employing lasers. More esoteric methods for performing this information transfer are being investigated by LEMONS and QUATE [5.215], who have developed an ultrasonic microscope, and by shadow casting techniques as described by HLINKA and SANDERS [5.216].

This section treats non-coherent optical scanners exclusively. Two important mechanisms which exist in the interface between the scanner and the computer are also treated, namely, 1) video amplifiers and pre-processing circuitry, and 2) digitizing circuitry for converting the analog pre-processed video signal into binary words.

5.4.1 Types of Scanners and Sources of Illumination

Historical information on the development of the television microscope has already been presented in Section 5.2. It should be noted that there are two primary types of optical scanners, i.e., image plane scanners and flying-spot scanners. In the image plane scanner the detector occupies a plane conjugate to the eyes of the microscopist. This is easily instrumented using most commercial microscopes which are designed for a trinocular tube for photomicrography (see Fig. 5.10).

In the flying-spot scanner the illuminating source is placed in the normal image plane of the microscope so that the optics image this source onto the specimen. The size of the source must be small and equal to that of a picture element. Hence, the name "spot". As the flying spot addresses the specimen, the light transmitted through the specimen is collected by the microscope condenser and impinges on a light detector.

In the 1950's and early 1960's before the advent of inexpensive and reliable television tubes, the primary image plane scanner, as exemplified by the Cytoanalyzer, was the Nipkow disc. Flying-spot scanners were devised from cathode ray tubes placed in the image plane whose illumination was directed down the microscope toward the specimen. The flying-spot scanner is best exemplified by CYDAC, as described by BOSTROM and HOLCOMB [5.217]. An excellent survey of scanners utilized in the mid 1960's is given in MANSBERG and OHRINGER [5.218]. A more recent survey is provided by BURN [5.219].

It should be mentioned that, because of the importance of fluorescence excitation and the creation of fluorescent images, much work has been done on scanners which excite fluorescence using violet illumination and form images at the fluorescence wavelength. Typical is the work of THIEME [5.220], PEARSE and ROST [5.221], ROST and PEARSE [5.222], and PLOEM et al. [5.223].

Recent scanners of all types (direct and fluorescent) have been placed under computer control for image transfer to a digital processor. Also the automatic transport of the specimen, including such operations as focusing and the location of objects of interest in the specimen plane, has been automated. In early systems, such as the Cytoanalyzer, focusing was done by means of an open-loop servo mechanism which mechani-

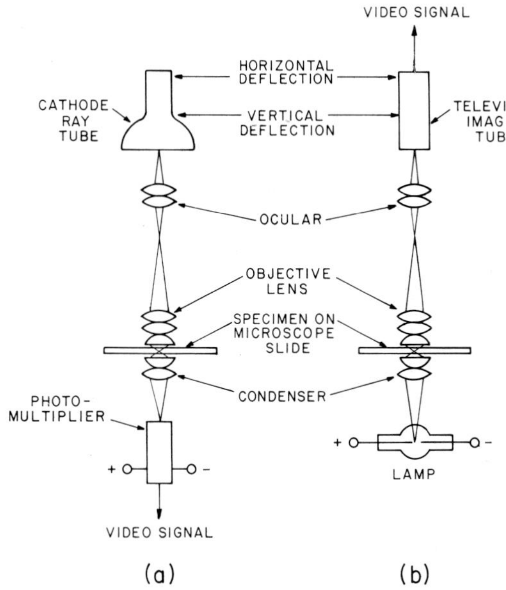

Fig. 5.10a and b. Schematic diagram comparing (a) the configuration of a flying spot scanner with (b) the configuration of an image plane scanner

cally sensed the position of the specimen and kept its surface in the focal plane of the objective lens. Later systems, such as CELLSCAN/GLOPR, used closed-loop focusing by means of detecting and maximizing the high temporal frequency content of the video signal. An alternative method of automatic focusing, as applied in CYDAC, is based on an analysis of the photometric histogram (see Subsection 5.5.1) as described by Kujoory et al. [5.224]. High-speed focusing is now incorporated in all commercial instruments and is capable of maintaining focus to within a few hundredths of a micrometer.

In blood smear analysis the CELLSCAN/GLOPR system was the first to include both automatic focusing and automatic search for finding leucocytes on the microscope slide. In this system a dedicated special purpose computer was used to monitor several segments of the moving microscope field of view and to select that segment in which the leucocyte appeared. The leucocyte was then centered electronically in the field of view of the scanner, which was transferred to the computer. This is described in detail by Preston [5.106].

In chromosome analysis high-speed metaphase cell detection using a laser was devised, as is described by Norgren [5.75], in a system which is presently employed by the Department of Radiation Health of the School of Medicine of the University of Pittsburgh. Other systems for the automatic location of metaphase cells, which have some selectivity

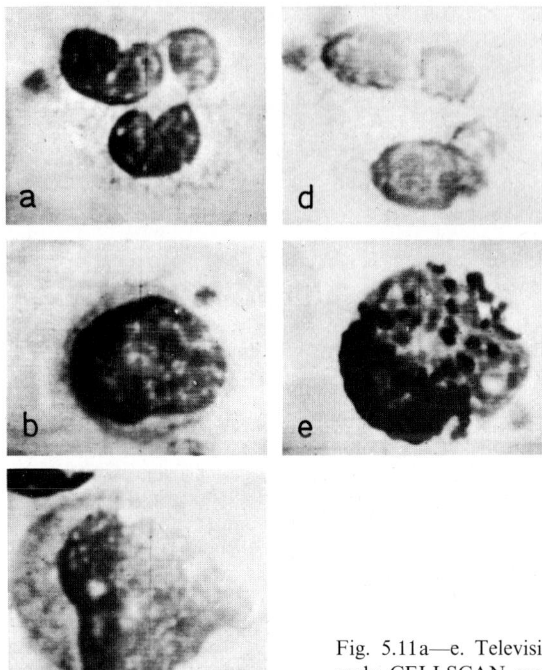

Fig. 5.11a—e. Television images generated by the early CELLSCAN system showing the five major types of human white blood cells

in the cells located, include the system at the Medical Research Council in Great Britain, as described by PATON [5.225] and by GREEN and CAMERON [5.226], and the system at Jet Propulsion Laboratory, as discussed in JOHNSON and GOFORTH [5.227].

Progress in the state of the art of video imaging is shown in Figs. 5.11 and 5.12. Figure 5.11 shows CELLSCAN images of leucocytes having a 0.1 μm scan-line spacing over a 20 by 20 μm field as recorded in 1966. Figure 5.12 shows an image generated recently by the Jet Propulsion Laboratory ALMS-2 system, where the field of view is about 500 by 500 μm with a 0.5 μm line spacing. In the first case the scanner used was a Dage slow-scan vidicon; in the latter case, it was a high-precision mirror scanner. As can be seen in both systems, superb images may be generated with carefully designed and optimized equipment. However, as stated below, no investigation was made as to whether the depth of digitization illustrated and the picture element spacing were justified in terms of the picture analysis to be carried out. This raises questions about image storage requirements requiring serious further study. It may be that workers in automatic cytology are taking no chances and are guilty of pictorial overkill.

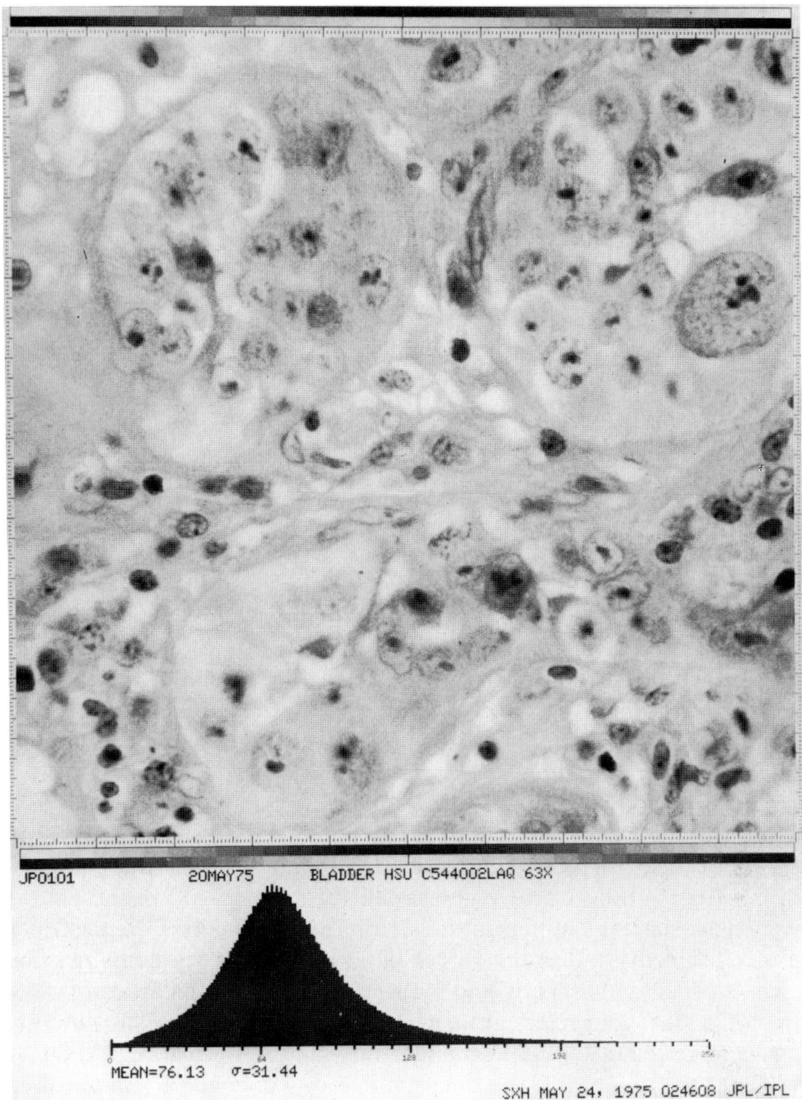

Fig. 5.12. Television image recently generated by the ALMS-2 system at the Jet Propulsion Laboratory using a mirror television scanner

5.4.2 Pre-Processing and Digitization

A typical cytological field may contain from 10 thousand to 1 million picture points. In imaging leucocytes, the typical field of view is 25 by 25 μm with resolution ranging from 0.1 to 0.5 μm. This leads to a computer memory requirement of from 4 to 64 kilowords (4 k to

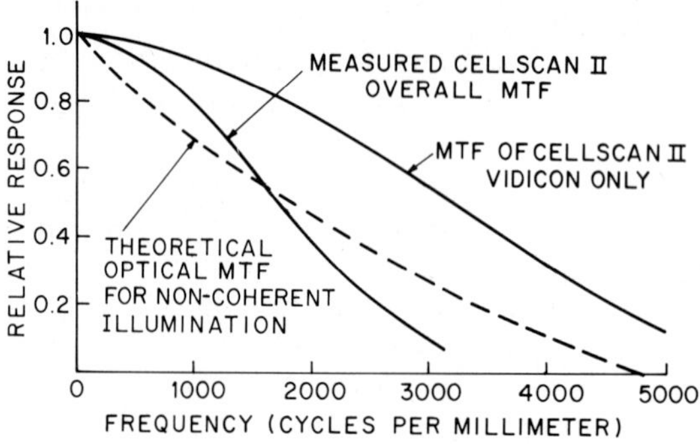

Fig. 5.13. The modulation transfer function of a Vidicon scanner which produced the images shown in Fig. 5.11. The modulation transfer function expresses contrast (relative response) in the image as a function of the fineness of image detail (cycles per millimeter). By electronically modifying the television signal the overall modulation transfer function can be made higher than the theoretical optical value

64 k words). For chromosome analysis the field of view must be enlarged to 80 by 80 μm so as to encompass a large metaphase cell. The highest possible resolution is usually required, leading to the need for storage of from 40 k to 640 k words. For certain tissue sections more than 1 million words must be recorded. This frequently leads to extreme demands on computer memory, great expense, and image transfer times too long for high-speed analysis. To combat these problems various methods of data compression have been investigated under the name of "pre-processing".

Three basic pre-processing techniques are utilized: 1) image enhancement or correction for the high spatial frequency fall-off of the optical imaging system, 2) minimization of the number of bits digitized per picture element so that the values of several picture elements may be packed per computer word, and 3) the use of an algorithm for combining the values of connected groups of picture elements in the original image in such a fashion that fundamental image information is not lost while, at the same time, the total number of words of storage required is reduced.

Image enhancement in pre-processing was originally introduced by PIKE [5.228] and SELZER [5.229] in biomedicine in the field of X-ray picture analysis. In most instruments for automatic cytology this correction for the modulation transfer function of the optics is done by tailoring the frequency response of the video amplifier. In television this is often called "aperture correction". An example is given in Fig. 5.13, which shows modification of the theoretical optical modulation transfer function (MTF) of the CELLSCAN television microscope by applying higher gain at high frequencies that at low frequencies in the video amplifier. Note also that Table 5.1 on scanner characteristics lists the MTF cut-offs of several scanners for purposes of cross reference. In most cases the MTF is carefully tailored in the scanner-to-computer link by analog video circuitry prior to digitization.

Surprisingly, the literature on digital cytologic picture analysis contains no references or studies relating to the evaluation of the quantizing noise introduced by digitization. Yet this has been thoroughly discussed in the general context of signal processing by such authors as BENNET [5.230], WIDROW [5.231], and BRUCE [5.232]. It appears that there is a general assumption that, since the human eye requires no more than six bits per picture element, digitization at this level is satisfactory for automatic cell image recognition. Many systems go somewhat beyond this and use eight bits per picture element because of the standard 8-bit byte length used in most general purpose digital computers. Also unused in the storage of cytologic images are the elegant methods of bandwidth compression as discussed by HUANG et al. [5.233]. This is probably because of the expense, complexity, and relatively low speed of mechanisms capable of bandwidth compression and the rapidly decreasing cost of computer memory.

Another remarkable omission in the automatic cytology literature is any discussion of work, such as that by GAVEN et al. [5.234], on the effects of the sampling interval (distance between picture elements digitized) on recognition success rate. Only one paper by WIED et al. [5.235] treats this problem. An unpublished paper by PRESTON (presented at the 1972 Engineering Foundation Conference on Automatic Cytology) treats this subject as regards the differentiation of two types of leucocytes. Preston's results are referred to in Fig. 5.14 but are too limited to be definitive. All that can be said at this point is that the high-speed systems that are imminent in the future will require a thorough trade-off analysis of these parameters (depth of digitization and sample spacing) in order to create economically effective units. The fact that this has not yet been done leaves a whole area of research wide open to study.

Finally, another almost totally unexplored area is that of local combination of picture point values in the process of digitization. NORGREN et al. [5.236] was able to obtain an order of magnitude reduction (from 40000 to 4096) in the number of computer words stored per image by using analog "white stretching" in the horizontal video signal and a vertical line combiner which made use of digital preprocessing. This is illustrated in Fig. 5.15. Although this was introduced in the CELLSCAN system in 1965, it was discarded in CELLSCAN/GLOPR in 1968 when inexpensive 16-bit minicomputers became available for use as image stores.

Table 5.1

SYSTEM	Cytoanalyzer I	Cytoanalyzer II	CYDAC	CELLSCAN	Cellscan/GLOPR
RESEARCHER	Tolles, W. E. (1956)[a]	Bostrum, R. C. (1963)[b]	Mendelsohn, M. L. (1968)[c]	Preston, K., Jr. (1962)[d]	Preston, K., Jr. (1972)[e]
AFFILIATION	Airborne Instr. Laboratory	Airborne Instr. Laboratory	University of Pa. Lawrence Livermore Lab.	Perkin-Elmer Corp.	Perkin-Elmer Corp.
LIGHT SOURCE	Tungsten	Tungsten	Cathode Ray Tube	Tungsten	Xenon Arc
Wavelength			503 nm	Green	540 nm and 570 nm
Waveband					± 50 nm
Flux/Pixel	15 μ Lumen				10 Nanowatts
OPTICS	45x Objective		100x Oil Objective	100x Oil Objective	100x Oil Objective
Focal Length	4.0 mm				
Numerical Aperture			1.25	1.3	1.3
Correction	Achromat		Planachromat	Achromat	Achromat
Spread Function	5 μm Circular Aperture	2 μm Circular Aperture			
Modulation Transfer Function			50% at 1150 cc/mm	50% at 1800 cc/mm	50% at 1800 cc/mm
SEARCH SCANNER	1800 rpm, 30-hole Nipkow Disc with 6467 PMT	1800 rpm, 48-Hole Nipkow Disc with PMT	Each Cell Manually Located	Each Cell Manually Located	20 Hz Mirror Scanner
Search Rate	50 mm^2/min.	25 mm^2/min.			0.06 mm^2/min.
Focus Method	Reference Surface	Reference Surface			Maximize High Video Freq.
Focus Response	N/A	N/A			1-2 Seconds
Stage Increment (X, Y, Z)	3 μm/line	1 μm/line			Continuous at 1.2 mm/min.
Increment Time (X, Y, Z)					N/A
Stage Travel (X, Y, Z)	10 x 20 mm				10 x 10 mm
Acquisition Time	N/A				30-60 Sec.
Bandwidth	0.25 MHz				18 kHz
IMAGE SCANNER	(Same as Search)	(Same as Search)	CRT with PMT Detector	DAGE Slow-Scan Vidicon	Same as search Scanner
Pixel Spacing	Continuous	Continuous	0.25 μm	0.1 μm	0.2 μm
Pixels/Second	900 Lines/Sec.	1240 Lines/Sec.	6,000	18,000	10,000
Bits/Pixel	Analog	Analog	8 Bits	1 Bit	8 Bits
Pixel Array Format	250 μm/Line		192 x 192	300 x 300	128 x 128
Frame Time			6.4 Sec.	5 Sec.	6 Sec.
Bandwidth			2.3 kHz	10 kHz	18 kHz
PROCESSOR	Special-Purpose	Special-Purpose	Off-line, Gen'l. Purpose	Special Purpose	Special Purpose (GLOPR) Gen'l. Purpose (Varian 620I)
Image Memory Capacity				90,000 Pixels	4096 Words
Measurements Made	4	5	72 Colorimetric 8 Morphologic	15	35
Measurement Time				10 Min.	40 Sec.
Recognition Time					
Processing Rate	2,000 Cells/Min.	1,000 Cells/Min.		4 Cells/Hr.	20-40 Cells/Hr.
Output	Abnormal and Total Cell Count	Abnormal and Total Cell Count			Data Sheet (1 Page)
SOFTWARE SYSTEM			CELLS		GLOL
DATA BANK			200 Cell Images		Measurements on 8,000 Cells 500 Cell Images

Table 5.1 (continued)

SYSTEM	(BIOSCAN)	(NICHHD)	SCANIT	ALMS	LARC	ACS 1000
RESEARCHER	Wald, N. (1968) [f]	Castleman, K. R. [g]	Neurath, P.W./Dew, B. (1974) [h]	Castleman, K. R. [i]	Megla, G. K. (1973) [j]	Powell, R. [k]
AFFILIATION	Univ. of Pittsburgh	Jet Propulsion Lab.	Tufts Med. Cntr./MIT	Jet Propulsion Lab.	Corning Electronics Co.	Honeywell Corp.
LIGHT SOURCE	Litton CRT		Xenon Arc	Tungsten		
Wavelength	P 36	Green	8 Wavebands Used	Variable by Filter	Orange, Blue/Green	
Waveband						
Flux/Pixel						
OPTICS	97 x Oil Objective	100x Oil Objective	100x Oil Objective	Selectable by Turret		
Focal Length	1.23	1.25	1.40			
Numerical Aperture Correction						
Spread Function		0.35 μm		10, 20, and 50 μm Apertures (in Plane of PMT)	5.0 μm Aperture (Search)	4.0 μm Aperture (Search) 0.5 μm Aperture (Focus)
Modulation Transfer Function		28% at 1000 cc/mm	30% at 500 cc/mm			
SEARCH SCANNER	Hughes 3076 mw Laser	Vidicon, 63x Obj.	8 μm Aperture Silicon Diode	Not Used in Search Mode		Mirror Scanner with 2 (Search/Focus) Photodiodes
Search Rate	300 mm² /Min.	6 mm² /min.	9 mm² /min.			38 mm² /min.
Focus Method	Focus Insensitive	Maximize High Video Freq.	Maximize High Video Freq.			
Focus Response	N/A	1 Sec.	2 Sec.		0.2 Sec.	0.1 Sec.
Stage Increment (X, Y, Z)				5 μm, 5 μm, 0.05 μm		
Increment Time (X, Y, Z)				5 ms., 5 ms., 1 ms. (max.)		
Stage Travel (X, Y, Z)						
Acquisition Time	0.4 Sec.	10-20 Sec.	0.33 Sec.		0.2 Sec.	0.15 Sec.
Bandwidth						
IMAGE SCANNER	Litton CRT and 4249 B Centronic PMT	(Same as Search)	Plumbicon	Plumbicon/Mirror Scanner and RCA 8645 PMT	Plumbicon	Standard Color Display for Operator
Pixel Spacing	0.1 μm	0.2 μm	0.2 μm	0.1 μm to 5 μm		
Pixels/Second	200	15,750	1,000,000	3300		N/A
Bits/Pixel	8 Bits	8 Bits	6 Bits	8 Bits		N/A
Pixel Array Format	512 x 512	480 x 525	256 x 256	Up to 1024 x 1024		N/A
Frame Time	4 Min.	17 Sec.	0.07 Sec.	Up to 5 Min.	0.13 Sec.	N/A
Bandwidth				2 kHz		4 MHz
PROCESSOR	Dig. Equip. Corp. KA10 + PDP 7	PDP 11/40	Nova 840	IBM 1130	PDP Minicomputer	Intel Microprocessor
Image Memory Capacity	2 Images (524 K Bytes)	$15 \cdot 10^6$ Bytes	$112 \cdot 10^6$ Bytes	512 K Disc		N/A
Measurements Made	12	6 (Homogeneous)/22 (Banded)	100	N/A	8	N/A
Measurement Time	3 Min.	58 Sec.	"Several Minutes"	N/A	0.3 Sec.	N/A
Recognition Time	5 Min.	21 Sec. (For 46 Chromosomes)		N/A	0.05 Sec.	N/A
Processing Rate	2 Karyograms/Hr.	8 Karyograms/Hr.	25 Cells/Hr. in "Data Collection" Mode	N/A	2 Cells/Sec. (max.) 2000 Cells/Hr. (Avg.)	N/A
Output	Karyogram (Calcomp)	Patient Report (2 Pages)		N/A	Printed Lab. Ticket	Printed Lab. Ticket
SOFTWARE SYSTEM		Mini-VICAR		VICAR (Available off-Line)		
DATA BANK	45 Cell Images	600 Cell Images	"Several Thousand"			

[a] Ref. [5.35], [b] Ref. [5.217], [c] Ref. [5.53], [d] K. PRESTON, JR.: Ann. N.Y. Acad. Sci. **97**, 482 (1962), [e] Ref. [5.312],
[f] Ref. [5.56], [g] Ref. [5.104], [h] Ref. [5.97], [i] Ref. [5.104], [j] Ref. [5.87], [k] Unpublished work.

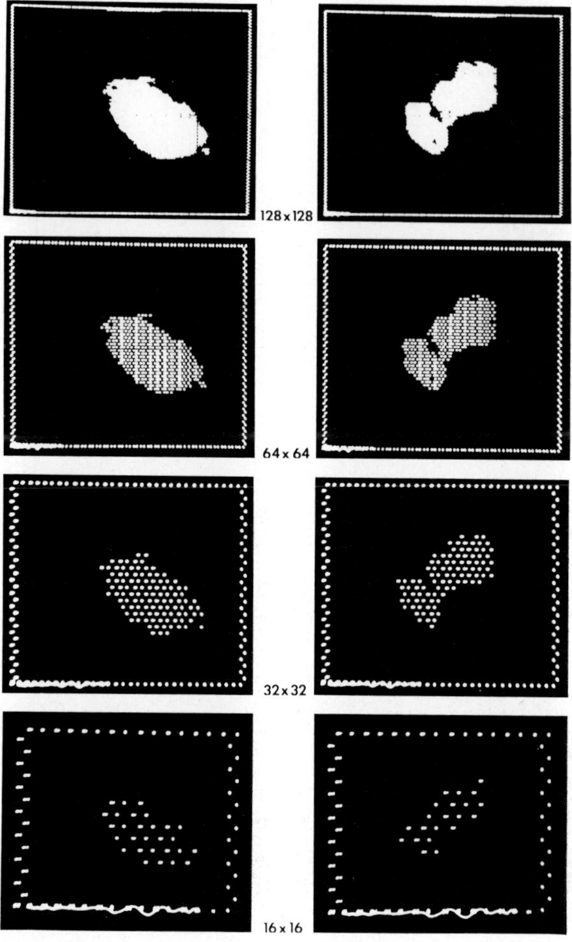

Fig. 5.14. Illustrations of the effect of varying the number of sample points per unit area as regards binary images of the nuclei of a lymphocyte (left-hand column) and a neutrophil (right-hand column). Using a population of 400 cells it was found that even with 16 by 16 it was possible to achieve 80% classification success rates using a sufficiently sophisticated analysis program

5.5 Automatic Recognition of Cytological Images

Previous sections have considered methods for preparing cytological specimens using carefully designed mechanical methods to separate individual cells and cytochemical treatments to tag them colorimetrically in such a fashion that their identity is most recognizable to the human eye and, it is hoped, to the machine. Previous sections have also

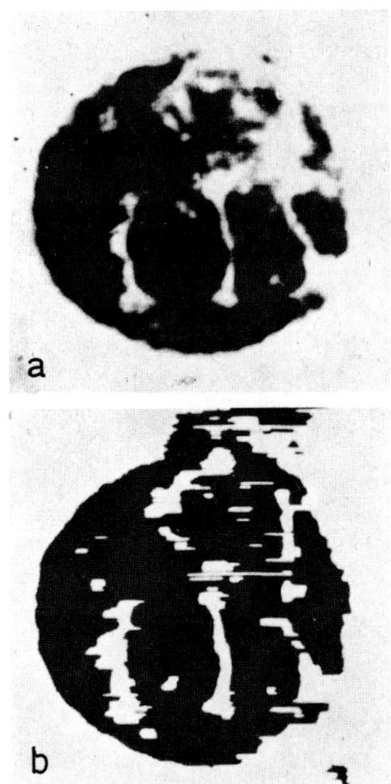

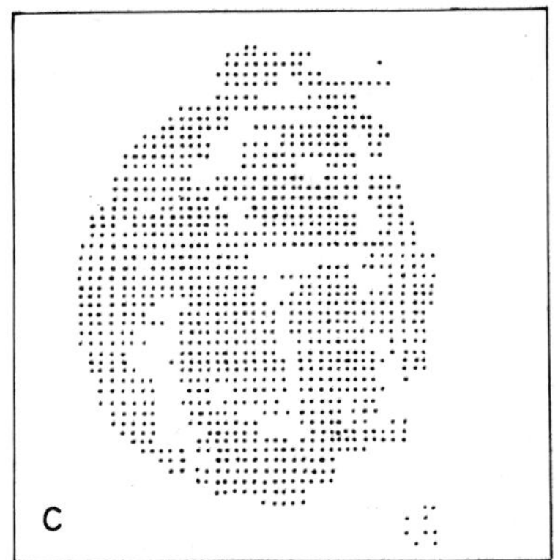

Fig. 5.15a—c. Illustration of the method of NORGREN [5.236] for using electronic preprocessing to "stretch" transmitting (white) regions in the analogue cell image so as to permit sampling of the 189 by 189 raster with only 63 by 63 sample points

mentioned the existence of methods for automatically transporting the microscope slide, maintaining the cells on it in focus, and locating cells at high speed. Electromagnetic radiation in the visible portion of the optical spectrum is then used to transfer information on the object, through the optics, into image space and, via the television microscope, preprocessor, and digitizer, into digital computer memory. This section describes techniques for processing the digital data array (or arrays) which represent the picture of the cell in computer memory for the purpose of extracting sufficient information to permit identification of the cell.

In general the process of cell image recognition is similar to object recognition in all pattern recognition problems. The process by which recognition takes place is diagrammed in Fig. 5.16. It consists of two major steps. The first is the generation of a measurement vector from the data array or arrays in memory. The second is the classification of this measurement vector. These two steps are treated separately in this section, with the emphasis on measurement techniques rather than on classification methods. It should be noted that measurement vectors

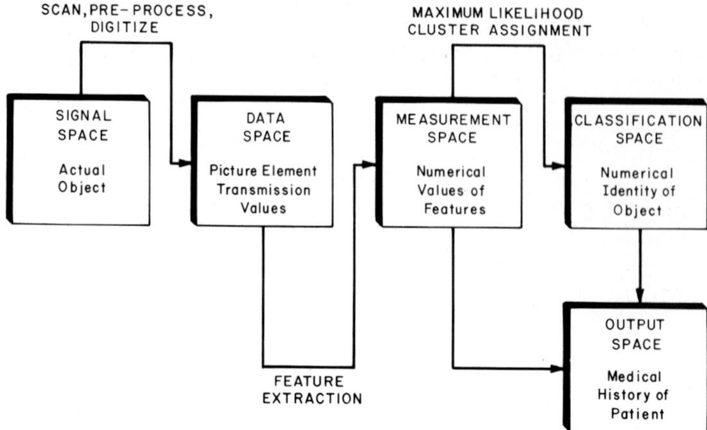

Fig. 5.16. Block diagram of the automatic recognition process whereby digital picture analysis is used to generate measurements from the data produced by scanning, preprocessing, and digitizing the cell image followed by measurement classification and storage of the final results as part of the patient's medical history

for a particular class of cell may be considerably different from each other; however, it is hoped that measurement vector clusters exist, so that the intraclass differences are smaller than interclass differences. If this is true, it is possible for the classifier to function by determining whether the measurement vector for an unknown cell being scrutinized is more likely to belong to a particular class (of the many possible classes) than to any other class.

Methods for classification, given a measurement vector, have been treated analytically in the literature and are well developed. One need only review the early work of FISHER [5.237] and the excellent text by RAO [5.238] to obtain a thorough working knowledge of classical classification techniques, except for hierarchical sorting. Some of the deficiencies of these techniques are reviewed by ORNSTEIN [5.239]. More recent approaches are reviewed below.

Unfortunately for workers in pattern recognition, both in automatic cytology and in other fields, no analytic approach to the generation of the most useful and appropriate measurement vector is available. At first one might think that the image itself might be used as this vector. However, since a typical cell image consists of from 10^4 to 10^6 picture points having at least 2^6 meaningful values per picture point, i.e., 6 bits of information, there can exist from $2^{6 \cdot 10^4}$ to $2^{6 \cdot 10^6}$ possible picture vectors. To map such a data array directly into classification space would require operations of a dimensionality so high as to make classification either meaningless or impossible. Thus one must perform the transformation from image space to measurement space with as great a reduction in dimensionality as possible. Since, in general, there are less than one hundred cell types in most cytological image analyzing problems, one would expect that the mapping into measurement space could reduce the information contained in the data array from the initial 60 thousand to 6 million bits to a much smaller number of bits. Current systems are usually capable of a reduction to between 100 to 1000 bits in measurement space, with the final reduction taking place in the classification process.

In the late 1950's and early 1960's a few workers attempted to develop analytic methods for generating and evaluating measurements. The work of UHR and VOSSLER [5.240] and LEWIS [5.241] might be reviewed by the interested reader primarily out of historical interest. However, the analytic approach to the formulation of measurements has been unsuccessful, and the world of measurement making for automatic classification is fraught with "adhockery"[1]. And, why not? Adhockery works. Analysis, so far at least, has not. Measurements are chosen intuitively by the experienced researcher and then tested by well-known classification algorithms. Thus measurements may be validated but not optimized.

Because of the variety of measurement-making methodologies, this section will include a generalized discussion of those methodologies which are categorized into certain broad general areas, namely 1) optical density histogramming, 2) stereologic analysis, 3) boundary and contour tracing, 4) counting and sizing, 5) shape analysis, 6) texture analysis, and 7) spatial frequency analysis. These seven subsections are then followed by a summary, followed in turn by subsections on classification methods and methods of implementation. Some illustrations of results are presented as part of this section. A broader discussion of results is reserved for Section 5.6. Computer languages for use in both measurement making and classification are left to the references; see also Subsection 5.5.10.

5.5.1 Optical Density Histograms

It has been found that the histogram of optical densities, i.e., the histogram of the frequency of occurrence of the picture point values, may be useful in the location of both nuclear and cytoplasmic boundaries, and even in direct cell identification. Figure 5.17 shows two television

[1] The author thanks J. FELDMAN of Stanford University (now at the University of Rochester) for this delightful new word.

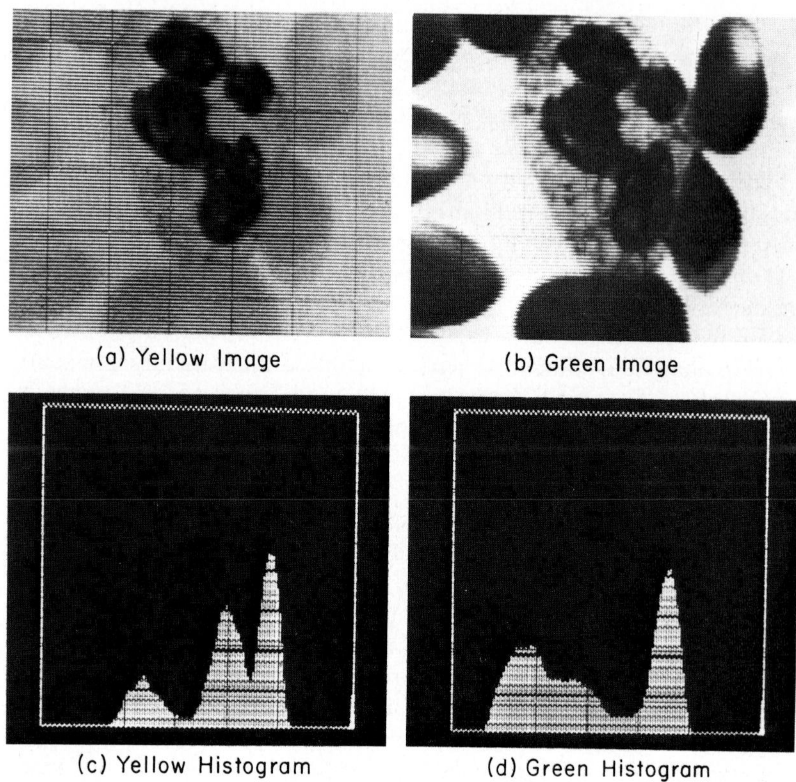

Fig. 5.17a—d. Television images generated by the CELLSCAN/GLOPR mirror scanner showing electronic images produced in the (a) yellow and (b) green and their associated optical transmission histograms. Note the tri-modality in (c) where the modes correspond to nucleus, cytoplasm and red cells, and background plasma, respectively

microscope images of the same human white blood cell illuminated in the yellow (left-hand view) and the green (right-hand view). The associated optical density histograms [shown on a scale which is logarithmic in frequency of occurrence, i.e., in the ordinate, and linear in optical transmission (the abscissa)] are given directly below each television image. The optical density histogram for the yellow shows three distinct peaks representing, from right to left, optical density values of the background, cytoplasm, and nucleus, respectively. The nuclear and cytoplasmic boundaries may readily be determined by thresholding the associated image at the nulls which occur between the background peak and the cytoplasm peak and between the cytoplasm peak and the nuclear peak, respectively, as discussed in Subsection 5.5.3.

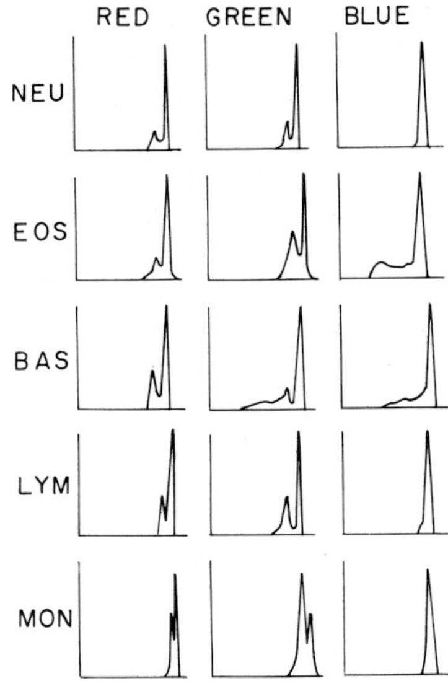

Fig. 5.18. Histograms generated by TYCKO [5.242] for the four commonly occurring types of human white blood cells stained using the tri-chrome stain of ORNSTEIN [5.200] and illuminated in the red, green, and blue

The multimodal optical density histogram may be operated upon by partitioning it into the sub-histograms associated with each mode and, for each sub-histogram, parameters such as the mean, mode, range, standard deviation, coefficient of deviation, skew, and kurtosis may be computed. These basic measurements and ratios of them (such as ratio of mean nuclear optical density to mean cytoplasmic optical density) have been used for blood cell classification without any further analysis of other image properties, as initially reported by PREWITT and MENDELSOHN [5.47]. TYCKO et al. [5.242] has expanded greatly on this theme by generating histograms from the trichrome stains of ORNSTEIN [5.200] as illustrated in Fig. 5.18.

BARTELS et al. [5.243–246] have used the diagram of optical density values, i.e., the histogram of probabilities of transitions from one picture element value to the value of its neighbor. The predecessor of this approach to classification came from the television and communications fields as exemplified by SCHAPHORST [5.247]. Here, the digram probabilities of television signals were computed for the purpose of coding television signals using delta modulation for bandwidth compression.

As can be seen from Fig. 5.18, there is a variation in the optical density histogram depending on the color used to illuminate the cell. Quantitative measurements of these multichromatic phenomena are reported by GELSEMA and POWELL [5.92], by CHENG [5.248], and by

TYCKO et al. [5.242]. Additional work relating to the statistics of the optical density histogram is given in KIRSCH [5.249]. There is still much open discussion as to whether the statistical properties of the picture element values (taken both directly and on a transitional basis) are of more or less use in automatic cell image identification than some of the morphological methods discussed below. Since optical density histograms are strongly dependent upon the cytochemical stains used and their methods of application, a definitive cross comparison of results (comparing colorimetric statistics and morphological statistics) is impossible. In general one expects intuitively that both types of measures are useful and must be investigated.

5.5.2 Stereology

What is now called "stereology" originated with the work of the geologist DELESSE in 1846, and of GLAGOLEFF [5.250]. Stereologic methods were first applied in biology by CHALKEY [5.251] and then extended to cytology by WEIBEL et al. [5.252]. Also see discussions by FREERE and WEIBEL [5.253] and WEIBEL et al. [5.70].

For two-dimensional pictures the basic stereologic equations are relatively simple, namely:

$$\begin{aligned} S_a &= 2N_a \\ A_a &= P_a \end{aligned} \tag{5.1}$$

where A_a is the area of the component a in the picture and S_a is its perimeter. The quantities N_a and P_a are best illustrated by referring to Fig. 5.19. This figure shows two stylized chromosomes in the field of view cut into horizontal slices. These one-dimensional slices occur since the picture scanner interrogates the field in a line-scanning mode. When the components of the field (the chromosomes) are of high contrast, the television signal consists of a series of pulses which occur whenever the scanner passes across a portion of a component. N_a is the number of pulses associated with the component a and P_a is the total number of picture points in the associated pulses. For the present the numbers associated with the pulses in Fig. 5.19 should be disregarded.

Let us call the pulses "intercepts" or "runs". An analysis of run lengths histogrammed collectively over the entire field of view for a cluster of chromosomes (having all possible orientations) will give some generalized information on the picture. For example, for the vertically or nearly vertically oriented chromosomes there will be a large number of runs whose length is characteristic of arm-width. For the horizontally

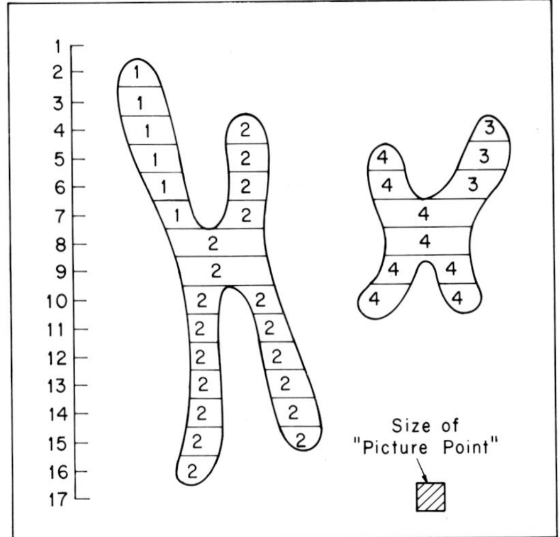

Fig. 5.19. Illustration of the "slicing" of a hypothetical binary image of two chromosome-like figures. The numerical labels relate to train-continuity object counting

oriented or nearly horizontally oriented chromosomes there will be a relatively smaller number of runs which relate to chromosome length. Therefore, from the overall run-length histogram one can qualitatively determine the smallest and largest characteristic distances across the components of the image. From the stereologic equations one can determine total area and an average perimeter-to-area ratio. This type of analysis is particularly useful in generating descriptions of simple, repetitive shapes such as non-overlapping bacterial colonies and also blood films consisting of non-overlapping red cells. More recently, run-length analysis has been applied to the problem of the quantification of "texture", as discussed in Subsection 5.5.6.

5.5.3 Counting and Sizing

Stereology is an elementary method of obtaining quantitative measurements on the morphology of components (objects) in the picture being analyzed. If the objects are all of similar shape, the area-to-perimeter ratio yields a size estimate and this, with total area, yields an estimate of the total numbers of objects in the picture. In digital pictures containing many cells of different shapes and sizes, or in a picture of one cell whose constituent parts must be counted and sized, stereologic methods do not suffice. However, if one could associate all intercepts belonging to a given component, it would be possible to determine: 1) the area of each component; 2) boundary points (i.e., the terminal points of each associated intercept); and 3) the total area of each component (i.e., the sum over all associated intercepts of the associated picture points). Early methods for associating intercepts were too simplistic to handle objects of arbitrary shape (see *Picture Processing—Stereology*). But, because these methods led to circuits which could perform at video rates, they were useful in high-speed counting systems for regular objects, as has been discussed by BERKLEY [5.4]. Here the method used in intercept association counting was to generate a particle count whenever an intercept was absent in the current scan line in a position where an intercept had appeared in the previous scan line. Thus, for the vertically oriented chromosomes of Fig. 5.19, a particle count of 4 would have been produced, corresponding to the 4 downward directed arm tips of both chromosomes.

A general method of intercept continuity counting which prevented such errors was devised in 1961 and patented by SCOTT and PRESTON [5.254]. Intercept continuity is implemented by assigning intercept numbers whenever the television scanner first intercepts the object. When portions of a single object coalesce, the initially assigned intercept numbers are merged. The left-hand chromosome in Fig. 5.19 is initially assigned the intercept numbers 1 and 2. When these are merged, the number 1 is erased from the intercept number memory. Similarly, the second chromosome is first assigned intercept numbers 3 and 4. Upon coalescence, the number 3 is erased from memory. A unique particle count independent of particle geometry is produced, when the entire picture has been scanned, by simply counting the number of assigned intercept numbers which remain in the assigned intercept number memory.

Intercept continuity object counting may be implemented at video data rates using relatively simple circuitry and can be used to produce object counts for such complex situations as multi-lobed objects, objects with voids, and objects contained within other objects. By utilizing additional memory which is addressed by using the intercept number itself, it is possible to totalize the area and to totalize the boundary points which are associated with each and every object in the picture.

5.5.4 Boundaries and Contours—Picture Segmentation

When the picture of a cell occurs at high contrast or unique modes exist in the optical density histogram, it is possible to select suitable thresholds to carry out picture segmentation, i.e., to find the boundaries of nucleus and cytoplasm and of their constituent parts (voids, vacuoles, etc.). Accurate boundary location is, of course, important so as to minimize errors in boundary determination which degrade the accuracy of area and perimeter measurements and confuse algorithms for shape analysis.

Since common use was made of the flying spot CRT scanner in early work in television microscopy, this provided workers with a scanner which could be programmed to depart from the standard raster-scan mode and be made to follow boundaries. Boundary-following scanners could be used to carry out extremely high speed video image processing. In the 1950's much effort was devoted to boundary-following techniques especially as applied to the problem of character recognition. The output of such scanners was the X and Y deflection waveforms generated as the boundary was followed. An analysis of such waveforms for shape recognition is given in Subsection 5.5.5. This is also discussed extensively by SPRICK [5.255, 256] and BROUILLETTE [5.257, 258].

Later FREEMAN [5.259, 260] developed a method of storing boundary information in a digital form. His technique was to store a list of boundary locations with the vector direction of the boundary at each location in a list structure which was specific to each object in the field of view. In a sense his was the digital equivalent to the earlier analog approaches.

All of this assumes that the boundary followed is at a well-defined level selected, for example, from the optical density histogram (see Fig. 5.17). In general there may not be a well-defined level. In this case the boundary between two regions is frequently defined by a locally high derivative, and boundary localization is often performed by taking the two-dimensional gradient or Laplacian. This is illustrated for a chromosome image in Fig. 5.20. Note that the Laplacian is bimodal and produces a light "outer boundary" and a dark "inner boundary". Also, see MENDELSOHN et al. [5.261] for a discussion of the problem of boundary detection in chromosome analysis. PREWITT [5.262] discusses "gradient tracking" as applied to finding nuclear and cytoplasmic boundaries in leucocytes. In other fields of biomedical image processing, gradient techniques have been applied and the reader may wish to refer to SELZER [5.263] who employs this method for the location of arterial walls in radiographs of the leg. HOU et al. [5.264] provides a general discussion in which a multiplicity of contours is used in describing biological forms. An illustration of contouring is shown in Fig. 5.21, taken from the field of autoradiography. The contour map of a cell has been generated (left-hand view). It contains regions of closely spaced contours where the dense autoradiographic particles (or "grains") occur. The regions may be located (right-hand view) in order to count the autoradiographic grains. The more general problems of finding

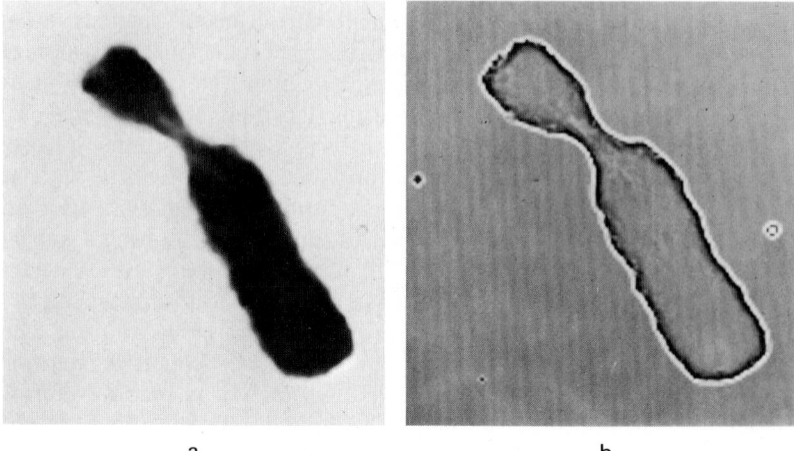

Fig. 5.20a and b. Electroradiographical image of a single chromosome shown in (a) with the results, shown in (b), of computing and displaying the Laplacian for the purpose of boundary elucidation

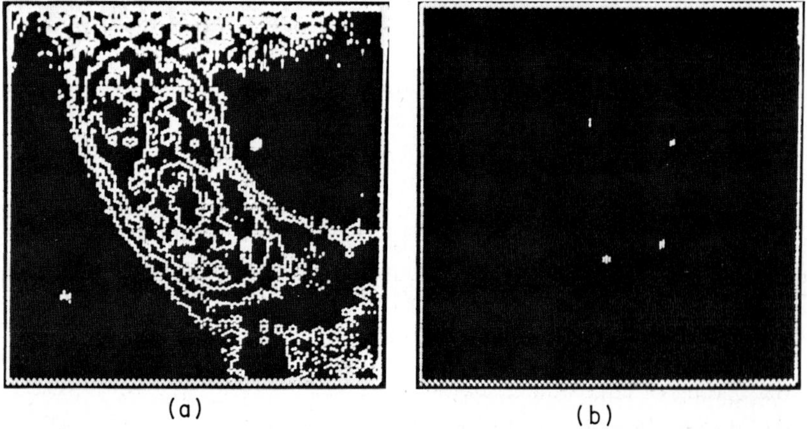

Fig. 5.21a and b. Illustration of the use of contouring to find the location of silver grains in autoradiography. Contours (isophotes) of the autoradiographed cell are shown in (a) with four grains located by marking points of high density shown in (b). In actuality the grains are located in a layer of photographic emulsion on top of the cell

boundaries in fields of view containing a multiplicity of cells have also been worked on; see YOUNG and PASKOWITZ [5.265].

CHIEN and FU [5.266] use a more complex method of boundary location in chromosome analysis employing an embedding metric and using a decision theoretic approach. They use the metric of FISCHLER and ELSCHLAGER [5.267] whose value at the point X is given by the equation

$$G(x)=\sum_{i=1}^{P}\sum_{j=1}^{i} g_{ij}(x_i, x_j) \tag{5.2}$$

where the kernel g_{ij} is the "global evaluation function", which is the cost of embedding the i-th component at the location x_i and the j-th component at location x_j in the image array of P elements. This approach has also been applied in radiological analysis by FU and CHIEN [5.268]. Another elegant and complex method for contouring has been developed by SKLANSKY [5.269], which to date has been applied only in the field of radiography. SKLANSKY extends the figure-of-merit approach originated by MONTANARI [5.270]. In this method the following objective function is maximized:

$$\underset{Z_1 \dots Z_N}{\mathrm{Max}} \sum_{i=1}^{N} G(Z_i)+\alpha \sum_{i=1}^{N-2} q(Z_i, Z_{i+1}, Z_{i+2})+\beta C(Z_{n-1}, Z_n, Z_1, Z_2) \tag{5.3}$$

where $G(Z_i)$ is the two-dimensional gradient modulus at the point i, $q(Z_i, Z_{i+1}, Z_{i+2})$ is defined as the "curvature" at the point Z_i, and $C(Z_{n-1}, Z_n, Z_1, Z_2)$ is the curve length variable. These quantities take on the values:

$$q=\begin{cases}1 & \text{if } Z_k,\ Z_{k+1},\ Z_{k+2} \text{ subtend an angle of } 135^\circ\\ 0 & \text{otherwise}\end{cases}$$

$$C=\begin{cases}1 & \text{if } Z_n=Z_1 \text{ or } Z_{n-1},\ Z_n,\ Z_2 \text{ subtend an angle} \geqq 135^\circ\\ 0 & \text{otherwise}\end{cases}$$

and $k(i, j)$ denotes a point in the I by J picture grid. SKLANSKY has successfully applied this method of boundary extraction to chest X-rays for the purpose of tumor detection. Other more complex approaches to the coding of edge and/or boundary information have been developed by workers in artificial intelligence. See for example, ALEXANDER and THALER [5.271], HUECKEL [5.272], and SARAGA and WAVISH [5.273]. Finally, ROSENFELD and THURSTON [5.274] have developed a computer algorithm for the location of the boundaries between regions of different texture.

5.5.5 Shape Analysis

Shape analysis uses the boundary of an object to extract morphological measures using a multiplicity of algorithms. These algorithms include the measurement of moments, the use of circumscribing rectangles and hexagons, projections and profiling, "shrinking" and skeletonizing, etc. These are discussed separately in the paragraphs below. Quality criteria for the selection of algorithms for applications of shape analysis to cytology are discussed in the experimental results reported in Subsection 5.5.8.

Trajectory Transforms

The boundary-following scanners of the 1950's produced as direct outputs the x and y trajectories of the boundary-following device. For example, when traversing the boundary of a circle of radius R at a uniform velocity v, the x and y coordinate trajectories are:

$$x = R\cos(vt/R), \qquad y = R\sin(vt/R). \tag{5.4}$$

The Fourier transform of either of the above trajectories may be used as a shape measurement. (In this case the Fourier transform yields a measure which is inversely related to the size of the object.) The Hankel transform may be taken by transforming the x and y coordinates into r and θ coordinates. High frequencies in either of these transforms indicate either small objects or rapid perturbations (irregularities) in the boundary. In general, the low frequency content relates to gross shape and the high frequencies to boundary irregularities.

Moments and Principal Axes

Using the boundary information alone or the picture data array values within the boundary, one may obtain further data on shape by means of the analysis of moments and principal axes of the object. This approach was pioneered by HU [5.275] and SAPP [5.276] in the field of character recognition and later applied to chromosome analysis by BUTLER et al. [5.277].

Optical Density Profiles

After determining moments and the principal axes, further measurements, which have been found useful particularly in the analysis of chromosomes, are obtained by projecting the picture data array (within the boundary) on these axes to form what are called "optical density profiles". (These not be confused with optical density histograms). In chromosome analysis these profiles are used for the location of the centromere. Differential (i.e., transitional) optical density profiles have also been applied successfully to shape analysis in radiology by HALL et al. [5.278], although this work has not yet been extended into digital

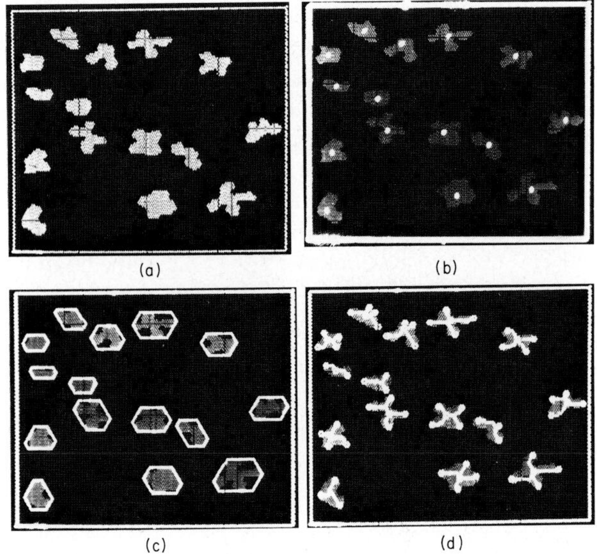

Fig. 5.22a—d. Illustration from PRESTON [5.106] showing: (a) a binary image of a chromosome spread, (b) centromeres located by an area reduction (shrinking) algorithm, (c) circumscribing hexagons erected for the purpose of locating arm tips, and (d) skeletons incorporating the centromeres and arm tips so located

picture processing for cytology. Various statistical measurements may be made on these profiles (both direct and differential), such as the location of the mean, mode, standard deviation, kurtosis, skew, etc.

Circumscribing Figures

In chromosome analysis we find the most prevalent use of a circumscribing polygon for shape analysis. This technique is most frequently used to locate arm tips (or ends) of individual chromosomes. Figure 5.22 shows how circumscribing hexagons may be used in locating arm tips as discussed by PRESTON [5.106]. This approach was originally pioneered by NEURATH and ENSLEIN [5.279]. Other discussions of the problem of centromere and arm tip location are provided by GALLUS et al. [5.280], RUTOVITZ [5.281], and REGOLIOSI [5.282] using this and other methods.

Shape Features

So far we have discussed many analytical measures of shape. Now we come to the adhockery, much of which sprang from early work in optical character recognition (OCR). Doyle [5.283] was the first great innovator, and his elaboration on methods of run length analysis (stereology) led to crude measurements of "concavities", "inlets", and "voids" which were not improved upon for a decade. It is worthwhile reviewing the early shape recognition techniques used in OCR such as the discussion by Greanias et al. [5.284] as well as later work by Kamentsky and Liu [5.285].

In cytology there has been work by Gallus [5.286], Gallus and Neurath [5.94], and Gallus and Regoliosi [5.287] on shape analysis relating to chromosomes. Shape recognition of the nucleus is discussed by Kiefer and Sandritter [5.288], Bourk and Tretiak [5.289], and Kiefer et al. [5.290]. Cowan and Wann [5.291] discuss both cell and nuclear sizing. Finally, Arcelli and Levialdi [5.292] extend the analysis to three dimensions for the purpose of neuron counting in slices of brain tissue.

Skeletonizing and Shrinking

The last method of shape analysis to be treated in this section is that of skeletonizing or shrinking. Shrinking or area reduction was first studied by Kirsch [5.293] at the National Bureau of Standards and Unger [5.294] at Bell Telephone Laboratories. A method of implementation was patented by Golay [5.295] and first implemented by Preston [5.296]. An example is shown in Fig. 5.23. It is found (strictly on an *ad hoc* basis) that reduced area techniques are useful in significantly improving cell recognition. This is discussed quantitatively by Preston [5.297].

A generalization of this method which created a true skeleton rather than the reduced area skeleton was devised by Blum [5.298]. Montanari [5.299] and Hilditch [5.300] have studied skeletonizing as applied to the analysis of chromosomes, and Levialdi and Montanari [5.301] have investigated area reduction and skeletonizing as applied to the analysis of chromosomes, while Levi [5.302] has investigated area reduction and skeletonizing techniques. Deutsch [5.303] has made a theoretical study of the subject. Rosenfeld [5.304] has analyzed the problem of connectivity in performing skeletonizing and shrinking operations. Prewitt [5.262] gives examples of both exoskeletons (exterior) and endoskeletons (interior) as applied to blood cell analysis. An example of the use of skeletonizing as an intermediate step in the analysis of chromosome images is shown in Fig. 5.22. One problem in skeletonizing is the difficult one of handling overlaps. Few people have worked in this area except for Tretiak [5.305], as regards chromosomes, and Voss [5.110], as regards overlapping red cells.

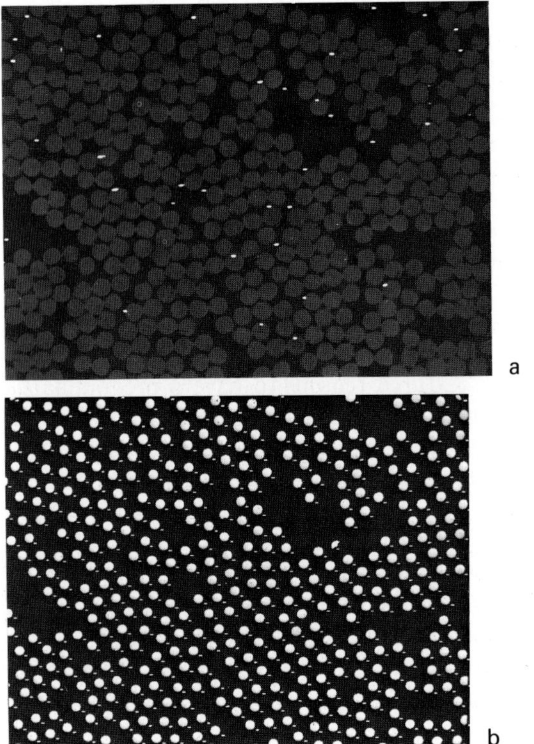

Fig. 5.23a and b. Illustration of the use of area reduction (shrinking) to create the correct count of a field of overlapping cells. The initial (incorrect) count is shown by the light markers in (a) and the correct count (after area reduction) is shown in (b). (Courtesy Image Analyzing Instruments Co., Ltd.)

5.5.6 Texture Analysis

In observing both the nucleus and cytoplasm of cells it is recognized that, in addition to variations in color and shape (configuration of the boundary), the identity of the cell is frequently determined by the internal structure of either the nucleus and/or the cytoplasm. Variations in chromatin structure within the nucleus and the presence of voids and vacuoles in both the nucleus and cytoplasm give significant clues as to cell type and age. Various workers, starting in the mid-1960's, have attempted to quantitatively describe internal structure. This area of research has been called "texture analysis".

Bacus [5.82] used a method of run-length analysis (see Subsection 5.5.2) to analyze changes in optical density within the nucleus. This is similar in all respects to the stereological techniques for analyzing clusters of objects (except that now the objects are smaller and are within the nuclear or cytoplasmic boundary).

It is also possible to apply the methods of Subsection 5.5.3 in counting and sizing to an analysis of internal structure, especially to voids and/or vacuoles. It is also of use to employ the probability density histograms (Subsection 5.5.1). In all cases these measures are restricted to a specific portion of the cell.

In addition texture may be described in terms of the spatial frequency content of the cell component being analyzed (see Subsection 5.5.7 below). This method of measurement consists of computing the two-dimensional autocorrelation function of the image or, equivalently, the Wiener spectrum (the spatial frequency power spectrum).

One would expect "fine" textured regions to have a predominance of high spatial frequencies, short run-lengths, and small internal object sizes. Conversely, "coarse" textured regions have the opposite of the above features. Specific studies relating to texture measurements of the interior structure of cells and cell constitutents are provided in KIRSCH [5.249], JAYARAMAMURTHY [5.306], and LIPKIN and LIPKIN [5.307]. VASTOLA et al. [5.308] discusses texture measurements as related to leukocytes, and CARVALKO [5.309] specifically treats differences between the nuclear textures of lymphocytes and monocytes.

In other areas of digital picture analysis, especially concerning studies in the automatic analysis of X-rays, texture measurements have also been developed. The reader is referred to the work of JAGOE and PATON [5.310] and to that of HALL et al. [5.311].

5.5.7 Spatial Frequency Analysis

As was mentioned in Subsection 5.5.6, the analysis of the spatial frequency content of cell constitutents (nucleus and cytoplasm) is part of texture analysis. Spatial frequency analysis is also useful in the overall analysis of cells both on microscope slides and in flow systems (Subsection 5.2.4). This is the reason for treating this subject in a separate section.

There is a particularly elegant relationship between the Wiener spectrum (spatial frequency spectrum) and the angular distribution of light scatter from a cell (or in general any particle) which is illuminated by coherent light. As is discussed in texts by both GOODMAN [5.156] and PRESTON [5.157], the light scatter distribution in the back focal plane of a coherently illuminated lens is given by the Fourier transform of the contents of the front focal plane, as follows:

$$F(w_x, w_y) = \int_{-A/2}^{+A/2} \int_{-A/2}^{+A/2} f(x, y) \exp[-j(w_x x + w_y y)]\, dx dy \qquad (5.5)$$

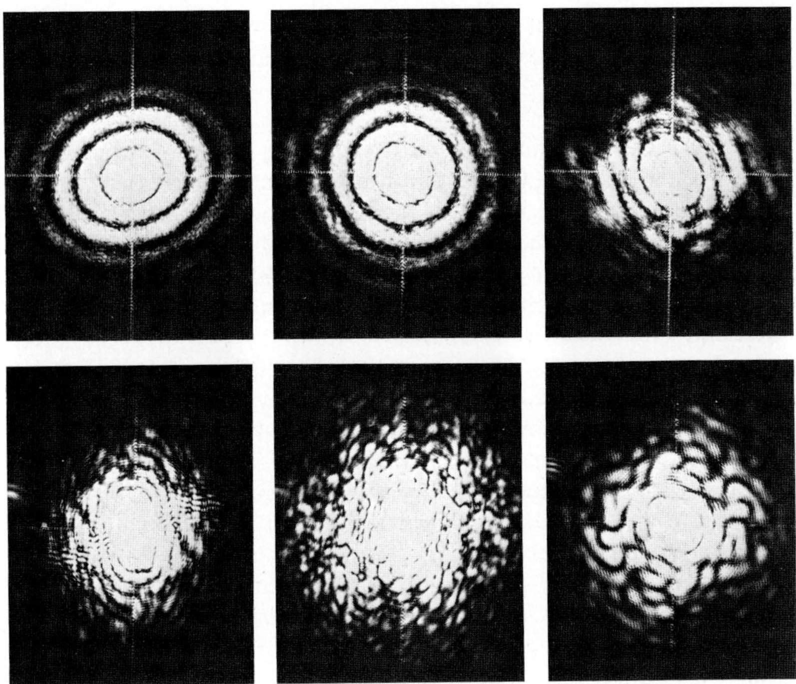

Fig. 5.24. Illustration of the use of Wiener spectra generated by optical scatter from normal epithelial cells, shown in the upper row, and malignant cells, shown in the lower row. (Courtesy Grumman Aircraft Co.)

where $f(x, y)$ is the light amplitude transmission of the field of view (cell or cells) in the front focal plane over a square region of dimensions A by A. Many workers in flow systems have taken advantage of this fact for obtaining the spatial frequency content of an object by adding laser-powered optical sensing systems to the flow environment. Recently this approach has also been applied to cells on the microscope slide.

Instruments for measuring angular scatter are treated by AUGHEY and BAUM [5.313]. Later several workers, such as BATEMAN [5.314], KOCH and EHRENFELD [5.315], and BERKMAN and WYATT [5.316], applied the techniques to studies of the size and shape of bacteria. At the same time other workers applied this technique to the identification of cells. BOWIE [5.317] worked on leukocyte identification using this method as did LURIE et al. [5.318] and RAUCH [5.319]. WARD et al. [5.320] have applied the same approach to the location and analysis of reticulocytes, and KOPP et al. [5.321] have performed an extensive study of its use in the detection of cancerous cells in cervical smears. As an example of this final application, Fig. 5.24 illustrates both normal and cancerous cells and their optically generated Wiener spectra. Although this is still preliminary and the results are uncertain, it holds the promise (especially in a flow environment) of very high speed and very simple cell analysis.

In flow system research the most significant work in using spatial frequency analysis for cell recognition by coherent optical methods has been by BRUNSTING [5.322, 323] and MULLANEY et al. [5.324–326]. FIEL [5.327] presents a general treatment of the approach and FIEL and SCHEINTAUB [5.328], FIEL et al. [5.329], and FIEL and MUNSON [5.330] present discussions of this technique applied to tissue cells from the spleen and liver as well as to viruses.

Analytical work on optical scatter and experimental comparisons is treated by BRYANT et al. [5.331], and papers by GRAVATT [5.155] and WYATT [5.332, 333] provide modern surveys. In general, it may be stated that light scatter and its analysis uses research which is not proven, but which, because of its speed, could provide dramatic results, especially in terms of the statistical analysis of large cytologic populations.

5.5.8 Summary

So far this section has concerned itself with measures which may be extracted from images of cells by digital picture analysis (plus some comments on purely optical methods of measurement extraction). It would be nice if there were a method of prediction whereby the utility of various measures for cell recognition could be determined. To date no such method has been devised nor is it likely to be devised. There is, however, some basis of comparison of measurement utility as regards the analysis of leukocytes. This is because significant statistical data have now been gathered in the multitude of both research and commercial programs mentioned above. It appears that, in the field of white blood cell analysis, only 5 to 10 carefully selected measurements are required to classify many cell types (see Subsect. 5.6.1). A thorough examination of the importance of various measures has not been published. Surprisingly, few workers have pooled their material so as to make cross comparisons possible.

Lists of measures with a priority ranking on utility are provided in Table 5.2. Comparing columns within this table it is noted that, depending on the research program, the method of measurement selection, and the implementation of measurement extraction, some projects prefer measures which are inconsistent with those preferred by others. Until it is possible to access the algorithms utilized and to evaluate the cell image transfer processes of the scanners involved, definitive decisions on optimum measurements for the analysis of leukocytes will not be established. This may take from five to ten years to accomplish depending upon the success of developing data bases and the degree to which the sharing of software routines and hardware specifications is carried out. For the analysis of cancer cells in cervical smears it will take even longer.

For the case of chromosome analysis a cross comparison study is already under way on a results basis only. Data on this exchange are given in Table 5.3 on p. 294.

5.5.9 Methods for Recognition and Classification

This chapter will not dwell in detail upon the many methods of classifying measurement data into clusters delineating identifiable cell types and sub-types. In general two approaches are taken. These are called "supervised" and "unsupervised".

Supervised Classification

Referring again to Fig. 5.16, it is seen that the classification step maps measurement (or "feature") space into classification space. When this mapping is "supervised", we mean that the user has used some a priori knowledge about measurement clusters upon which to base the classification. The classical techniques of HOTELLING [5.334] and FISHER [5.237] may then be used to contruct weighting functions which, when applied to the measurements obtained from an unknown cell, determine its most likely classification. To perform such a classification the class coefficients c_i are computed from the C linear equations given by:

$$c_i = \sum_{j=1}^{j=M} w_{ij} m_j \tag{5.6}$$

where the m_j are the values of the M measurements and (w_{ij}) is a matrix of weights. The highest-values component of the vector c_i determines the class to which the vector m_j belongs. Some references which summarize the classical approach are HIGHLEYMAN [5.335], KARUSH [5.336], and SEBESTYEN [5.337]. The general area of analysis to which these refer is called "decision theory".

Early in the 1960's several researchers in decision theory as applied to biological samples (including cells) discussed the fact that the classical method was inadequate when the measurements made were not readily interpretable as points in a Euclidean space. TANIMOTO and LOOMIS [5.338], TANIMOTO [5.339], and ORNSTEIN [5.239, 340] investigated this problem and proposed a metric which could be used to

Table 5.2

Researcher(s):	Prewitt, J. M. S. (1972)[a]	Preston, K., Jr. (1970)[b]	Neurath/Powell[c]	Bacus, J. W. (1970)[d]
Affiliation:	Nat'l. Insts. of Health	Perkin-Elmer Corp.	Tufts/CERN	Univ. of Illinois
Material Preparation:	Manual Smear; Gallo-Cyanin Chromalum and Naphthol Yellow S Stain	Spun Smear Wright-Giemsa Stain	Manual Smear Wright-Giemsa Stain	Manual Smear Wright-Giemsa Stain
Transducer:	TV Microscope (CYDAC)	TV Microscope (Cellscan)	TV Photograph Scanner	TV Microscope
Illumination:	503 nm (Monochrome)	540 nm (Green) 570 nm (Yellow)	Wratten 44 (Green) Wratten 22 (Yellow)	Wratten 44 (Green) Wratten 15 (Yellow)
Pixel Spacing:	0.25 Micrometer	0.20 Micrometer	0.10 Micrometer	0.33 Micrometer
Pixel Range:	8 Bits	8 Bits	6 Bits	8 Bits
Total Number of Measurements:	80	35	46	8
Types of Measurements:				
1. Optical Density Histogram:	Mean (N, C, T) Mode (N, C, T) Inter-Mean Distance (N-C, N-T) Inter-Mode Distance (N-C, N-T) Absorbance (N, C, T) Mean Ratios (N/C, N/T) Absorbance Ratios (N/C, N/T)	Mode (N, C-2 Colors) Mode Ratios (N/C-2 Colors)	Mean (N, C, T-2 Colors) Absorbance (N, C, T-2 Colors) Mean Ratios (N/C, N/T-2 Colors) Contrast• Color Angle•	Mean Color Angle (C)• Mean Ratio (N/C)
2. Stereologic:			Average Run Length Run Length Histogram Total Pixels in Runs Average Run Color Average Run Density (The above taken at High and Low Thresholds for N, C, T)	
3. Morphologic:	Area (N, C, T) Perimeter (N, T) Diameter (N, T) Asphericity (N, T) (Area/Perim.^{2}) Area Ratios (N/C, N/T)	Area (N, C, T) Perimeter (N, T) Area Dense Nuclear Chromatin Perimeter Dense Nuclear Chromatin	Area (N, C, T) Perimeter (N) Asphericity (N) (Perim.^{2} / Area) Eccentricity (N) (Centroid/C. G.)	Area (N) Perimeter (N) Asphericity (N) (Area/Perim.^{2})

	Diameter Ratios (N/C, N/T) Asphericity Ratios (N/C, N/T)	Number of Lobes (N) Number of Small, Med., Large Concavities (N) Area Nuclear Bridges Area Nuclear Inlets Reduced Area (N) Reduced Perim. (N) Area Nuclear Skeleton Ends in Skeleton Area Nuclear Voids Area Cytoplasmic Vacuoles	Number of Lobes (N) Average Chain Value (N) [5.259]	
4. Textural:	Density Moments (N, C, T) (Variance, Skew, Kurtosis) Entropy (N, C, T)	Fine structure of Dense Nuclear Chromatin, Nuclear Voids, Cyto- plasmic Vacuoles, Number of Clumps of Dense Nuclear Chromatin	Color Variance (N, C)	Variance Color Angle (C) Nuclear Area Having Given Edge Derivative Cytoplasm Area with Given Color Angle Derivative
Best Measurements:	Mean (T), Variance (T), Absorbance (T), Absorbance Ratio (N/T)	Area (T) Reduced Area (N) Reduced Perimeter (N) Fine Structure (N) Vacuoles (C) Mode (N, C-2 Colors)	Color Contrast Angle Mean (N, T) Area (T) Mean (T-Yellow) Asphericity (N) Color Variance (N) The Above are best for a 6-Class Analysis. When a 16-Class Analysis was carried out, the Best were: Mean (N-Yellow) Color Angle (C, T) Color Contrast Magnitude Mean (N, T) Mean (T-Green) Contrast (N/C-Green)	N/A (All Meas. Used)

LEGEND

N = Nucleus
C = Cytoplasm
T = Total Cell

* Color Angle and Contrast Diagram as Below:

Color Contrast Angle (N/C)
Contrast (N/C) (Magnitude of Vector)
Density at Color 2
C
N
Color Angle (N)
Density at Color 1

Footnotes:
a [5.262]
b [5.13]
c Unpublished work
d [5.82]

determine the similarity between an unknown set of measurements m_i and several standard measurement sets r_{ij} representing different types of cells as given below:

$$\sigma_j = \frac{\sum_i \min(m_i, r_{ij})}{\sum_i \max(m_i, r_{ij})} \tag{5.7}$$

where $\min(m_i, r_{ij})$ stands for the smaller of the numbers in the pair m_i, r_{ij} and $\max(m_i, r_{ij})$ for the larger.

Other workers realized that (5.7) is a set of linear expressions which could be solved by linear programming. This technique was originally developed for optimizing problems in process control and transportation as described by MANGASARIAN [5.341] and later extended by NELSON and LEVY [5.342] and MAR [5.343]. Linear programming was applied in cytology to the recognition of leucocytes by PRESTON [5.312].

Other methods for pattern recognition (primarily relating to cytology) are given in the treatments of KOLERS and EDEN [5.344], LEDLEY [5.345], LIPKIN et al. [5.346], and SHAPIRO et al. [5.347]. Also of interest is an early tutorial survey by MATTSON [5.348]. For the diligent reader who wishes to be exposed to a thorough modern review of the classification problem, the annotated compendium by BULLOCK [5.349] is recommended as one of the most comprehensive to date. The fact remains, however, that the most sophisticated of the more recent methods have not provided any significant breakthrough in measurement classification in comparison to the more classical approaches of some 40 years ago.

Non-supervised Classification

In the case of "non-supervised" measurement analysis, one assumes that there is no a priori knowledge as to the inherent structure of a large set of measurements taken from images of many classes of cells. The problem of non-supervised classification or "learning" is to determine in this measurement set whether there are significant modes or "clusters". This non-supervised approach to measurement analysis is of the most basic and fundamental nature. Using this approach, one can discover structures in data which are data dependent rather than dependent upon somewhat subjective and arbitrary a priori human classifications. This approach to data analysis is treated extensively in the pattern recognition literature. Only a few of these papers relate directly to cytology. BARTELS et al. [5.350] are the major authors on the problem of data structures in cytology and discuss non-supervised or "self-learning" methods. The use of various statistical and decision theoretic methods for detecting clusters (especially as related to detecting the presence of abnormal cells within a population of normals) is treated by BHATTACHARYA et al. [5.351, 352]. RUTOVITZ [5.353] has written on data structures in regards to chromosome analysis.

In some cases it has been found that the use of computers having graphic displays permits measurement clustering on an interactive basis whereby measurement points are displayed visually to the researcher who is performing data analysis. Such interactive cluster mapping schemes are described in papers by PATRICK and FISCHER [5.354] and by SAMMON [5.355, 356]. The efficacy of this method is frequently improved by means of simultaneous computer trials at clustering using such techniques as the ISODATA program originated by BALL and HALL [5.357]. This interactive approach has given rise to a series of computer systems for use in both general pattern recognition and for cell recognition which are described in Subsection 5.5.10.

5.5.10 Implementation

Two types of measurement analysis systems for use in what is called "interactive pattern recognition" have been implemented. The first type is a measurement analysis facility for use in interactive clustering of measurements taken from any general data source. The second type is a complete system for research in cytology automation which incorporates a television microscope scanner, a portfolio of measurement generating programs, and additional programs for measurement clustering under both supervised and non-supervised decision regimes.

The earliest systems to be implemented in the United States were of the first type. These handled measurements from a generalized data source and sought to analyze them. The best example of this implementation is OLPARS, described by SAMMON [5.356]. This system is in general use by the United States Air Force at the Rome Air Development Center (RADC). Other similar systems are CYCLOPS, described by MARRILL [5.358], in use at Bolt, Beranek, and Newman; and the system of HALL et al. [5.359] at RADC.

One of the earliest systems devoted to scanning, measurement making, and on-line analysis of biomedical images is described by HODES [5.360] at the National Institutes of Health. A more recent system is that of JOHNSTON [5.361] at the M.D. Anderson Hospital, Houston. Some systems use on-line measurement making and off-line data analysis, such as the CYDAC system of PREWITT and MENDELSOHN [5.362, 363], the CELLSCAN/GLOPR system of PRESTON [5.364], the TICAS system of WIED et al. [5.64], and the Medical Research Council system in Great Britain for chromosome analysis, which is discussed by HILDITCH [5.365].

Many of the above systems rely on the utility of general-purpose pattern recognition or measurement-making languages which have been developed for picture processing. The earliest of these languages was PAX, which was originated as part of the ILLIAC III program; it is described by SNIVELY and BUTT [5.366]. A more recent version of PAX has evolved at the University of Maryland and is presented by JOHNSTON [5.367]. STONE [5.368] wrote CHROMO for the analysis of chromosome images at Lawrence Livermore Laboratory; PRESTON [5.369] developed GLOL for general use in cytology; while LEMKIN et al. [5.100] describe languages in use at the National Institutes of Health. Another more general language for artificial intelligence is SAIL, which is described by FELDMAN [5.370] at Stanford University.

5.6 Results to Date—Summary of Progress

This chapter has traced digital picture processing in cytology from the era of the Cytoanalyzer program, which commenced in the mid-1950's. Now, some twenty years later, automatic picture analysis equipment is only just beginning to enter the commercial market-place. It is a monumental job just to read the literature which documents progress in this field (even if the reader's attention is restricted to the references cited in this chapter). Therefore it is important that the author present the key milestones which have been achieved in the three major areas of application: 1) blood smear analysis, 2) cervical smear analysis, and 3) chromosome analysis. This is done below, while the following section attempts to extrapolate the past 20 or 30 years of progress into the future.

5.6.1 Blood Smear Analysis

As has already been discussed, the earliest work which reported successful machine implementation of the routine visual recognition of the commonly occurring types of white blood cells in blood smears on microscope slides was that of INGRAM et al. [5.46] and PREWITT and MENDELSOHN [5.47]. In both cases television pictures of small samples (100–300 cells) were machine analyzed. The results, although hopeful, were statistically inconclusive. The next major advance in this area occurred in work reported by INGRAM and PRESTON [5.13] wherein complete (but slow) automation was achieved with the implementation of a computer-controlled television microscope which focused, searched, located, measured, and classified leukocytes at the rate of approximately

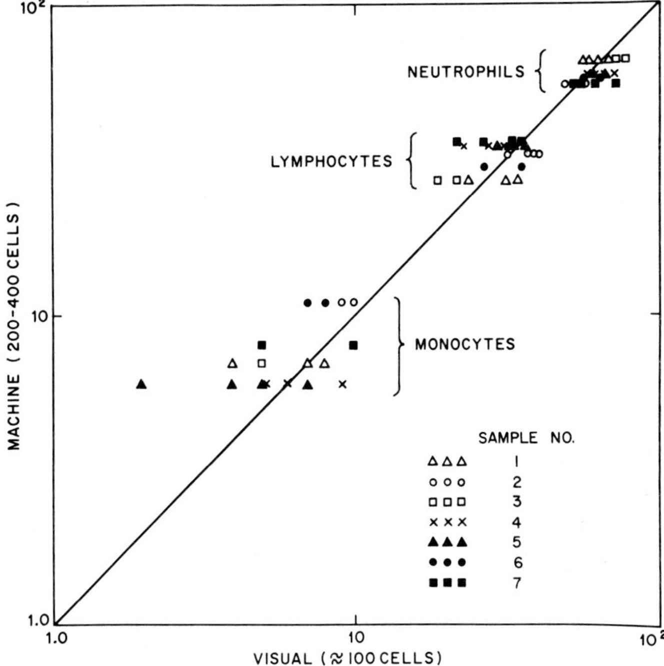

Fig. 5.25. Results of the first reduction to practice of a fully automated system for recognizing three major types of white blood cells. Seven blood smears (from seven different subjects) were automatically scanned by the machine and visually examined by four separate clinical technologists. Points lying on the diagonal indicate exact agreement between machine and human observer. In general the disagreement in counts among technologists is as great or greater than that between the machine count and the average result of the technologists taken together

one per minute. This permitted the researchers involved to make continuous runs (over about a twelve hour period) wherein a few hundred leukocytes per microscope slide were automatically recognized and sorted without human intervention. Thousands of cells were processed in this manner so that statistical results were obtained as graphed in Fig. 5.25.

Figure 5.25 reports on an experiment, covering the three major categories of cells, where machine classification is compared with visual counts on the same blood smears (but not on identical blood cells) by four hospital technologists. The disagreement in counts among technologists is as great as or greater than that between the machine count and the average results of the technologists, indicating that statistical "noise" rather than machine error prevented exact agreement.

		MACHINE					
		NEU	LYM	MON	EOS	BAS	OTHER
HUMAN	NEU	9887	19	25	44	3	–
	LYM	34	3872	38	7	13	–
	MON	59	48	1481	7	1	–
	EOS	54	0	0	750	0	–
	BAS	20	19	14	8	249	–
	OTHER	38	92	63	12	6	1339
		NEU	LYM	MON	EOS	BAS	OTHER
	MYE	19	3	33	7	0	190
	PRO	0	0	0	1	0	101
	NRC	17	72	6	0	2	663
	BLA	1	14	18	4	3	289
	PLA	1	3	6	0	1	86

Fig. 5.26. Confusion matrix plotting some of the early data taken using the Hematrak white blood cell differential counter produced by Geometric Data Corp. (SmithKline, Inc.). (For an explanation of the abbreviations see Fig. 5.27)

This accomplishment and associated work of MILLER [5.371], BACUS [5.82], YOUNG [5.49], and CHENG and LEDLEY [5.48], led to the commercial developments which have been discussed in Subsection 5.2.3. The work of MILLER was commercialized by the Geometric Data Corp. (a subsidiary of Smith, Kline, and French) and that of Bacus by the Corning Electronics Corp. (a subsidiary of Corning Glass Works).

Now that commercialization has begun, research results are no longer published, and a true quantitative picture of the status of digital picture analysis as related to the performance of commercial instruments is not available. Algorithms are not disclosed. Only a few results are available in the literature. Some data on the performance of a typical commercial instrument (one or two years out of date) are given in Fig. 5.26. The significance of these graphs is that they cover a large sample (almost 20000 cells!). Good statistics on performance are available. Success rates are shown on a cell by cell basis (unlike the presentation of Fig. 5.25) and exact false positive and false negative rates may be computed. It is difficult to evaluate errors, because one must judge who is the better classifier: machine or human?

Figure 5.26 shows an overall false positive rate of 3%. Some misclassification problems are still evident (at least as of the date of the preparation of Fig. 5.26) and are delineated in detail in the lower portion of the figure. These misclassifications may be, from a strict medical point of view, serious (such as the classification of 13% of the nucleated red blood cells as leukocytes). However, with the rapid progress being made and with the already obsolete nature of the data presented in Fig. 5.26, we are encouraged to predict performance improvements

BAN	BAND
LYM	LYMPHOCYTE
NEU	NEUTROPHIL
MON	MONOCYTE
EOS	EOSINOPHIL
BAS	BASOPHIL
MYE	MYELOCYTE
PRO	PROMYELOCYTE
NRC	NUCLEATED RED CELL
BLA	BLAST
PLA	PLASMACYTE
MET	METAMYELOCYTE
MYB	MYELOBLAST
LYB	LYMPHOBLAST
LYA	ATYPICAL LYMPHOCYTE
LYI	IMMUNO-LYMPHOCYTE

Fig. 5.27. Abbreviations for the types of leukocytes used in the confusion matrices given in Figs. 5.26 and 5.28

		MACHINE														
		BAS	EOS	LYM	MON	NEU	BAN	MET	MYE	PRO	MYB	LYB	LYA	LYI	NRC	PLA
HUMAN	BAS	40	0	1	0	1	0	1	2	0	0	5	0	0	0	0
	EOS	0	38	0	0	0	0	0	0	0	0	0	0	0	0	0
	LYM	1	0	31	4	0	0	0	0	0	0	0	6	0	0	0
	MON	0	0	0	29	0	0	2	3	0	0	1	3	0	0	0
	NEU	0	0	0	0	25	8	2	0	0	0	0	0	0	0	0
	BAN	0	0	0	0	6	32	2	0	0	0	0	0	0	4	0
	MET	0	0	0	6	0	0	18	7	0	0	0	0	0	1	0
	MYE	5	0	0	1	0	0	11	16	0	0	5	1	0	0	0
	PRO	0	0	0	1	0	0	2	5	6	8	2	1	1	0	2
	MYB	0	0	0	7	0	0	0	0	1	25	3	5	3	0	1
	LYB	4	0	1	0	0	0	0	1	1	2	23	1	2	1	0
	LYA	0	0	1	3	0	0	0	0	0	1	3	22	1	1	7
	LYI	0	0	0	1	0	0	0	0	1	2	1	3	10	0	3
	NRC	0	0	1	0	2	2	2	0	0	1	6	3	1	111	1
	PLA	0	0	0	2	0	0	0	0	0	0	0	0	2	0	27

Fig. 5.28. Confusion matrix produced by NEURATH [5.373] in which images (taken from photographs) of 15 types of leukocytes were machine classified. The results were obtained for a population of approximately 650 individual cells

which will lead to a thorough medical validation of digital picture analysis in automatic microscopy as applied to blood smear analysis.

As the commercial work proceeds, it is worth noting that there has been a simultaneous government-financed program in the automation of the analysis of blood smears at the New England Medical Center (Boston) which is described by NEURATH et al. [5.372], NEURATH and BRENNER [5.91], NEURATH [5.373], and NEURATH et al. [5.374]. Results are given in Fig. 5.28. In the early stages of this work a parallel program,

HUMAN \ MACHINE	SSQ	INT	CYT	PAR	MME	IME	NKD	KED	DYS	CIS	INC	OTHER
SSQ	98	2	0	0	0	0	0	0	0	0	0	0
INT	1	89	0	0	9	0	0	0	0	0	0	1
CYT	0	0	99	0	0	0	0	0	0	1	0	0
PAR	0	0	0	67	9	6	3	0	3	1	0	11
MME	0	4	0	0	93	1	0	0	1	0	0	1
IME	0	0	0	14	26	52	0	0	2	0	0	6
NKD	0	0	0	5	9	4	58	3	4	0	2	10
KED	0	0	0	1	22	1	5	55	1	0	0	15
DYS	0	0	0	1	0	2	2	2	59	23	5	6
CIS	0	0	0	0	0	0	0	0	25	73	2	0
INC	0	0	0	0	0	0	3	0	15	16	57	9

Fig. 5.29. Confusion matrix for the 11 types of ectocervical cells examined by BARTELS [5.410]. Results are given for approximately 1100 cells. (For an explanation of the abbreviations see Fig. 5.30)

SSQ - SUPERFICIAL SQUAMOUS CELLS
INT - INTERMEDIARY CELLS
CYT - CYTOLYTIC CELLS
PAR - PARABASAL CELLS
MME - MATURE METAPLASTIC CELLS
IME - IMMATURE METAPLASTIC CELLS
NKD - NON-KERATINIZED DYSPLASTIC CELLS
KED - KERATINIZED DYSPLASTIC CELLS
DYS - DYSPLASTIC IN METAPLASIA CELLS
CIS - CARCINOMA IN SITU CELLS
INC - INVASIVE CARCINOMA CELLS

Fig. 5.30. Abbreviations for the ectocervical cell types referred to in the confusion matrix shown in Fig. 5.29

using the same digitized pictures of cells but variations in computer measurement techniques, was conducted in Geneva and reported by GELSEMA and POWELL [5.92] of CERN. Also reporting on this work is BRENNER et al. [5.375].

A still broader research program on the analysis of cells is being conducted by BARTELS et al. [5.376], JARKNOWSKI et al. [5.377], OLSON et al. [5.378], KWAN and NORMAN [5.379], and ZAJICEK et al. [5.380]. These workers are concentrating on lymphocytes taken both from blood and from various tissue sections and aspirates. This is an extensive research program wherein tens of thousands of digitized leukocyte images have been recorded. These images are archived on magnetic tape. Large amounts of measurement data on images of both normally occurring and rarely occurring cells are being amassed. A matrix indicating the extent of the work of Bartels is shown in Fig. 5.29. Papers on this program are being published monthly in *Acta Cytologica.*

Other research is continuing in this field at the National Institutes of Health as reported by LIPKIN [5.381] and by LIPKIN and LIPKIN [5.307]. An excellent discussion of the general nature of digital picture analysis as it relates to leukocytes is presented by PREWITT [5.262]. A discussion of some of the problems relating to validity and validation of work in this area—especially as regards commercial instruments—and the legalistic and regulatory aspects of their use is provided in Section 5.7.

5.6.2 Chromosome Analysis

Digital picture processing in chromosome analysis differs from blood smear analysis in that both colorimetry and texture are unimportant. There is no requirement to separate the constituents of the objects studied (in comparison with the necessity of separating nucleus from cytoplasm, from vacuoles, etc., in blood cells). In chromosome picture analysis boundary determinations are the most important initial task. After the boundary of each chromosome is located, the centromeres and the arm tips are located, and the integrated optical density over the entire chromosome and over each arm is also computed.

Much of the difficulty in automating the analysis of chromosome spreads resides in the initial location of suitable spreads on the microscope slide. This is not an easy task, as there are only on the order of 10–100 spreads per cm^2, and their individual size is typically 50–100 μm in diameter. Complete automation of the analysis of chromosomes (sample preparation, automatic location, and automatic analysis) has not been accomplished to date, although a few systems may achieve this goal eventually.

The pioneering work in the automation of chromosome analysis in the United States has been by MENDELSOHN et al. [5.51, 53–55], with additional work being reported by MENDELSOHN and MAYALL [5.382]. Other pioneers in the United States have been NEURATH et el. [5.60, 61, 383], LEDLEY and RUDDLE [5.12], LEDLEY et al. [5.384], and CASTLEMAN and WALL [5.104]. More recent work on a somewhat different scale has been undertaken by AGGARWAL and FU [5.385, 386] and is also reported by AGGARWAL [5.387]. Early work at the Lawrence Livermore Laboratory is reported by STONE et al. [5.388] and by STONE [5.368]; see also KLINGER and KOCHMAN [5.389].

The major effort in automatic karyotyping overseas is that of RUTOVITZ [5.57, 58, 281], using a software system which has been described by HILDITCH and RUTOVITZ [5.390] and HILDITCH [5.365].

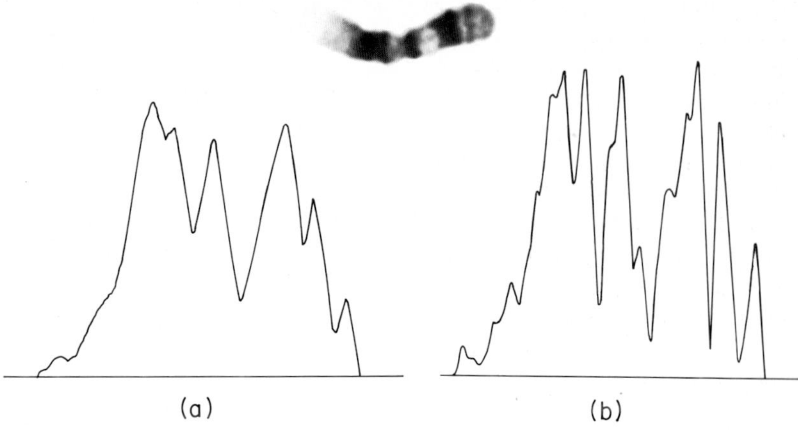

Fig. 5.31a and b. Illustration of a banded chromosome with (a) a photometric trace along the chromosome arms showing the position and amplitude of the bands; (b) a second trace produced by applying an inverse Gaussian frequency filter to trace (a) as performed by THOMAS. (Case-Western Reserve University, unpublished)

On the European continent work on chromosome analysis has been carried out in Italy by GALLUS et al. [5.280, 391]. As a whole these programs are similar in nature. There has been much interchange of data and software as well as cross comparison of results, as commented on in Section 5.7.

In the early 1970's karyotyping was revolutionized by the introduction of Caspersson's approach to cytochemical staining, which elucidates the so-called "banding" patterns, as discussed in Subsection 5.3.2. This change in the appearance of the chromosomes has required a revision in the digital picture processing approach which has been taken. A typical trace taken along the chromosome arms showing the bands is given in Fig. 5.31. Each trace may be expanded in a Fourier series and the Fourier coefficients utilized as a feature vector to classify each chromosome. This work is still relatively recent, much of it is unreported, but the reader is referred to the presentation by CASPERSSON [5.171] with an associated discussion of the photometric problems involved by LUNDSTEEN and PHILIP [5.176].

An approach to the automation of chromosome analysis for other than automatic karyotyping is given by WALD and PRESTON [5.56]. Here automatic location of chromosome spreads is carried out by optical scatter (Wiener spectrum techniques). The analysis of the metaphase spreads, once located, is carried out for the purpose of elucidating radiation damage; see LI [5.392] and WALD and FATORA [5.393].

The output of the system is not the karyogram but a tabulation of abnormal chromosome configurations. Typical abnormal configurations are, for example, chromosomes with two centromeres (called "dicentric"), chromosomes having a circular or "ring" configuration, and chromosome fragments.

A typical modern chromosome analysis system is that of CASTLEMAN at the Jet Propulsion Laboratory, being developed under contract with the National Institute of Child Health and Human Development. The system of Castleman incorporates a minicomputer (PDP 11/45) with an automatic microscope using a television imaging tube (the Plumbicon) as a scanner. Parallel operations are carried out wherein, as one metaphase spread is being located on the microscope slide, the spread previously located (which has also been stored in computer memory) is being processed. Simultaneously, the computer generated karyogram of the still previously processed spread is being recorded by a hard-copy output device. An entire cycle for these three simultaneous operations takes approximately eight minutes. The system, therefore, will produce about eight karyograms per hour. Since approximately fifteen karyograms are presently required per patient, this system could handle from ten to twenty patients per day (about 3000 to 6000 per year). With 500 to 1000 such systems in operation, all U.S. newborns could be karyogrammed.

Typical outputs generated by the Jet Propulsion Laboratory system are shown in Figs. 5.32 and 5.33. The basic steps required to produce this output are, first, to locate each chromosome and isolate its boundary. For overlapping chromosomes or touching chromosomes the Jet Propulsion Laboratory system provides for operator intervention (via a joy-stick and cursor) to permit editing of the image. The karyogram is then carried out by centromere location and arm length and arm area measurement in the traditional fashion. Software for carrying out the karyogram of banded chromosomes has now also been incorporated.

5.6.3 Cervical Smear Analysis

All major work in the automation of cervical smear analysis in the United States halted after the demise of the Cytoanalyzer program. Exploratory efforts were conducted in certain laboratories, but as a whole, this particular problem of digital picture analysis in cytology was avoided. Recently the National Cancer Institute has contracted with the University of Chicago for further work in the application of digital picture analysis to cervical smears. Other contracts have been placed with several other organizations as regards sample preparation.

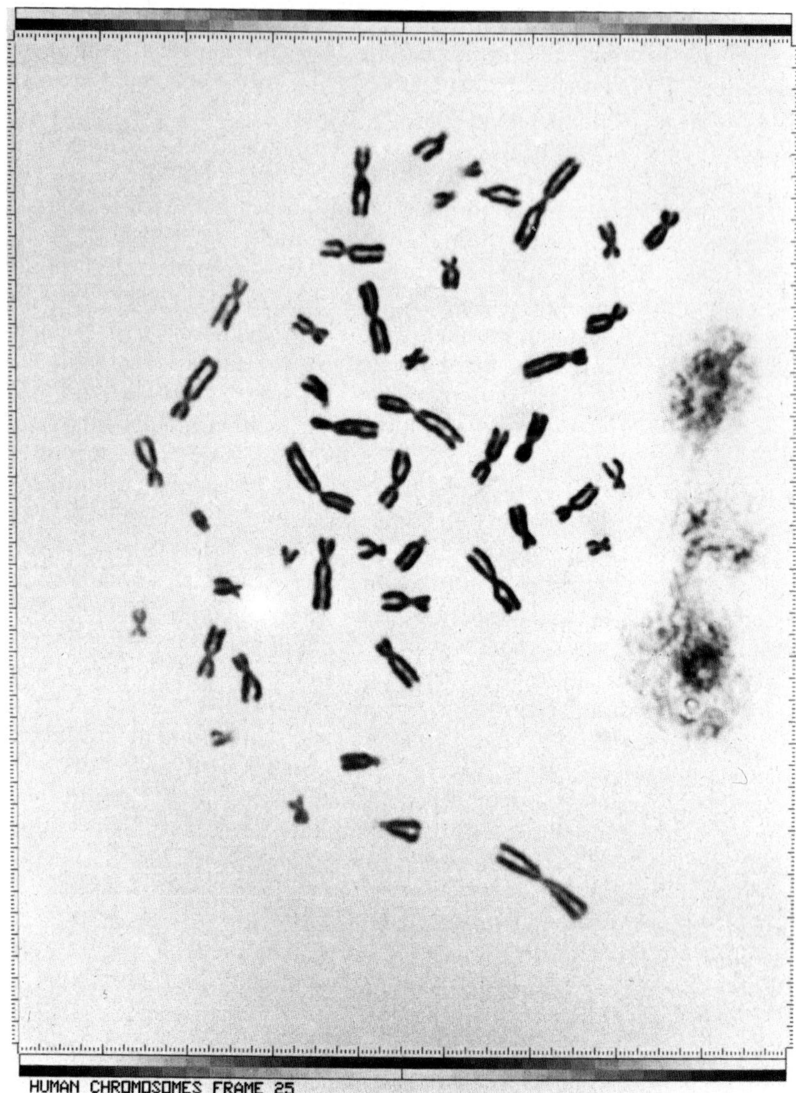

Fig. 5.32. Illustration of the Jet Propulsion Laboratory computer display of a chromosome spread prior to analysis

This has introduced a major new effort in the field. At the same time two corporations in Japan (Hitachi Corp. and Toshiba Corp.) have commenced the development of instruments for eventual clinical use. Results have not yet appeared in the open literature.

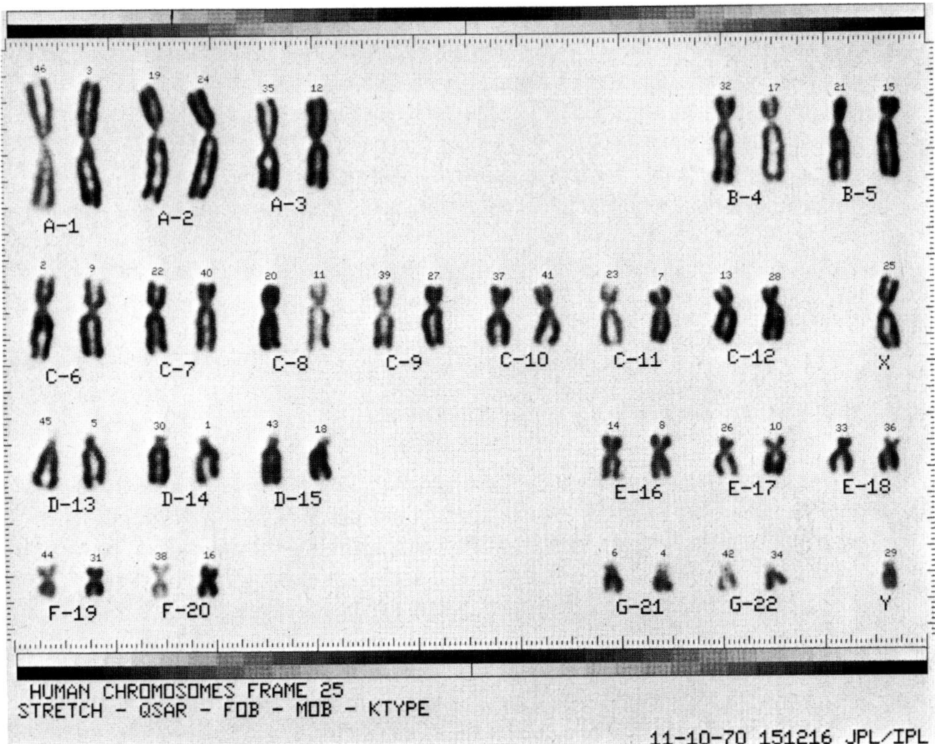

Fig. 5.33. Illustration of the Jet Propulsion Laboratory computer display after automatically performing the analysis of the chromosome spread shown in Fig. 5.32

Simultaneously there has been extensive work using flow systems (see Subsection 5.2.4) for the same purpose. Such systems are now in operation at the ERDA laboratories at Los Alamos and Livermore, with one such system having been delivered to the National Cancer Institute. It is hoped that these new efforts (part of the National Cancer Program) may lead to automatic systems for the detection of cervical cancer. Few if any results are in the published literature and, at the date that this chapter was written, success is still speculative.

The status of the Cytoanalyzer program as of the late 1950's is documented in HORVATH et al. [5.394] and in BOSTRUM et al. [5.395]. The programs which started in the mid- and late 1960's include work by ROSENBERG [5.43], the CESAR system of DAWSON [5.396], and the Cytoscreen system of HUSAIN and HENDERSON [5.397] and HUSAIN et al. [5.398]. Now, more than twenty years after the publication of the initial work of MELLORS [5.33], researchers in this field are still working on the problem of quantifying the nuclear size and optical density of cells in cervical scrapings. Other work worth reading is that by GUSBERG et al. [5.399], BAHR and WIED [5.400], BIBBO et al. [5.401], and BARTELS and WIED [5.402]. Another approach using slit scan fluorimetry is that of WHEELESS [5.183].

Finally there is the work using the optical Fourier transform as illustrated in Subsection 5.5.7, as conducted by KOPP et al. [5.321] of the Grumman Aircraft Corp.

From the results of all of the above work there is literally no way of judging how far we are from full automation.

5.7 Future Trends and Expectations

Although enormous advances have taken place in the application of digital picture analysis in cytology, the field has just reached the point where the results of three decades of research are beginning to bear fruit. Successful automation in cytology is still restricted to blood smear analysis. But automation of blood smear analysis represents one of the most exciting technical accomplishments in pattern recognition achieved in this century. This is the first time that naturally occurring objects (as opposed to man-made symbols) have been recognized by a machine—and at the rate of thousands per hour!

Present commercial equipment, which is just now being marketed on a trial basis, is still questionable in cost effectiveness and must be proven as to its medical qualifications. Using the cost justification approach taken by Geometric Data Corp. [5.403], which assumes that a technologist can do 40 smears/hour by machine and 10 per hour manually and is paid $12.25/hour, a labor cost saving of $.92 per smear is indicated. For an installation averaging 120 smears per day (365 days per year) this yields an annual saving of $40296 and a 2.5 year payback on a $90000 machine (with a $7200 annual service contract). However, the Chicago Hospital Council [5.404] determined that a technologist could read 20 smears per hour at $2.50/hour (see Table 5.4). Even with inflation this would indicate no cost saving through automation.

The debate on cost data goes on indefinitely, as was indicated at the 1970 Engineering Foundation Research Conference on Multiphasic Health Testing in Davos (see PRESTON and RUTOVITZ [5.14]), where the Automatic Cytology Workshop estimated the cost of blood smear examinations at $1.00 per smear, which is closer to the Geometric Data Corp. estimate than to that made by the Chicago Hospital Council.

Table 5.4

Time cost study—Blood smear	Hr.
1. Sort request slips, log data, mark slide	0.01293
2. Stain Smear	0.09695
3. Travel in laboratory during staining	0.00895
4. Clean slide, take to reading area	0.00147
5. Set up microscope	0.02893
6. Load slide	0.00941
7. Read and totalize WBCD	0.02964
8. Transcribe results in log	0.00285
9. Record laboratory slip	0.00590
10. Clean up	0.01796

Total time: 0.21499 Hrs; Cost (at $ 2.50/hr.[a]): $ 0.54

Batch operations (for a practical batch size of 10) permit a proportional reduction in tasks 3, 4, 5, 10 and parts of task 2. The result is:

Total time (per smear): 0.07564 Cost: $ 0.17

Even in batch operations the time required for reading the smear consumes only 40% of the total time. The balance of the time is consumed as follows:

Data logging operations	28%
Staining operations	13%
Slide loading (on microscope)	12%
Set-up and clean up	7%

[a] *Clinical Laboratory Study*, Chicago Hospital Council, 1965.

Whatever the cost-effectiveness considerations may be, it is still indisputable that the blood smear examination as presently conducted manually is statistically invalid. An average human being has 10^{10} white blood cells instantaneously in the peripheral blood stream. The present examination catalogues only 10^2 of these cells. Yet this examination is used to determine the rate of occurrence of cell types such as the basophil, which appear at the rate of one per several hundred cells. For such types of cells (which are normal yet rare), one would require the examination of at least 10^4 cells before even achieving 20 to 25% accuracy in determining their rate of occurrence in the blood stream.

The above considerations lead to the prediction that the future, albeit far distant, will require automated television microscopes with transport velocities and data processing rates at least two orders of magnitude beyond what is now becoming commercially available. The development of such machines will take research laboratories and commercial industry another one or two decades. Some predictions on possible physical limitations to achieving this goal are given below.

5.7.1 Physical Limitations on Cytological Picture Processing Rates

The basic information carrier which transfers data from the television microscope to the computer is electromagnetic energy in the optical spectrum. The basic unit of electromagnetic energy is the photon which, in the optical spectrum, has an energy of approximately 0.3 electron volts. As scanning rates are increased in order to increase the rate at which cell images are transferred to the computer, the major physical barrier to further increases in speed will be shot noise. This noise occurs due to the fact that, over the time interval during which the value of a picture element is transferred by the scanner to the computer, only a finite number of photons is gathered by the television system. The number of photons available is directly dependent upon the brilliance of the illuminating light source. Using the most intense (noncoherent) microscope illuminators (such as the zirconium arc), one has available only a few tens of nanowatts per optical resolution element. This fact leads to the physical bounds on picture transfer rates shown in Fig. 5.34. The rate (in terms of cells transferred per minute) is dependent upon the fourth power of the picture element size as given in the following equation.

$$R_{\text{scan}} = \frac{(d)^4\, \varepsilon(\lambda)\, \eta(\lambda)\, (\text{N.A.})^2}{(S/N)^2\, D^2} \int I(\lambda)\, d\lambda$$

d = Effective detector diameter (related to object plane)
D = Object diameter
$I(\lambda)$ = Irradiance
N.A. = Numerical aperture in object plane
S/N = Acceptable peak signal to RMS noise
$\varepsilon(\lambda)$ = Optical transfer efficiency
$\eta(\lambda)$ = Quantum efficiency
λ = Wavelength.

The fourth power dependence is due to the fact that, as the picture element size increases, the number of photons per picture element increases as the square of the picture element diameter. At the same time the number of picture elements per cell decreases as the square of the picture element diamter. This leads to the fourth power law given above (assuming that the detector diameter equals the picture element diameter).

Figure 5.34 shows how the physical limitation imposed by shot noise limits scan rates as a function of various types of light handling devices: 1) scanned point detectors (such as the photomultiplier or semiconductor diode), 2) light integrating devices (such as the television imaging tube),

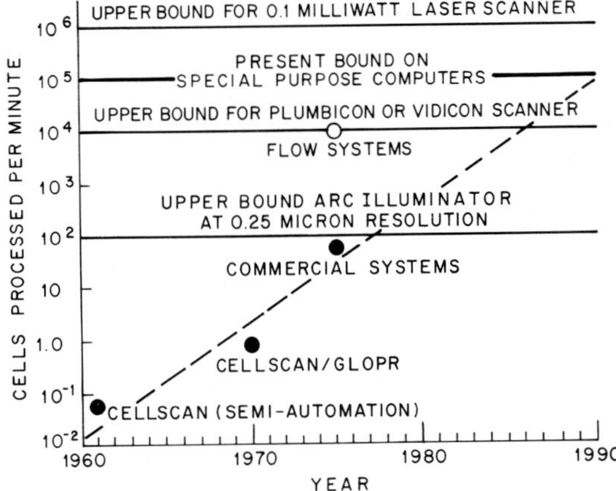

Fig. 5.34. System performance (as a function of the source of optical energy and of the advance of technology) for various types of scanning and digital picture processing systems as related to the automatic white blood cell differential count

and 3) coherent emitting devices (lasers). As can be seen, at high resolution, the point detector reaches an upper bound of approximately 100 cells per minute at 0.25 micrometer resolution (with $S/N=100$). Light integrating devices and scanning lasers have the potential capability of operating orders of magnitude faster. As the reliability of lasers increases, the author believes that the competition between light integrating scanners and laser scanners will be won by the laser. This is because light integrating devices have a narrower dynamic range and experience the phenomenon of retaining a residual image after readout (called "image lag"). The laser still suffers somewhat from high cost and poor reliability. But note that for some years we have seen commercial cytology instruments, such as those of Bio/Physics Systems, Inc., which employ lasers as their standard optical power source. Thus, as lasers decrease in cost and reach the point where the color of their emission may be rapidly altered or "tuned", they will begin to predominate in automatic television microscopes.

5.7.2 Standards and Validation

With the introduction of commercial cytologic picture analyzing systems into the clinical laboratory environment, it will be necessary to establish standards for calibration and methods for validating performance. This is an area which surprisingly has received little consideration over the years. In an excellent series of experiments Bacus [5.405]

1. CABOT RINGS
2. ELLIPTOCYTES
3. HEMOGLOBIN C CRYSTALS
4. HOWELL-JOLLY BODIES
5. IMMATURE NUCLEATED CELLS
6. FRAGMENTED RED BLOOD CELLS
7. ROULEAUX
8. POLYMACROCYTES
9. SIDEROCYTES
10. MALARIA PARASITES
11. SCHUFFNER'S DOTS
12. MALARIA CRESCENT SHAPED GAMETOCYTES
13. MAURER'S DOTS
14. MALARIAL DOUBLE RINGS 1-3 TROPHOZOITES
15. OTHER MALARIAL FORMS
16. POLYCHROMATOPHILIA, SLIGHT
17. POLYCHROMATOPHILIA, MARKED
18. BASOPHILIC STIPPLING, SLIGHT
19. BASOPHILIC STIPPLING, MARKED
20. SPHEROCYTES, FEW
21. SPHEROCYTES, NUMEROUS
22. TARGET CELLS, FEW
23. TARGET CELLS, MANY
24. SICKLE CELLS, POINTED
25. SICKLE CELLS, BLUNT END
26. BURR CELLS OR SPINOCYTES, FEW
27. BURR CELLS OR SPINOCYTES, MANY

Fig. 5.35. Red cell descriptive terms (glossary numbers 1—27) taken from the publication of the Center for Disease Control. U.S. Department of Health, Education and Welfare

has investigated the variability of the human technologist in the visual classification of human leukocytes. Other unreported experiments have investigated similar variability in both chromosome analysis and the analysis of cervical smears. The Center for Disease Control (Atlanta, Georgia) is also concerned with performance testing of clinical laboratories in the performance of cytological determinations in hematology. For this purpose hundreds of standard slides are prepared per year which are sent to participating laboratories in a self-testing and evaluation program. There is no equivalent program in either chromosome or cervical smear analysis. The Center for Disease Control is also responsible for the standardization of terminology. Figures 5.35 and 5.36 give the current cytological vocabulary as agreed upon by this Federal institution.

Unfortunately, automation of sample preparation and staining is not currently used in preparing standard slides. Few workers have examined the influence of sample preparation on automatic cell recognition. In the field of cervical cytology DIACUMAKOS [5.41] prepared microscope slides for testing the Cytoanalyzer using microsurgical instruments to extract selected cells from cervical scrapings. DECAMPOS

35. ALDER'S GRANULATION
36. BASOPHILIC BANDS
37. BASOPHILIC METAMYELOCYTES
38. BASOPHILIC MYELOCYTES
39. HYPOSEGMENTATION (NEUTROPHILS)
40. AUER RODS
41. POLYPLOIDY, DIPLOID CELLS, TETRAPLOID CELLS, ETC.
42. EOSINOPHILIC BANDS
43. EOSINOPHILIC METAMYELOCYTES
44. EOSINOPHILIC MYELOCYTES
45. IMMATURE CELLS UNIDENTIFIED
46. MITOSIS
47. MONOCYTES, ATYPICAL
48. MONOCYTES, VACUOLATED
49. NEUTROPHILS, VACUOLATED
50. PLASMOCYTES
51. STEM CELLS
52. TART CELLS
53. MARKED LEUKOCYTOSIS
54. TOXIC GRANULATION, SLIGHT
55. TOXIC GRANULATION, MARKED
56. SMUDGE CELLS, FEW
57. SMUDGE CELLS, MANY
58. HYPERSEGMENTATION OF NEUTROPHILS, FEW
59. HYPERSEGMENTATION OF NEUTROPHILS, MANY
60. MACROPOLYCYTES
61. ATYPICAL LYMPHOCYTES OCCASIONAL (0-5%)
62. ATYPICAL LYMPHOCYTES PLASMACYTOID TYPE (6-20% of total)
63. MAJORITY OF THE LYMPHOCYTES ATYPICAL
64. DOEHLE (RNA) BODIES

Fig. 5.36. List of descriptive terms for white blood cells (glossary numbers 35—64) taken from the publication of the Center for Disease Control, U.S. Department of Health, Education and Welfare

et al. [5.198] have studied the influence of various stains on pattern recognition of cell nuclei as regards size and shape measures. Workers using fluid flow systems have worked on standards in a different manner. In this regard FULWYLER et al. [5.406] have examined the problem of the accurate generation of microspheres. Such workers as BRUNSTING and MULLANEY [5.407], COOKE and KERKER [5.408], and PHILLIPS et al. [5.409] have analytically predicted the light scattering properties of such spheres. These spheres are then introduced in the flow stream for the purposes of calibrating measurements obtained by both optical and electronic flow sensors.

In the future these meager steps toward both analytic and subjective standardization will have to be considerably improved. Serious thought will have to be given to data banking not only the traditional cell classification table but also the total machine output for the

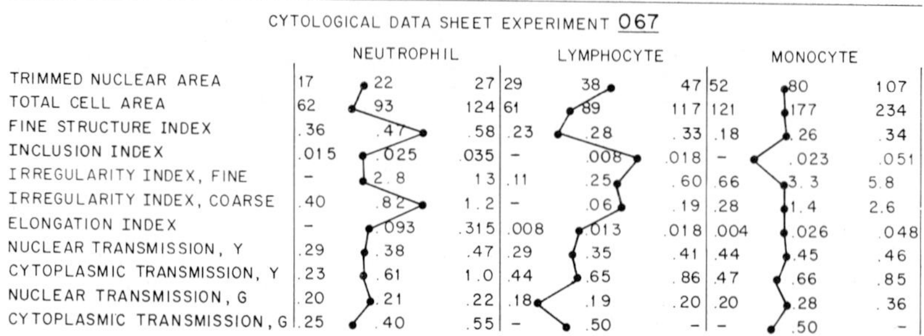

CYTOLOGICAL DATA SHEET EXPERIMENT 067

	NEUTROPHIL			LYMPHOCYTE			MONOCYTE		
TRIMMED NUCLEAR AREA	17	22	27	29	38	47	52	80	107
TOTAL CELL AREA	62	93	124	61	89	117	121	177	234
FINE STRUCTURE INDEX	.36	.47	.58	.23	.28	.33	.18	.26	.34
INCLUSION INDEX	.015	.025	.035	–	.008	.018	–	.023	.051
IRREGULARITY INDEX, FINE	–	2.8	13	.11	.25	.60	.66	3.3	5.8
IRREGULARITY INDEX, COARSE	.40	.82	1.2	–	.06	.19	.28	1.4	2.6
ELONGATION INDEX	–	.093	.315	.008	.013	.018	.004	.026	.048
NUCLEAR TRANSMISSION, Y	.29	.38	.47	.29	.35	.41	.44	.45	.46
CYTOPLASMIC TRANSMISSION, Y	.23	.61	1.0	.44	.65	.86	.47	.66	.85
NUCLEAR TRANSMISSION, G	.20	.21	.22	.18	.19	.20	.20	.28	.36
CYTOPLASMIC TRANSMISSION, G	.25	.40	.55	–	.50	–	–	.50	–

Fig. 5.37. Cytological data sheet (for CELLSCAN experiment No. 067) showing a patient profile as compared with norms developed from a small population of patients (see Fig. 5.25)

individual patient, i.e., raw measurement data on such fundamental qualities as nuclear and cytoplasmic size, shape, and color on a per-patient basis. As reported by MELLECKER [5.15], PRESTON has made the sole attempt in this direction by introducing the cytological patient profile as illustrated in Fig. 5.37.

Finally, such agencies as the Food and Drug Administration must determine under the Freedom of Information Act whether the individual patient has the right not only to the measurement data produced by automatic cell analyzers but also to a description of the algorithms which are applied to analyze the pictures of his or her blood cells. An open disclosure of these algorithms and cross comparison between them and results obtained with various commercial and research instruments on a standard cell population would be of extreme utility. Finally, the National Bureau of Standards should play a role in the standardization of stains for use in automatic cytology.

These are the trends towards the future. Although the rate of success in developing techniques for automatic picture analysis in cytology is slow, the derivative is positive, and we have great expectations for continued progress.

References

5.1 D. SENCER: Statement to Senate Subcommittee on Anti-Trust and Monopoly (Committee on the Judiciary), Washington, D.C. (1967)

5.2 A. K. PARPART: Science **113**, 483—484 (1951)

5.3 L. E. FLORY: Proc. IRE **47**, 1889—1894 (1959)

5.4 C. BERKLEY: "Electronic Counting, Sorting and Sizing"; Proc. 2nd Intern. Conf. Med. Electron. (Iliffe and Sons, Ltd., London 1959)

5.5 P. O. B. MONTGOMERY (ed.): Ann. New York Acad. Sci. **97**, 329—526 (1962)

5.6 D. M. RAMSEY (ed.): *Image Processing in Biological Science* (Univ. Calif. Press 1968)

5.7 W. E. Tolles (ed.): Ann. New York Acad. Sci. **157**, 1—530 (1969)
5.8 W. E. Bushor: Electronics **34**, 65—71 (1961)
5.9 B. Shackel: Med. Electron. Biol. Engrg. **1**, 35—50 (1963)
5.10 G. N. Eaves, in: *Computers and Biomedical Research*, Vol. 1 (Academic Press, New York 1967) pp. 112—123
5.11 P. W. Neurath: Med. Res. Engrg. 9—15 (June—July 1971)
5.12 R. S. Ledley, F. H. Ruddle: Sci. Am. **213**, 40—46 (1966)
5.13 M. Ingram, K. Preston, Jr.: Sci. Am. **223**, 72 (1970)
5.14 K. Preston, Jr., D. Rutovitz: Automatic Cytology. In: *Automated Multiphasic Health Testing*, ed. by C. Berkley (Engineering Foundation, New York 1971)
5.15 J. Mellecker: Biocharacterist **2**, 127 (1972)
5.16 B. Mayall (ed.): J. Histochem. Cytochem. **22**, 451—765 (1974)
5.17 D. M. D. Evans (ed.): *Cytology Automation* (E. and S. Livingston, Edinburgh and London 1970)
5.18 G. L. Wied, B. Bahr, P. H. Bartels: Automated Analysis of Cellular Images by TICAS. In: *Automated Cell Identification and Cell Sorting*, ed. by G. L. Wied, G. F. Bahr (Academic Press, New York 1970) pp. 195–360
5.19 J. Z. Young, F. Roberts: Nature **167**, 231 (1952)
J. Z. Young, F. Roberts: Nature **169**, 963 (1952)
5.20 W. H. Walton: Nature **169**, 518 (1952)
5.21 P. G. W. Hawksley, J. H. Blackett, E. W. Meyer, A. E. Fitzsimmons: Brit. J. Appl. Phys. **5**, 165—173 (1954)
5.22 D. Causley, J. Z. Young: Research **8**, 430—434 (1953)
5.23 R. Barer: J. Opt. Soc. Am. **47**, 545 (1957)
5.24 E. G. Ramberg: IRE Trans. ME-**12**, 58 (1958)
5.25 A. W. Pollister, M. J. Moses: J. Gen. Physiol. **32**, 567 (1949)
5.26 M. J. Moses: Exp. Cell Res. **2**, 75 (1952)
5.27 T. L. Skeggs, Jr.: Am. J. Clin. Pathol. **28**, 311 (1957)
5.28 G. Z. Williams: IRE Trans. ME-**6**, 68—74 (1959)
5.29 P. O. B. Montgomery, W. A. Bonner: Proc. IRE **47**, 1889 (1959)
5.30 H. N. Coates: Proc. IEE **103-B**, 479 (1956)
5.31 W. K. Taylor: IEE Rpt. 3154, 447 (1959)
5.32 H. P. Mansberg: Science **126**, 823 (1957)
5.33 R. C. Mellors, A. Glassmann, G. N. Papanicolaou: Cancer **5**, 458—468 (1952)
5.34 W. E. Tolles: Trans. New York Acad. Sci. **17**, 250 (1955)
5.35 W. E. Tolles, R. C. Bostrom: Ann. New York Acad. Sci. **63**, 1211—1218 (1956)
5.36 R. Barer, J. Wardley: Nature **192**, 1060 (1961)
5.37 V. K. Zworykin, F. L. Hatke: Science **126**, 805 (1957)
5.38 J. C. Pruitt, S. C. Ingraham, II., R. F. Kaiser, A. W. Hilberg: "Preparation of Vaginal-Cervical Material for Automatic Scanning—Preliminary Results with the Cytoanalyzer"; Trans. 6th Ann. Mtg. Intern. Soc. Cytol. Council (1959)
5.39 R. C. Bostrom, H. S. Sawyer, W. E. Tolles: Proc. IRE **47**, 1895—1900 (1959)
5.40 E. G. Diacumakos, E. Day, M. J. Kopac: Acta Cytol. **6**, 238 (1962)
5.41 E. G. Diacumakos, E. Day, M. J. Kopac: Ann. New York Acad. Sci. **97**, 329 (1962)
5.42 E. G. Diacumakos, E. Day, M. J. Kopac: Acta Cytol. **6**, 423 (1962)
5.43 S. A. Rosenberg, K. S. Ledeen, T. Kline: Science **163**, 1065—1067 (1965)
5.44 A. I. Spriggs, R. A. Diamond, E. W. Meyer: Lancet **1**, 359 (1968)
5.45 N. F. Izzo, W. Coles: Electronics **35**, 52—57 (1962)
5.46 M. Ingram, P. E. Norgren, K. Preston, Jr., in: *Image Processing in Biological Science*, ed. by D. M. Ramsey (University of California Press 1968) pp. 97—117
5.47 J. M. S. Prewitt, M. L. Mendelsohn: Ann. New York Acad. Sci. **128**, 1035–1053 (1966)

5.48 G.C. Cheng, R.S. Ledley, in: Proc. of 13th Ann. Tech. Symp. of SPIE (1968) p. 393
5.49 I.T. Young: Automated Leukocyte Recognition. PhD dissertation, Massachusetts Institute of Technology (1969)
5.50 E.M. Nadel: Acta Cytolog. **9**, 203 (1965)
5.51 M.L. Mendelsohn, W.A. Kolman, R.C. Bostrum: Ann. New York Acad. Sci. **115**, 998 (1964)
5.52 M.L. Mendelsohn, T.J. Conway, D.A. Hungerford, W.A. Kolman, B.H. Perry, J.M.S. Prewitt: Cytogenetics **5**, 223 (1964)
5.53 M.L. Mendelsohn, B.H. Mayall, J.M.S. Prewitt, R.C. Bostrum, W.G. Holcomb, in: *Advances in Optical and Electron-Microscopy*, ed. by V. Cosslett, R. Barer (Academic Press, New York 1968) pp. 77—150
5.54 M.L. Mendelsohn, B.H. Mayall, J.M.S. Prewitt, in: *Image Processing in Biological Science*, ed. by D.M. Ramsey (University of California Press 1969) pp. 119—136
5.55 M.L. Mendelsohn, D.A. Hungerford, B.H. Mayall, B.H. Perry, T.J. Conway, J.M.S. Prewitt: Ann. New York Acad. Sci. **157**, 376—392 (1969)
5.56 N. Wald, K. Preston, Jr.: Automatic Screening of Metaphase Spreads for Chromosome Analysis; in: *Image Processing in Biological Science*, ed. by D. Ramsey (University of California Press 1968) pp. 9–34
5.57 D. Rutovitz, in: *Human Radiation Cytogenetics* (North-Holland Publ. Co., Amsterdam 1967) p. 589
5.58 D. Rutovitz: Brit. Med. Bull. **24**, 260—267 (1968)
5.59 R.S. Ledley, L.S. Rotolo, T.J. Golab, J.D. Jacobsen, M.D. Ginsberg, J.B. Wilson, in: *Optical and Electro-Optical Inf. Proc.* (MIT Press, Cambridge 1965) pp. 591—613
5.60 P.W. Neurath, A. Falek, B.L. Bablouzian, T.H. Warms, R.C. Serbagi: Ann. New York Acad. Sci. **128**, 1013—1028 (1966)
5.61 P.W. Neurath, D.H. Brand, E.D. Schreiner: Ann. New York Acad. Sci. **157**, 324—338 (1969)
5.62 C.W. Gilbert: Nature **212**, 1437—1440 (1966)
5.63 G. Gallus, N. Montanaro, G.A. Maccacaro: An Application of Pattern Recognition in Biology; Proc. 14th Cong. for Electronics, Rome (1967)
5.64 G.L. Wied, P.H. Bartels, G.F. Gunter, D.G. Oldfield: Acta Cytol. **12**, 180—204 (1968)
5.65 P.G. Stein, L.E. Lipkin, N.H. Shapiro: Science **166**, 328—333 (1969)
5.66 C. Beadle, in: *Advances in Optical and Electron Microscopy*, ed. by V. Coslett, R. Barer (Academic Press, New York 1971) p. 361
5.67 C. Fisher: Microscope **19**, 20 (1971)
5.68 C. Zimmer: ZEISS-Inform. **17**, 126—131 (1969)
5.69 W. Muller: Leitz Sci. Tech. Inform. **1**, 101—116 (1974)
5.70 E.R. Weibel, C. Fisher, J. Gahm, A. Schaefer: J. Microscopy **95**, 367—391 (1971)
5.71 R.R.A. Morton, in: *Stereology and Quantitative Metallography* (American Society of Testing Materials 1972) pp. 81—94
5.72 S.S. Nelson, O.E.A. Boldman, W.A. Shurcliff: Ann. New York Acad. Sci. **97**, 290—297 (1962)
5.73 H.P. Mansberg, J.M. Segarra: Ann. New York Acad. Sci. **99**, 309—322 (1962)
5.74 L.L. Salomon, G.H. Musgrave: Ann. New York Acad. Sci. **157**, 168—182 (1969)
5.75 P.E. Norgren: Ann. New York Acad. Sci. **157**, 514—524 (1969)
5.76 C.A. Soffen: Ann. New York Acad. Sci. **157**, 159—167 (1969)
5.77 F.W. Oberst, J.W. Crook, S.F. Swain, F.P. Ward, W.S. Koon, W.P. Musselman: "Toxic Effects of High Concentrations of Bromo-Benzylnitrile Vapor in Various Animal Species"; 8th Ann. Mtg. Soc. of Toxicology, Williamsburg, Virginia (1969)

5.78 L.E. Mawdesley-Thomas, P. Healey: "The Use of an Automated Image Analyzer in the Quantitative Histologic Evaluation of Toxicologic Material with Particular Reference to the Respiratory Tract"; 8th Ann. Mtg. Soc. of Toxicology, Williamsburg, Virginia (1969)

5.79 L.E. Mawdesley-Thomas, P. Healey: Am. Rev. Respir. Disease **100**, 231—233 (1969)

5.80 L.E. Mawdesley-Thomas, P. Healey: Science **163**, 1200 (1969)

5.81 M.N. Miller: "Design and Clinical Results of Hematrak—An Automated Differential Counter"; Geometric Data Corp. (1973)

5.82 J.W. Bacus: "An Automated Classification of the Peripheral Blood Leukocytes by Means of Digital Image Processing", PhD dissertation, Dept. Physiol., Univ. of Ill., Chicago (1970)

5.83 D.A. Cotter: Am. J. Med. Tech. **39**, 383—389 (1973)

5.84 C.N. Carter, D.A. Cotter: "Leukocyte Nucleus Shape Evaluation—Systems and Methods", presentation at La. State Univ., Baton Rouge (1974)

5.85 M.N. Miller: "Leukocyte Classification by Morphological Criteria", Engrg. Found. Conf. on Automatic Cytology, Saxtons River (1972)

5.86 G.K. Megla: "The LARC Analyzer and its Acceptability", Engrg. Found. Conf. on Automatic Cytology, Saxtons River (1972)

5.87 G.K. Megla: Acta Cytol. **17**, 3—14 (1973)

5.88 E.T. Smithline: US Patent No. 3, 315, 229 (1967)

5.89 A.E. Brain: US Patent No. 3, 824, 383 (1974)

5.90 B.H. Mayall: Computer **7**, 5—81 (1974)

5.91 P.W. Neurath, J.F. Brenner: "A Computer Classification of White Blood Cells into 15 Classes Using the Tufts PIQUANT System", Proc. 25th Ann. Conf. on Engrg. in Med. and Biol., Bal Harbor, Florida (1972)

5.92 E.S. Gelsema, B.W. Powell: CERN DD/72/74 (1972)

5.93 E.S. Gelsema, B.W. Powell, P.L. Berrini: Lect. Notes, Econ. Math. Syst. **83**, 237 (1973)

5.94 G. Gallus, P.W. Neurath: Phys. Med. Biol. **15**, 435—455 (1970)

5.95 P.W. Neurath, G. Gallus, W.D. Selles: J. Assoc. Adv. Med. Instr. **4**, 6—14 (1970)

5.96 D.K. Green, P.W. Neurath: J. Histochem. Cytochem. **22**, 531—535 (1974)

5.97 B. Dew, T. King, D. Mighdoll: J. Histochem. Cytochem. **22**, 685 (1974)

5.98 I.T. Young, in: *Automated Cell Identification and Cell Sorting*, ed. by G.L. Wied, G.F. Bahr (Academic Press, New York 1970) pp. 187—194

5.99 I.T. Young: IEEE Trans. BME-**19**, 291—298 (1972)

5.100 P. Lemkin, G. Carman, L. Lipkin, G. Shapiro, M. Shultz, P. Kaiser: J. Histochem. Cytochem. **22**, 725 (1974)

5.101 G. Carman, P. Lemkin, L. Lipkin, B. Shapiro, M. Shultz, P. Kaiser: J. Histochem. Cytochem. **22**, 732 (1974)

5.102 M. Bibbo, P.H. Bartels, G.F. Bahr, A.B.P. Ng, J.W. Reagan, D.L. Richards, G.L. Wied: Acta Cytol. **14**, 574—582 (1970)

5.103 P.J. Golab: Proc. 23rd ACEMB (1970) p. 316

5.104 K.R. Castleman, R.J. Wall, in: *Chromosome Identification*, ed. by T. Caspersson, L. Zech (Academic Press, New York 1973) pp. 77—84

5.105 R.J.P. Le Go: Medinfo 74, Commissariat L'Energie Atomique, France (1974)

5.106 K. Preston, Jr., in: *Computer Techniques in Biomedicine and Medicine* (Auerbach 1973) pp. 295—331

5.107 D.F. Wann, T.A. Woolsey, M.L. Dierker, W.M. Cowan: IEEE Trans. BME-**20**, 233—247 (1973)

5.108 D.F. Wann, J.L. Price, W.M. Cowan, M.A. Agulnek: Brain Research **80**, 1—28 (1974)

5.109 J. Zajicek, P. H. Bartels, G. F. Bahr, M. Bibbo, P. A. Jacobsson, G. L. Wied: Acta Cytol. **17**, 179—187 (1973)
5.110 K. von Voss: Exp. Path. **9**, 131—139 (1974)
5.111 H. M. Aus, K. Koschel, V. ter Meulen, P. H. Bartels: Proc. Intern. Computer Symp., Davos, Switzerland (North-Holland/American Elsevier, Amsterdam, 1973) p. 407
5.112 M. Eden, J. E. Green, A. Sun: Bull. Org. Mond. Sante/Bull. Wld. Hlth. Org. **48**, 211—218 (1973)
5.113 V. ter Meulen, P. H. Bartels, G. F. Bahr: Acta Cytol. **16**, 454—463 (1972)
5.114 H. M. Aus, V. ter Meulen, M. Kaekell, W. Scholz, K. Koschel: J. Histochem. Cytochem. **22**, 577—582 (1974)
5.115 K. Christov, G. Kiefer, R. Kiefer, W. Sandritter: "Changes in the Nuclear Structure During Thyroid Carcinogenesis in Rats—An Image Analysis Study", Inst. of Path., Univ. of Freiburg (1974)
5.116 D. H. Moore: J. Histochem. Cytochem. **22**, 663—667 (1974)
5.117 P. J. Crosland-Taylor: Nature **171**, 37 (1953)
5.118 W. Coulter: "High-Speed Automatic Blood Cell Counter and Cell Size Analyzer", Proc. Nat. Electron. Conf. (1956) pp. 1034—1042
5.119 L. A. Kamentsky, M. R. Melamed, H. Derman: Science **150**, 630 (1965)
5.120 H. R. Ansley, L. Ornstein: "Enzyme Histochemistry and Differential White Cell Counts on the Technicon Hemalog D", Technicon Intern. Cong. (1970)
5.121 H. R. Ansley, L. Ornstein; in: *Advances in Automated Analysis*, Vol. 1 (Thurman Assoc., Miami 1971) p. 437
5.122 H. R. Ansley, L. Ornstein: US Patent No. 3, 741, 875 (1973)
5.123 A. M. Saunders, W. Groner, J. Kusnetz: A Rapid Automated System for Differentiating and Counting White Blood Cells; in: *Advances in Automated Analysis*, Vol. 1, Technicon Intern. Cong., Tarrytown, New York (1971)
5.124 H. P. Mansberg: "High Speed, High Precision Leukocyte Differential Counting by Cytochemical Electro-Optical Detection", 26th Ann. Conf. on Engrg. in Med. and Biol. (1973)
5.125 R. V. Pierre, M. B. O'Sullivan: Evaluation of the Technicon Hemalog D Automated Differential Machine; in: *Advances in Automated Analysis*, Vol. 3, Technicon Intern. Cong., Tarrytown, New York (1973)
5.126 H. P. Mansberg, A. M. Sanders, W. Groner: J. Histochem. Cytochem. **22**, 711—724 (1974)
5.127 A. M. Saunders, F. Scott: J. Histochem. Cytochem. **22**, 707—710 (1974)
5.128 A. M. Saunders: "A Rapid Automated System for Differentiation and Counting White Blood Cells"; in: *Advances in Automated Analysis*, Technicon Intern. Cong. (1970)
5.129 A. M. Saunders: Clin. Chem. **18**, 783 (1972)
5.130 A. M. Saunders: "Hemalog D—Recent Development"; in: *Advances in Automated Analysis*, Vol. 3, Technicon Intern. Cong., Tarrytown, New York (1973)
5.131 R. G. Sweet: Rev. Sci. Instr. **36**, 131 (1965)
5.132 M. J. Fulwyler: Science **150**, 910 (1965)
5.133 M. J. Fulwyler, R. B. Glasscock, R. D. Hiebert: Rev. Sci. Instr. **40**, 42 (1969)
5.134 M. A. Van Dilla, M. J. Fulwyler: Acta Cytol. **15**, 98 (1971)
5.135 M. A. Van Dilla, L. L. Steinmetz, D. T. Davis, R. N. Calvert, J. W. Gray: IEEE Trans. NS-**12**, 714 (1974)
5.136 J. A. Steinkamp, M. J. Fulwyler, J. R. Coulter, R. D. Hiebert, J. L. Horney, P. F. Mullaney: Rev. Sci. Instr. **44**, 1301 (1973)
5.137 D. J. Arndt-Jovin, T. M. Jovin: J. Histochem. Cytochem. **22**, 622—625 (1974)
5.138 L. A. Kamentsky, M. R. Melamed: Proc. IEEE **57**, 2007—2016 (1965)

5.139 L. KAMENTSKY, M. R. MELAMED: Science **156**, 1364 (1967)
5.140 L. A. KAMENTSKY, M. R. MELAMED: Ann. New York Acad. Sci. **157**, 310 (1969)
5.141 L. A. KAMENTSKY, M. R. MELAMED: Proc. IEEE **57**, 2007 (1969)
5.142 L. A. KAMENTSKY: The Rapid Cell Spectrophotometer; in: *Automated Cell Identification and Cell Sorting*, ed. by G. L. WIED, G. F. BAHR (Academic Press, New York 1970)
5.143 G. C. FINKEL, S. GRAND, M. P. EHRLICH, R. DECOTE: J. Assoc. Adv. Med. Instr. **4**, 1—6 (1970)
5.144 G. C. SALZMAN: Biophys. Soc. Abstr. **17**, 302 (1973)
5.145 R. C. LEIF, in: *Automated Cell Identification and Sorting*, ed. by G. L. WIED, G. F. BAHR (Academic Press, New York 1970) pp. 21—96, 131—159
5.146 R. C. LEIF, R. A. THOMAS: Clin. Chem. **19**, 853 (1973)
5.147 J. L. LADINSKY, H. W. GRUCHOW, M. S. STANLEY, L. INHORN: Am. J. Clin. Path. **54**, 254—465 (1970)
5.148 R. A. THOMAS, B. F. CAMERON, R. C. LEIF: J. Histochem. Cytochem. **22**, 626—641 (1974)
5.149 N. B. GROVER, J. NAAMAN, S. BEN-SASSON, J. DOLJANSKI: Biophys. J. **9**, 1398 (1969)
5.150 D. C. GOLIBERSUCH: Rpt. 72-CRD 334, General Electric Technical Information Series (1972)
5.151 W. DITTRICH, W. GOHDE: Z. Naturforsch. **B-24**, 360 (1969)
5.152 D. M. HOLM, L. S. CRAM: Exp. Cell Res. **80**, 105 (1973)
5.153 X. YATAGANAS, B. D. CLARKSON: J. Histochem. Cytochem. **22**, 651—659 (1974)
5.154 M. KERKER: *The Scattering of Light* (Academic Press, New York 1969)
5.155 C. C. GRAVATT: Appl. Spectrosc. **25**, 509 (1971)
5.156 J. W. GOODMAN: *Introduction to Fourier Optics* (McGraw-Hill, New York 1968)
5.157 K. PRESTON, JR.: *Coherent Optical Computers* (McGraw-Hill, New York 1972)
5.158 L. W. DIGGS, D. STURM, A. BELL: "The Morphology of Blood Cells in Wright Stained Smears of Peripheral Blood and Bone Marrow"; in: *What's New* (Abbott Laboratories 1954)
5.159 M. M. WINTROBE: *Clinical Hematology* (Lea and Febiger, Philadelphia 1956)
5.160 T. A. HARPER: *The Peripheral Blood Film* (Butterworth, London 1968)
5.161 J. B. MIALE: *Laboratory Medicine Hematology* (C. V. Mosby Co., St. Louis 1967)
5.162 US Army: Tech. Manual 8-227-4 (1963)
5.163 K. PRESTON, P. E. NORGREN: US Patent No. 3, 577, 267 (1971)
5.164 J. J. STAUNTON: US Patent No. 3, 705, 048 (1972)
5.165 M. INGRAM, F. M. MINTER: Am. J. Clin. Pathol. **51**, 214 (1969)
5.166 C. W. ROGERS: "Sample Preparation—A Quantitative Evaluation of Different Techniques", Engrg. Found. Conf. on Automatic Cytology, Saxtons River (1972)
5.167 C. W. ROGERS: Am. J. Med. Tech. **39**, 435—442 (1973)
5.168 J. W. BACUS: J. Histochem. Cytochem. **22**, 506 (1974)
5.169 D. A. HUNGERFORD: Observations on the Morphology and Behavior of Normal Human Chromosomes; in: *Mammalian Cytogenetics and Related Problems in Radiobiology*, ed. by C. PAVAN (Pergamon Press, New York 1964);
5.170 T. CASPERSSON, G. LOMAKKA, R. RIGLER, JR.: Acta Histochem. (Jena) **6**, 123 (1965)
5.171 T. CASPERSSON, G. LOMAKKA, A. MILLER: Hereditas **67**, 103 (1971);
T. CASPERSSON, G. LOMAKKA, J. LORZECH: Hereditas **67**, 89 (1971)
5.172 T. CASPERSSON, J. LINDSTEN, G. LOMAKKA, A. MOLLER, L. ZECH: Intern. Rev. Ext. Pathol. **11**, 2 (1973)
5.173 W. R. BREG, D. A. MILLER, P. W. ALLDERDICE, O. J. MILLER: Nature **236**, 76 (1974)
5.174 W. R. BREG, D. A. MILLER, P. W. ALLDERDICE, O. J. MILLER: Am. J. Dis. Child. **123**, 561 (1974)
5.175 A. T. SUMNER, H. J. EVANS: Exp. Cell Res. **81**, 223 (1973)

5.176 C. Lundsteen, J. Philip: J. Med. Genetics **11**, 196—200 (1974)
5.177 G. N. Papanicolaou: *Atlas of Exfoliative Cytology* (Harvard University Press 1954)
5.178 H. J. Davis: Acta Cytol. **6**, 459 (1962)
5.179 J. S. Parry, B. K. Cleary, A. R. Williams, D. M. Evans: Acta Cytol. **15**, 1963 (1971)
5.180 J. H. Tucker, G. A. Gresham: J. Obs. Gyn. Brit. Comm. **78**, 947 (1971)
5.181 G. F. Bahr, P. H. Bartels, M. Bibbo, M. deNicolas, G. L. Wied: Acta Cytol. **17**, 106—112 (1973)
5.182 L. L. Wheeless, M. A. Onderdonk: J. Histochem. Cytochem. **22**, 522—525 (1974)
5.183 L. L. Wheeless, S. F. Patten: Acta Cytol. **17**, 333—391 (1973)
5.184 L. R. Adams: J. Histochem. Cytochem. **22**, 492—494 (1974)
5.185 R. Feulgen, H. Rossenbech: Z. Physiol. Chem. **135**, 203 (1924)
5.186 A. W. Pollister, H. Ris: Cold Spring Harbor Symp. Quant. Biol. **12**, 147 (1947)
5.187 H. S. DiStefano: Chromosoma **3**, 282 (1948)
5.188 A. Deitch: Lab. Invest. **4**, 324 (1955)
5.189 A. Deitch: J. Histochem. Cytochem. **12**, 451—461 (1964)
5.190 A. DeCapoa, D. A. Miller, O. J. Miller: Nature **220**, 264 (1968)
5.191 N. Bohm, E. Sprenger: Histochemie **16**, 100 (1968)
5.192 T. T. Trujillo, M. A. Van Dilla: Acta Cytol. **16**, 26 (1972)
5.193 S. A. Latt: J. Histochem. Cytochem. **22**, 478—491 (1974)
5.194 J. F. Golden, S. S. West: J. Histochem. Cytochem. **22**, 495—505 (1974)
5.195 E. Davidson: Adv. Automated Anal. **1**, 447 (1971)
5.196 R. Kiefer, G. Kiefer, U. Wolf: "Feulgen Hydrolysis Kinetics in Euchromatin and Heterochromatin", Inst. of Path., Univ. of Freiburg (1972)
5.197 J. W. Gray: J. Histochem. Cytochem. **22**, 642—650 (1974)
5.198 F. B. DeCampos, G. W. Moore, G. Schulter: Acta Cytol. **17**, 510—521 (1973)
5.199 J. E. Gill, M. M. Jotz: J. Histochem. Cytochem. **22**, 470—477 (1974)
5.200 L. Ornstein, H. R. Ansley: J. Histochem. Cytochem. **22**, 453 (1974)
5.201 W. Sandritter, G. Kiefer, G. Schluter, W. Moore: Beitr. Path. **151**, 87—96 (1974)
5.202 J. A. Steinkamp, H. A. Chrissman: J. Histochem. Cytochem. **22**, 616—621 (1974)
5.203 V. Bremeskow, P. Kaden, C. Mittermayer: Eur. J. Cancer **6**, 379 (1970)
5.204 R. L. Henry, R. M. Nalbandian, B. M. Nichols, P. L. Wolf, F. R. Camp, Jr., N. F. Conte, in: *Advances in Automated Analysis*, Vol. 1. Technicon Intern. Cong. (Thurman Assoc., Miami 1971) pp. 471, 476
5.205 B. J. Davis, L. Ornstein: J. Histochem. Cytochem. **7**, 297 (1959)
5.206 F. Sweetman, H. R. Ansley, L. Ornstein: Fed. Proc. **31**, 630 (1972)
5.207 H. Gilbert, L. Ornstein: Blood **46**, 279—286 (1975)
5.208 M. Atamer, W. Groner: Adv. Automated Anal. **3**, 33 (1973)
5.209 L. R. Adams, L. A. Kamentsky: Acta Cytol. **15**, 289 (1971)
5.210 M. R. Melamed, L. R. Adams, F. Traganos, L. A. Kamentsky: J. Histochem. Cytochem. **22**, 526—530 (1974)
5.211 J. A. Steinkamp, A. Romero, M. A. Van Dilla: Acta Cytol. **17**, 113 (1973)
5.212 J. J. Sheppard: *Human Color Perception: A Critical Study of the Experimental Foundation* (American Elsevier Publishing Co., New York 1968)
5.213 G. I. Kaufman, D. E. Wasserman, J. F. Nester: J. Assoc. Adv. Med. Instr. **6**, 230—234 (1972)
5.214 P. Davidovits, M. D. Egger: Appl. Opt. **10**, 1615—1619 (1971)
5.215 R. A. Lemons, C. F. Quate: Appl. Phys. Letters **24** (4), 163—165 (1974)
5.216 J. Hlinka, F. K. Sanders: Trans. New York Acad. Sci. **132**, 675—687 (1970)
5.217 R. C. Bostrom, W. G. Holcomb: IEEE Intern. Conf. Rec., Pt. 9, 110 (1963)
5.218 H. P. Mansberg, P. Ohringer: Ann. New York Acad. Sci. **157**, 5—37 (1969)
5.219 G. P. Burn: Med. Biol. Engrg. **9**, 87—95 (1971)
5.220 G. A. Thieme: Acta Physiol. Scand. **67**, 514 (1966)

5.221 A. A. PEARSE, F. W. D. ROST: J. Microsc. **89**, 321 (1969)
5.222 F. W. D. ROST, A. G. E. PEARSE: J. Microsc. **94**, 93—105 (1971)
5.223 J. S. PLOEM, J. A. DESTERKE, J. BONNET, H. WASMUND: J. Histochem. Cytochem. **22**, 668—677 (1974)
5.224 M. A. KUJOORY, B. H. MAYALL, M. L. MENDELSOHN: IEEE Trans. BE-**20**, 126 (1973)
5.225 K. A. PATON: Ann. Hum. Genet. (London) **35**, 67—77 (1971)
5.226 D. K. GREEN, J. CAMERON: Metaphase Cell Finding by Machine. IEEE (in press)
5.227 E. T. JOHNSON, L. J. GOFORTH: J. Histochem. Cytochem. **22**, 536—545 (1974)
5.228 W. S. PIKE: Ann. New York Acad. Sci. **97**, 395—407 (1962)
5.229 R. H. SELZER: NASA Tech. Rpt. No. 32-1028 (1966)
5.230 W. BENNETT: "Spectra of Quantized Signals", Bell Sys. Tech. J. **27**, 446—472 (1948)
5.231 B. WIDROW: IRE Trans. CT-**3** (4), 266—276 (1956)
5.232 J. D. BRUCE: MIT/RLE Tech. Rpt. No. 429 (1965)
5.233 T. S. HUANG, W. F. SCHREIBER, O. J. TRETIAK: Proc. IEEE **59**, 1586—1609 (1971)
5.234 J. V. GAVEN, JR., J. TAVITIAN, A. HARAKEDIAN: Phot. Sci. Engrg. **14**, 16—20 (1970)
5.235 G. L. WIED, M. BIBBO, G. F. BAHR, P. H. BARTELS: Acta Cytol. **14**, 136—141 (1970)
5.236 P. E. NORGEN, M. INGRAM, K. PRESTON, JR., in: *Digest 7th Intern. Conf. Med. Biol. Engrg.*, Stockholm (1967) p. 122
5.237 R. A. FISHER: Ann. Eugenics **7**, 179—188 (1936)
5.238 C. R. RAO: *Advanced Statistical Methods in Biometric Research* (Wiley & Sons, New York 1952)
5.239 L. ORNSTEIN: "Pattern Recognition, Morphology, and the Generation of Hypotheses", Am. Assoc. Adv. Sci. Conf. (1960)
5.240 L. UHR, C. VOSSLER: Proc. Western Joint Comp. Conf. (1961), p. 555
5.241 P. M. LEWIS: General Electric Rpt. 61-RL-2796 E (1961)
5.242 D. H. TYCKO, S. ANBALAGAN, H. C. LIU: Tech. Rpt. No. 32, Dept. of Comp. Sci., State Univ. of New York at Stony Brook (1974)
5.243 P. H. BARTELS, G. F. BAHR, G. L. WIED: Acta Cytol. **13**, 210—217 (1969)
5.244 P. H. BARTELS, M. BIBBO, G. F. BAHR, J. TAYLOR, JR., G. L. WIED: Acta Cytol. **17** 449—453 (1973)
5.245 P. H. BARTELS, M. BIBBO, J. TAYLOR, JR., G. L. WIED: Acta Cytol. **18**, 165—169 (1974)
5.246 P. H. BARTELS, G. F. BAHR, W. S. JETER, G. B. OLSON, J. TAYLOR, JR., G. L. WIED: J. Cytochem. Histochem. **22**, 69—79 (1974)
5.247 R. A. SCHAPHORST: US Dept. of Commerce Clearinghouse Document No. AD 322669 (1961)
5.248 G. C. CHENG: J. Histochem. Cytochem. **22**, 517—520 (1974)
5.249 R. A. KIRSCH: NBS Rpt. No. 10173 (1969)
5.250 A. A. GLAGOLEFF: Trans. Inst. Econ. Min. **59**, 1 (1933)
5.251 H. W. CHALKEY: J. Nat. Cancer Inst. **4**, 47 (1943)
5.252 E. R. WEIBEL, G. S. KISTLER, W. F. SCHERLE: J. Cell Biol. **30**, 23 (1966)
5.253 R. H. FREERE, E. R. WEIBEL: J. Roy. Microscop. Soc. **87**, 25—34 (1967)
5.254 L. B. SCOTT, K. PRESTON, JR.: US Patent No. 3, 408, 485 (1968)
5.255 W. SPRICK: US Patent No. 2, 738, 499 (1956)
5.256 W. SPRICK, K. GANZHORN: IEE Rpt. 3154 (1959)
5.257 J. W. BROULLETTE, JR: US Patent No. 2, 986, 643 (1961)
5.258 J. W. BROULLETTE, JR: Proc. 4th Nat. Conv. Mil. Electron. (1961) p. 1790
5.259 H. FREEMAN: IEEE Trans. EC-**10**, 260—268 (1961)
5.260 H. FREEMAN: Proc. Nat. Electron. Conf. **17**, 421 (1961)
5.261 M. L. MENDELSOHN, B. H. MAYALL, B. H. PERRY, in: *Advances in Medical Physics*, ed. by J. S. LAUGHLIN, R. W. WEBSTER (1971) p. 327
5.262 J. M. S. PREWITT: *Adv. Comput.* **12**, 285—414 (1972)

5.263 R.H.SELZER: Radio. Clinics of N. Am. **7**, 461—471 (1969)
5.264 C.Y.HOU, S.A.VERESS, J.W.PROTHERO: IEEE Trans. BME-**17**, 106—108 (1970)
5.265 I.T.YOUNG, L.I.PASKOWITZ: "Localization of Cellular Structures", NIH Grant GM 14940-07 (1974)
5.266 Y.P.CHIEN, K.S.FU, Comput. Graph. Image Proc. **3**, 125—140 (1974)
5.267 M.A.FISCHLER, R.A.ELSCHLAGER: IEEE Trans. C-**22**, 67—92 (1973)
5.268 K.S.FU, Y.P.CHIEN: Rpt. TR-EE-73-6, School of Elec. Engrg., Purdue University (1973)
5.269 J.SKLANSKY, D.BALLARD, H.WECHSLER: "Computer Aided Analysis of Radiographs", US Public Health Service Grant GM-17632, University of California, Irvine (1973)
5.270 U.MONTANARI: Comm. Assoc. Comp. Mach. **14**, 335—345 (1971)
5.271 J.C.ALEXANDER, A.I.THALER: J. Assoc. Comp. Mach. **18**, 105—112 (1971)
5.272 M.H.HUECKEL: J. Assoc. Comp. Mach. **18**, 113—125 (1971)
5.273 P.SARAGA, P.R.WAVISH: Electron. Lett. **7**, 736—738 (1971)
5.274 A.ROSENFELD, M.THURSTON: IEEE Trans. C-**20**, 562—569 (1971)
5.275 M.K.HU: Proc. IRE **49**, 1428 (1961)
5.276 E.R.SAPP: US Dept. of Commerce Clearinghouse Document AD 289676 (1962)
5.277 J.W.BUTLER, M.K.BUTLER, A.STROUD, in: *Data Acquisition and Processing in Biology and Medicine*, Vol. 3, ed. by K.ENSLEIN (Pergamon Press, New York 1966) pp. 261—275
5.278 E.L.HALL, W.D.CRAWFORD, JR., K.PRESTON, JR., F.E.ROBERTS: Proc. 1st Intern. Joint Conf. on Pattern Recog. (1973) pp. 77—87
5.279 P.W.NEURATH, K.ENSLEIN: Rpt. NYO-3748-12, US Atomic Energy Commission (1968)
5.280 G.GALLUS, N.MONTANARO, G.A.MACCACARO: Comput. Biomed. Res. **2**, 187 (1968)
5.281 D.RUTOVITZ: Machine Intelligence **5**, 435 (1970)
5.282 G.REGOLIOSI: Mong. Centro G. Zambon. **5**, 136 (1970)
5.283 W.DOYLE: Rpt. No. 54-12, Lincoln Laboratory (1959)
5.284 E.C.GREANIAS, C.J.HOPPEL, M.KLOOMOK, J.S.OSBORNE: IBM J. Res. Develop. **1**, 8—18 (1957)
5.285 L.A.KAMENTSKY, C.N.LIU: IBM J. Res. Develop. **7**, 2—13 (1963)
5.286 G.GALLUS: Appl. Biomed. Calcolo Elettronico **2**, 95 (1968)
5.287 G.GALLUS, G.REGOLIOSI: J. Histochem. Cytochem. **22**, 546—553 (1974)
5.288 G.KIEFER, W.SANDRITTER: "Objectivization of the Microscopic Image of the Cell Nucleus", Inst. of Path., Univ. of Freiburg (1970)
5.289 T.BOURK, O.J.TRETIAK: "Computer Measurement of Neutrophil Nucleus Shape", Engrg. Found. Conf. on Automatic Cytology (Asilomar 1973)
5.290 G.KIEFER, R.KIEFER, G.W.MOORE, R.SALM, W.SANDRITTER: J. Histochem. Cytochem. **22**, 569—576 (1974)
5.291 W.M.COWAN, D.F.WANN: J. Microscopy **99**, 331—349 (1973)
5.292 C.ARCELLI, S.LEVIALDI: Acta Cybernetica (Hungary) **5**, 65—68 (1973)
5.293 R.A.KIRSCH, L.C.RAY, L.CAHN, G.H.URBAN: Proc. Eastern Joint Comp. Conf. (1957) p. 221
5.294 S.H.UNGER: Proc. IRE **46**, 1744—1750 (1958)
5.295 M.J.E.GOLAY: US Patent No. 3, 214, 574 (1965)
5.296 K.PRESTON, JR.: Proc. Western Joint Comp. Conf. (1961) p. 173
5.297 K.PRESTON, JR., in: *Computers and Automata* (Polytechnic Press, 1972) pp. 609—623
5.298 H.A.BLUM: "A Transformation for Extracting New Descriptors of Shape", Symp. on Models for the Percept. of Speech and Visual Form, Boston (1964)

5.299 U. MONTANARI: J. Assoc. Comp. Mach. **15**, 600—624 (1968)
5.300 C. J. HILDITCH, in: *Automatic Interpretation and Classification of Images*, ed. by A. GRASSELLI (Academic Press, New York 1969) p. 363
5.301 G. LEVIALDI, U. MONTANARI: Inf. Control. **17**, 62 (1970)
5.302 S. LEVIALDI: Comm. Assoc. Comp. Mach. **15**, 7—10 (1972)
5.303 E. S. DEUTSCH, in: *Computer Processing in Communications* (Polytechnic Press, 1969) pp. 221—234
5.304 A. ROSENFELD: J. Assoc. Comp. Mach. **17**, 146—160 (1970)
5.305 O. J. TRETIAK: Ann. New York Acad. Sci. **157**, 365—375 (1969)
5.306 S. N. JAYARAMAMURTHY: "Computer Methods for Analysis and Synthesis of Visual Texture", Dept. of Computer Science, Univ. of Illinois (1973)
5.307 B. S. LIPKIN, L. E. LIPKIN: J. Histochem. Cytochem. **22**, 583—593 (1974)
5.308 E. F. VASTOLA, R. A. HERTZBERG, P. W. NEURATH, J. F. BRENNER: Acta Cytol. **18**, 231 (1974)
5.309 J. CARVALKO, JR.: IEEE Trans. C-**21**, 1430—1433 (1972)
5.310 R. JAGOE, K. PATON: IEEE Trans. C-**25**, 95—97 (1976)
5.311 E. L. HALL, W. D. CRAWFORD, JR., K. PRESTON, JR., F. E. ROBERTS, in: Proc. 1st Intern. Joint Conf. Pattern Recognition, 1973, pp. 77—87
5.312 K. PRESTON, JR.: Biomed. Engrg. **7**, 226 (1972)
5.313 W. H. AUGHEY, F. J. BAUM: J. Opt. Soc. Am. **44**, 833 (1954)
5.314 J. B. BATEMAN: J. Colloid Interface Sci. **27**, 458 (1968)
5.315 A. L. KOCH, E. EHRENFELD: Biochem. Biophys. Acta **165**, 262 (1968)
5.316 R. M. BERKMAN, P. J. WYATT: Appl. Microbiol. **20**, 510 (1970)
5.317 J. E. BOWIE: Proc. 23rd ACEMB (1970), p. 256
5.318 O. B. LURIE, R. E. BYKOV, E. P. POPECHITELEV, V. I. ULYANOV: "Automatic Recognition of Leukocyte Cells in Blood Preparations", Proc. 8th Conf. on Engrg. in Medicine and Biology (1969)
5.319 D. J. RAUCH: "Digital Matched Filter Detection and Discrimination of a Class of Lymphocytes", 5th Hawaii Conf. on Sys. Sci., Univ. of Hawaii (1972)
5.320 J. E. WARD, III., F. P. CARLSON, J. D. HEYWOOD: IEEE Trans. BME-**21**, 12—20 (1974)
5.321 R. E. KOPP, J. LISA, J. MENDELSOHN, B. PERMICK, H. STONE, R. WOHLERS: J. Histochem. Cytochem. **22**, 598—604 (1974)
5.322 A. BRUNSTING: "Light Scattering from Mammalian Cells", PhD dissertation, University of New Mexico (1972)
5.323 A. BRUNSTING: J. Histochem. Cytochem. **22**, 607—615 (1974)
5.324 P. F. MULLANEY, M. A. VAN DILLA, J. R. COULTER, P. N. DEAN: Rev. Sci. Instr. **40**, 1029—1032 (1969)
5.325 P. F. MULLANEY, P. N. DEAN: Appl. Opt. **8**, 2361 (1969)
5.326 P. F. MULLANEY, J. A. STEINKAMP, H. A. CRISSMAN, L. S. CRAM, D. M. HOLM: Laser Flow Microphotometer for Rapid Analysis and Sorting of Individual Mammalian Cells; in: *Laser Applications in Medicine and Biology*, ed. by M. L. WOLBARSHT (Plenum Publishing Corp., New York 1974)
5.327 R. J. FIEL: Exp. Cell Res. **59**, 413 (1970)
5.328 R. J. FIEL, H. M. SCHEINTAUB: Arch. Biochem. Biophys. **158**, 164 (1973)
5.329 R. J. FIEL, E. H. MARK, B. R. MUNSON: Arch. Biophys. **141**, 547 (1970)
5.330 R. J. FIEL, B. R. MUNSON: Exp. Cell Res. **59**, 421 (1970)
5.331 F. D. BRYANT, P. LATIMER, B. A. SEIBER: Arch. Biochem. Biophys. **135**, 109 (1969)
5.332 P. J. WYATT: J. Colloid Interface Sci. **39**, 479 (1972)
5.333 P. J. WYATT, in: *Methods in Microbiology*, ed. by J. R. MORRIS, D. W. RIBBONS (Academic Press, New York 1973) p. 183
5.334 H. HOTELLING: J. Educ. Psychol. **24**, 417—441 and 498—520 (1933)

5.335 W.K. HIGHLEYMAN: Bell Sys. Tech. J. **41**, 723 (1962)
5.336 W. KARUSH: US Dept. of Commerce Clearinghouse Document No. AD 410789 (1963)
5.337 G.S. SEBESTYEN: *Decision Making Processes in Pattern Recognition* (MacMillan, New York 1962)
5.338 T.T. TANIMOTO, R.G. LOOMIS: "A Taxonomy Program for the IBM 704", IBM Program Library, New York (1960)
5.339 T.T. TANIMOTO: Ann. New York Acad. Sci. **23**, 576—578 (1961)
5.340 L. ORNSTEIN: J. Mt. Sinai Hosp. **32**, 837 (1965)
5.341 O.L. MANGASARIAN: Oper. Res. **13**, 44 (1956)
5.342 G.D. NELSON, D.M. LEVY: IEEE Trans. SSC-**4**, 145—151 (1968)
5.343 W.C. MARTIN: IEEE Trans. SMC-**1**, 87—88 (1971)
5.344 P.A. KOLERS, M. EDEN: *Recognizing Patterns* (MIT Press, Cambridge 1968)
5.345 R.S. LEDLEY: Science **146**, 216—223 (1964)
5.346 L.E. LIPKIN, W.C. WATT, R.A. KIRSCH: Ann. New York Acad. Sci. **129**, 984—1012 (1966)
5.347 B. SHAPIRO, P. LEMKIN, L. LIPKIN: J. Histochem. Cytochem. **22**, 741 (1974)
5.348 R.L. MATTSON: US Dept. of Commerce Clearinghouse Document No. AD 246244 (1960)
5.349 B. BULLOCK: Res. Rpt. 444, Hughes Aircraft Corp. (1971)
5.350 P.H. BARTELS, G.F. BAHR, D.W. CALHOUN, G.L. WIED: Acta Cytol. **14**, 313—324 (1970);
P.H. BARTELS, G.F. BAHR, J.C. BELLAMY, M. BIBBO, D.L. RICHARDS, G.L. WIED: Acta Cytol. **14**, 486—494 (1970)
5.351 P.K. BHATTACHARYA, P.H. BARTELS, G.F. BAHR, G.L. WIED: Acta Cytol. **15**, 533—544 (1971)
5.352 P.K. BHATTACHARYA, P.H. BARTELS, J. TAYLOR, JR., G.L. WIED: Acta Cytol. **17**, 538—552 (1973)
5.353 D. RUTOVITZ, in: *Pictorial Pattern Recognition*, ed. by G.C. CHENG, R.S. LEDLEY, D.C. POLLOCK, A. ROSENFELD (Thompson Book Co., 1968) pp. 106—133
5.354 E.A. PATRICK, F.P. FISCHER: IEEE Trans. C-**18**, 987—991 (1969)
5.355 J.W. SAMMON, JR.: IEEE Trans. C-**18**, 401—409 (1969)
5.356 J.W. SAMMON, JR.: IEEE Trans. C-**19**, 594—616 (1970)
5.357 G.H. BALL, D.J. HALL: Behav. Sci. **12**, 153—155 (1957)
G.H. BALL, D.J. HALL: "ISODATA, an Iterative Method of Clustering Multivariate Data", Stanford Research Institute (1968)
5.358 T. MARRILL, A.K. HARTLEY, T.G. EVANS, B.H. BLOOM, D.M.R. PARK, T.P. HART, D.L. DARLEY: Rpt. No. AFCRL-69-0323, US Air Force (1969)
5.359 D.J. HALL, D. WOLF, G.H. BALL: Rpt. No. RADC-TR-68-572, US Air Force (1969)
5.360 L. HODES: Comm. Assoc. Comp. Mach. **13**, 279—283 (1970)
5.361 D.A. JOHNSTON: "An Interactive Classification System", M.D. Anderson Hospital and Tumor Institute, University of Texas at Houston
D.A. JOHNSTON: "A General Purpose Image Digitizing System", M.D. Anderson Hospital and Tumor Institute, University of Texas at Houston
5.362 J.M.S. PREWITT, M.L. MENDELSOHN: Ann. New York Acad. Sci. **128**, 1035 (1966)
5.363 J.M.S. PREWITT, M.L. MENDELSOHN: A General Approach to Image Analysis by Parameter Extraction, in: *The Use of Computers in Radiology*, ed. by G.S. LODWICK (University of Missouri, 1966)
5.364 K. PRESTON, JR.: "Automatic Computer Identification of Human Leukocytes", Proc. 2nd Intern. Tutorial on Quant. Cytochem., University of Chicago (1968)
5.365 C.J. HILDITCH, in: *Human Population Genetics*, ed. by P.A. JACOBS (University of Edinburgh Press 1970) pp. 198—235

5.366 J.W. SNIVELY, JR., E.B. BUTT: Tech. Rpt. 68-67, University of Maryland (1968)
5.367 E.G. JOHNSTON, in: *Picture Processing and Psychopictorics*, ed. by B.S. LIPKIN, A. ROSENFELD (Academic Press, New York 1970) pp. 427—512
5.368 S.P. STONE: Rpt. TID-45000 UC, University of California at Livermore (1967)
5.369 K. PRESTON, JR., in: *Advances in Optical and Electron Microscopy*, ed. by V. COSSLETT, R. BARER (Academic Press, New York 1973) pp. 43—93
5.370 J.A. FELDMAN, J.R. LOW, D.C. SWINEHART, R.H. TAYLOR: Proc. FJCC **41**, 1193 (1972)
5.371 M.N. MILLER: "Design and Clinical Results of Hematrak—An Automated Differential Counter", Geometric Data Corp. (1973)
5.372 P.W. NEURATH, G. GALLUS, E. VASTOLA: Phys. Med. Biol. **17**, 716 (1972)
5.373 P.W. NEURATH: "White Cell Differential", Engrg. Found. Conf. on Automatic Cytology (Asilomar 1973)
5.374 P.W. NEURATH, J.F. BRENNER, W.D. SELLES: "Computer Identification of White Blood Cells", presentation at Intern. Computing Symp., Davos (1973)
5.375 J.F. BRENNER, E.S. GELSEMA, T.F. NECHOLES, P.W. NEURATH, W.D. SELLES, E. VASTOLA: J. Histochem. Cytochem. **22**, 797 (1974)
5.376 P.H. BARTELS, G.F. BAHR, J. GRIEP, H. RAPPAPORT, G.L. WIED: Acta Cytol. **13**, 557—568 (1969)
5.377 T.L. JARKNOWSKI, G.F. BAHR, G.L. WIED, J.C. BELLAMY, P.H. BARTELS: Acta Cytol. **15**, 147—153 (1971)
5.378 G.B. OLSON, G.L. WIED, P.H. BARTELS: Acta Cytol. **17**, 454—461 (1973)
5.379 D.K. KWAN, A. NORMAN: Acta Cytol. **18** (3), 189—191 (1974)
5.380 J. ZAJICEK, P.H. BARTELS, G.F. BAHR, M. BIBBO, G.L. WIED: Acta Cytol. **16**, 284—296 (1972)
5.381 B.S. LIPKIN: "Effective Partial Representations of White Blood Cell Images", Engrg. Found. Conf. on Automatic Cytology, Saxtons River (1972)
5.382 M.L. MENDELSOHN, B.H. MAYALL, in: *Human Chromosome Methodology*, ed. by J. YUNIS (Academic Press, New York 1974) pp. 311—346
5.383 P.W. NEURATH, B. KESS, D.A. LOW: Comput. Biol. Med. **2**, 181 (1972)
5.384 R.S. LEDLEY, H.A. LUBS, F.H. RUDDLE: Comp. Biol. Med. **2**, 107 (1972)
5.385 R.K. AGGARWAL, K.S. FU: Rpt. TR-EE 73-12, School of Elec. Engrg., Purdue University (1973)
5.386 R.K. AGGARWAL, K.S. FU: Rpt. TR-EE 74-35, School of Elec. Engrg., Purdue University (1974)
5.387 R.K. AGGARWAL, K.S. FU: J. Histochem. Cytochem. **22**, 561—568 (1974)
5.388 S.P. STONE, J.L. LITTLEPAGE, B.R. CLEGG: Proc. Seminar Soc. Photo Optical Eng. (1967) pp. 157—171
5.389 A. KLINGER, A. KOCHMAN, N. ALEXANDRIDIS: IEEE Trans. C-**20**, 1014 (1971)
5.390 C.J. HILDITCH, D. RUTOVITZ: Ann. New York Acad. Sci. **157**, 339 (1969)
5.391 G. GALLUS, A. MORABITO, G. REGOLIOSI: Proc. IFAC Symp., Bruxelles (1971) p. 111
5.392 C.C. LI, S. SHIPKOVITZ, D.W. TSAO, F.K. SUN: Proc. of 1st Intern. Symp. on Computers and Chinese Input/Output Systems (Academia Sinica, Nankang 1973) pp. 347—371
5.393 N. WALD, S.R. FATORA: "Biomedical Significance of Human Chromosome Aberrations: A Justification of Automated Image Processing", Proc. IEEE Conf. Sys. Man, and Cyber., Dallas (1974)
5.394 W.J. HORVATH, W.E. TOLLES, R.C. BOSTROM: Trans. 1st Intern. Cancer Cytol. Congr. (1956) pp. 371—397
5.395 R.C. BOSTRUM, H.S. SAWYER, W.E. TOLLES: "The Cytoanalyzer—An Automatic Pre-Screening Instrument for Cancer Detection", Proc. 2nd Intern. Conf. on Med. Elec., Paris (1959)

5.396 I. M. P. DAWSON, C. P. HEANLEY, A. C. HEBER-PEREY, J. K. TYLKO: J. Clin. Path. **20**, 724—730 (1967)
5.397 O. A. N. HUSAIN, M. W. HENDERSON, in: *Cytology Automation* (E. and S. Livingston, Ltd., Edinburgh and London 1970) pp. 214—227
5.398 O. A. N. HUSAIN, E. F. D. MACKENZIE, R. W. B. ALLEN: "A Critical Analysis of the Size and Density of Nuclei in Cervical Smears" (submitted)
5.399 S. B. GUSBERG, C. J. COHEN, K. YANNOPOULIS, S. GRAND, P. KARDON, G. C. FINKEL: Obstet. Gynecology **34**, 284—287 (1969)
5.400 G. F. BAHR, G. L. WIED: Multiphasic Health **1**, 67—82 (1970)
5.401 M. BIBBO, P. H. BARTELS, G. F. BAHR, J. TAYLOR, JR., G. L. WIED: Acta Cytol. **17**, 340—350 (1973)
5.402 P. H. BARTELS, G. L. WIED: J. Histochem. Cytochem. **22**, 660—662 (1974)
5.403 Geometric Data Corp.: "How to Buy a Differential Counter" (1974)
5.404 Chicago Hospital Council: "Clinical Laboratory Study" (1965)
5.405 J. W. BACUS: Am. J. Clin. Path. **59**, 223 (1973)
5.406 M. J. FULWYLER, J. D. PERRINGS, L. S. CRAM: Rev. Sci. Instr. **44**, 204 (1973)
5.407 A. BRUNSTING, P. F. MULLANEY: Appl. Opt. **11**, 675 (1972)
5.408 D. D. COOKE, M. KERKER: J. Colloid Interface Sci. **42**, 150 (1973)
5.409 D. T. PHILLIPS, P. J. WYATT, R. M. BERKMAN: J. Colloid Interface Sci. **34**, 159 (1970)
5.410 P. H. BARTELS, G. L. WIED, in: Proc. Intern. Opt. Comp. Conf. (1975) pp. 38—43

Table 5.3. (Referred to on p. 261). Investigators on chromosome aberration analysis (World Health Organization Rpt. RHL/WP/72.5, 1972)

Laboratory (see Notes)	*Within laboratory overall standard deviations*			
	Arm length (μm)	Centromeric index (length)	Area	Centromeric index (area)
1.	0.27	0.023	0.31	0.023
2.	0.29	0.032	—	—
3.	0.27	0.032	0.33	0.029
4.	0.30	0.047	0.35	0.046
5.	—	—	0.42	0.045
6.	0.45	0.063	—	—
Between laboratories comparative standard deviations				
3.–1.	0.14	0.024	0.22	0.022
3.–2.	0.15	0.028	—	—
3.–4.	0.18	0.039	0.28	0.037
3.–5.	—	—	0.31	0.026
3.–6.	0.30	0.052	—	—
2.–1.	0.13	0.025	—	—
2.–4.	0.19	0.039	—	—
4.–1.	0.19	0.034	0.25	0.032
4.–5.	—	—	0.25	0.033
5.–1.	—	—	0.30	0.027

Notes: 1. Jet Propulsion Laboratory (K. R. CASTLEMAN)
2. University of Colorado (H. A. LUBS)
3. Tufts New England Medical Center (P. W. NEURATH)
4. National Biomedical Research Institute (R. S. LEDLEY)
5. Medical Research Council (D. RUTOVITZ)
6. National Research Council (T. KASVAND)

6. Picture Analysis in Character Recognition

J. R. ULLMANN

With 13 Figures

The automatic recognition of printed and written characters is a highly evolved art that has a large commercial literature. The present chapter is intended to give an impression of the state of the art of digital picture analysis in commercial character recognition systems, and is not much concerned with the research literature that records the use of character recognition as a demonstration of basic techniques of pattern recognition.

After a very brief discussion of scanners, this chapter considers reasons and techniques for thresholding so that the many-level grey scale is changed to a two-level grey scale. It is usual to apply denoising or smoothing techniques before segmenting a line of text into individual characters and standardizing or normalizing characters for position and size. The simplest technique of recognition is template matching, which is commonly implemented by hardware techniques of digital correlation. For coping with variations in shapes of characters, Boolean techniques are generally superior to template matching. Because of its versatility, microprogramming provides an attractive method of implementing Boolean techniques. Distortion or variation in shape is a particularly acute problem in the recognition of handprinted characters, and because connectivity remains unchanged by distortion, topological properties may be useful for distinguishing between characters.

Although character recognition may appear to be one of the most elementary application areas in the field of pattern recognition, it still conspicuously lacks commercially useful theory, and practical successes are limited.

6.1 Introduction

Character recognition has a rich literature of patents that records a great deal of ingenuity and practical experience accumulated over at least forty years. Among all the ideas and techniques that have been developed for use in character recognition, there may perhaps be some that can be adapted for use in other applications of digital picture analysis.

Character recognition remains a challenging and tantalizing field because the big research and development effort that has gone into it has not solved all its commercially urgent and intellectually interesting problems. For example, there is still a wide gap between human and machine error rates in the recognition of isolated letters and numerals that have been freely written without conformity to rules of constraint. This gap is particularly provocative because it still exists after a very big effort has been made to close it, and also because it presents a relatively simple problem in visual pattern recognition.

As a method of entering data into an automatic data processing system, optical character recognition has to compete with other methods such as keyboard, keypunch, and optical mark reading [6.1, 49]. Detailed system design factors determine whether, in a particular data processing application, it is best to use optical character recognition, or a combined ("multimedia") keyboard-to-disc and optical character recognition system, or some other method of data entry. The following paragraphs mention just a few applications where optical character recognition has actually been used.

There may be at least one million names and addresses in the list of customers, subscribers, or clients of a mail-order company, publisher of a periodical, or insurance company. To keep one million addresses up to date it may be necessary to record about 25000 changes of address per week, and optical character recognition has been used for this purpose [6.28].

When an item is sold in a large department store, the salesman may record data that identify the item sold, the salesman, and the price. Character recognition has been used for reading these data into a computer system that determines, for instance, when the stocks of a particular item should be replenished from the warehouse, or whether sales personnel should be transferred to other parts of the store where more sales are occurring at a particular time. The data can also be used for accounting and auditing and other purposes listed in [Ref. 6.28, pp. 21–23].

Credit card companies use character recognition to enter records of credit card sales transactions into a computer system which, for instance, debits the purchaser's account. Character recognition has also been used for reading ticket and coupon numbers on passenger flight tickets [6.28].

In England in about 1971 the Eastern Electricity Board had more than 2200000 electricity consumers, and each day there were about 36000 meter readings [6.21]. Character recognition was used to read the consumer reference number on each meter-reading slip, and automatic optical mark reading was used to read the actual meter reading from the slip into a computer system that printed invoices. When an invoice was paid, the payment advice document was read by character recognition so that the computer system could automatically record payment in the consumer file. Optical character recognition has also been used by insurance companies for reading reference numbers on invoice payment advice documents.

One method of entering text into an automatic type-setting system is to type the text using a special high-quality typewriter, and to read this typed text by means of character recognition. An important advantage of this method is that it allows proof-reading corrections to be made on the typescript before the type is finally set [6.28]. Corrections are more costly after the type has been set. Furthermore, typists are more readily available, and generally have a greater throughput, than keypunch operators.

In Post Offices, availability of labor for manual sorting may fall short of the demand that arises from the ever-increasing volume of mail [6.53]. Postal code character recognition systems have been developed under the auspices of, for example, the United States and the Japanese Post Offices [6.53, 68, 103].

In a commercial character recognition system, the total cost of automatic document transport, optical scanner, format control, output buffering, computer control of manual correction of reject errors, and so on, may be higher than the cost of the part of the system that does the automatic picture analysis and recognition [6.115]. In the following pages we will be concerned only with some aspects of character recognition that may perhaps interest a reader of a book on digital picture analysis. Useful chapters on topics such as paper and printing for character recognition, font design, and document transport are included in a handbook published by the British Computer Society [6.21].

6.2 Hardware for Character Recognition

6.2.1 Scanners

Almost every known technique of opto-electronic transduction has been considered for use in character recognition. One technique is to focus an image of a character onto a two-dimensional array of photodetectors that constitute a retina (e.g. [6.14, 113, 135]). The document moves in the horizontal direction that is parallel to a line of characters, and this motion eventually brings a character into correct horizontal alignment with the retina. To allow for vertical misalignment the retina is typically at least three times the height of the tallest character to be recognized. A possible advantage of a two-dimensional retina is that analogue video information is simultaneously available from all the photodetectors. A disadvantage is the cost of employing a separate video amplifier for each photodetector.

A less costly technique is to use just one single vertical column of photodetectors, possibly solid-state photodiodes (e.g. [6.102, 151]). The processing speed depends on the document speed and this is the same for a single column of cells as it is for a two-dimensional retina [6.115]. References [6.164, 173] disclose techniques for compensating for differences in the sensitivity of cells within a column.

Mechanical scanners are devices in which an aperture is, in effect, mechanically scanned over an image of a character, and the light transmitted through the aperture is measured photoelectrically (e.g. [6.7, 130, 132, 172]). Mechanical scanners are reliable and fairly cheap, but noisy and relatively slow. Whereas a mechanical scanner moves an aperture over an image, the electronic image dissectors described in [6.51, 54] electronically shift an image over an array of detector devices. Each such device measures the intensity over a small area of the image, and this technique has in common with a retina the possible advantage of providing parallel analogue video data from a two-dimensional array of small areas of a pattern. These image dissectors [6.51, 54] have the further advantage that an image can be electronically deflected over the detector array at high speed without the mechanical (possibly mirror) movement that a retina technique requires.

When photocell arrays, mechanical or electronic image dissectors, or vidicons are employed, a character is flooded with light while being read. In a further class of techniques, only a single small area of the document is illuminated at any instant, and a measure of the total reflected light is taken to be a measure of the brightness of a small area of the document. For instance, in a cathode-ray tube (CRT) flying spot scanner (e.g. [6.120, 158]), a bright spot on the face of a CRT is imaged onto the document and the total light diffusely reflected from the document is measured by one or more photomultipliers. This technique generally has finer resolution and higher scanning speed than a vidicon [6.21]. However, the CRT provides only a weak source of light, and extraneous light must be excluded by enclosing the scanner in a light-tight box.

When the flying spot originates not from a CRT but from a laser, a light-tight box may be unnecessary because the spot is very bright. In a laser scanner (e.g. [6.21, 33]), a rotating mirror sweeps the bright spot repeatedly along a single line, and the motion of the document ensures that two-dimensional video information is obtained from a photomultiplier that measures the total light reflected from the document. Instead of using a CRT or a laser, the IBM 3886 [6.81] uses a light-emitting diode array scanner (cf. [6.168]) in which successive diodes in turn illuminate small areas of a document with infrared light. Compared with laser and light-emitting diode array scanners, CRT flying spot scanners have the advantage that the entire raster can be electronically deflected over a wide area, and this may be helpful if the position of a character on a document is not even roughly known in advance, for instance in reading addresses on envelopes in the mail.

Whatever the type of scanner used, the output typically consists of a discrete collection of grey-scale values. Each such value is roughly the average grey-scale value of a small area of the document, and this small area is typically 130 μm in diameter. If the small areas were smaller, irrelevant noisy detail would be detected, and if they were larger, significant detail might be lost. The small areas usually form a rectangular array, and the distance between the centers of horizontally or vertically neighboring areas is also typically 130 μm. We will sometimes refer to the grey-scale value derived from a particular small area as a *pattern element*, whether or not this value has been binarized.

6.2.2 The Use of Special Purpose Hardware in Character Recognition

Character recognition systems can be classified into three main categories according to how much special purpose hardware they use. By far the largest number of commercial character recognition systems belongs to the category in which the entire recognition process is carried out in special purpose hardware. Judging from recent patents, the trend is for this hardware to become more purely digital. A particularly important development is the use of microprogramming, and we will return to this later.

There is a further category in which recognition is achieved by the cooperative effort of a computer and special purpose hardware. Special purpose hardware may be used, for instance, to produce a thinned skeleton of a character or to measure the perimeter of a character, while the higher-level decision making is done in a general purpose computer [6.79]. In a Scan-Data system [6.127] recognition is normally done in special purpose hardware, and as usual a computer is employed for such purposes as format control and editing of reject errors. However, if the reject rate rises, perhaps due to faulty printing, the computer stores data derived from specimens of troublesome characters, and subsequently uses this to assist the hardware in recognition. Here the computer provides a line of defense against high reject rates.

There is a third category of character recognition systems in which, except for scanning and binarizing, the entire process is carried out in a general purpose computer, e.g. [6.31, 75]. In some applications the economy arising from the low cost of minicomputers may not be outweighed by the relatively low speed of recognition. Furthermore, when the speed of recognition is low it may be appropriate to use relatively simple and cheap document transport.

The categorization of character recognition systems according to their use of hardware versus software will not be pursued in the following pages. Instead we will consider some of the main ideas of character recognition, regardless of whether they lead us into hardware or software.

6.3 Binarization

6.3.1 The Requirement for Binarization

Binarization is the process of deeming any given grey-scale value black or white according to whether it lies on the blacker or whiter side of a threshold that is known as the binarizing threshold. Instead of calling the two binary levels *black* and *white* we sometimes call them 1 and 0.

An obvious advantage of binarization is that it reduces an analogue or multi-level grey scale to a binary scale, and it is generally cheaper to process binary pictorial data. Economy is not the only reason why binarization is almost always employed in commercial character recognition equipment. For instance, BEISER et al. [6.14] have patented a recognition system in which both binary and analogue data are used, and binarization substantially increases the cost of the system. A tentative explanation of the other important practical reason for binarization is as follows.

A character is essentially a distribution of black on white, or of one color on another, or of one texture on another. In other words, a character is essentially binary, and any departure from a binary definition of a character is unnecessary for the purposes of communication between writer and reader. Ideally only two levels, which we can call *black* and *white*, should be necessary for the recognition of a character. In practice, due to the imperfections of printing, paper and scanners, the blackness of black and whiteness of white may vary considerably from document to document or indeed within the area occupied by a single character. This variation may be regarded as distortion of the entire grey scale [6.65]. If there were no such distortion, then we could ideally find a fixed binarizing threshold θ at which characters could be binarized without appreciable change in their shape. In practice, at a particular point (x, y) in a particular character, distortion of the grey scale takes θ to a generally different value θ_{xy}. Binarizing the distorted grey scale at threshold θ_{xy} should yield the same result as would be obtained by binarizing the non-distorted grey scale at threshold θ. By determining θ_{xy} and using this as the binarizing threshold we would annul the effect of distortion of the grey scale. Ideally the result would be the same as would be obtained by normalizing the grey scale to remove the distortion. Just as normalizing for size and position of a character may reduce the cost of the subsequent process of recognition, so also the cost and complexity may be reduced by annulling the effect of distortion of the grey scale. Unfortunately it is notoriously difficult to determine a suitable (generally locally varying) binarizing threshold. We will now briefly consider some of the principal techniques that have been used.

6.3.2 Measurement of Limb-Width

Local operator techniques [6.58, 10, 122] can be used to measure approximately the limb-width of a character that has been binarized at an arbitrary threshold. If the limb-width is found to be less than an ideal value, then the character can be binarized again at a whiter threshold so as to increase the limb-width. On the other hand, if the limb-width is greater than the ideal value, then the character can be binarized again at a blacker threshold [6.78, 131, 122]. One of the disadvantages of this technique is that it applies the same threshold to all points in a given character, and to remedy this BARTZ [6.10] has combined this technique with a contrast determining technique.

6.3.3 Contrast Determination

In contrast determining techniques it is usual first to process the video data so as to bring to a fixed predetermined level the analogue signal that corresponds to background white [6.10, 64]. The binarizing threshold is then set at a level that depends on a global measurement of blackest black [6.27, 63, 66] or on the average of all grey-scale values blacker than a predetermined threshold [6.10]. HALL and SEGAR [6.63], and BARTZ [6.10] also took account of a local measure of black, so that the binarizing threshold may be different at different points within a given character, to accord with local variations of the blackness of print.

Over regions close to black/white edges of characters, WESZKA et al. [6.157] have obtained histograms that show the numbers of pattern elements at each discrete grey level. The binarizing threshold can be placed in the valley between the two peaks that correspond to black and white in a histogram. This technique avoids the questionable implicit assumption [6.10, 27, 63] that the whitest level that should be deemed black depends predictably on the blackest level of black or on the mean of the black peak in the grey-level histogram. On the other hand, a histogramming technique is relatively costly to implement. Furthermore, to obtain a respectable histogram it may be necessary to take grey-level data from a fairly large area of a document. At different points within this area the distortion of the grey scale may be different, and it may be undesirable to ignore such differences.

6.3.4 Fixed Local Contrast Requirement

In a *Laplacian* binarizing technique (e.g. [6.14, 30, 67, 50]) a pattern element is deemed black if it is more than a prescribed amount blacker than the average grey level of a set of neighboring points. BRUST [6.23] deemed a point black if it was more than a prescribed amount blacker than at least one of a set of neighboring points (cf. [6.169]). In these techniques the prescribed amount is usually fixed in advance, and it is not a function of a measurement of the blackness of black.

6.4 Smoothing

To smooth or de-noise a binary pattern, DINNEEN [6.44] proposed a technique in which a new binary pattern was constructed from a noisy one according to the following procedure. A square window was centered on the (i, j)-th bit in the noisy pattern, and the (i, j)-th bit in the new pattern was set to 1 or 0 according to whether the number of 1's within the window did or did not exceed a threshold. This was repeated for every bit, that is, for all i, j.

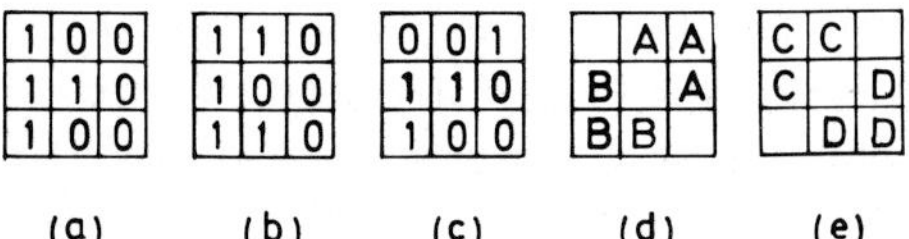

Fig. 6.1a–e. Examples for discussion of smoothing

In Fig. 6.1a the central element may be a bump on the side of a character limb, and in Fig. 6.1b the central element may be a notch or void. By applying a threshold of 5 to the 3 by 3 window we can remove the bump in Fig. 6.1a and smooth out the notch in Fig. 6.1b. Unfortunately we will also remove the central element in Fig. 6.1c, which may render recognition more difficult. In an attempt to remove only black elements that are likely to be due to noise, UNGER [6.148] made the (i, j)-th bit in the new pattern 0 unless at least one of the bits labelled A and at least one of the bits labelled B in Fig. 6.1d is 1 or at least one of the bits labelled C and at least one of the bits labelled D in Fig. 6.1e is 1. This (presumably correctly) removes the central 1 from Fig. 6.1a but does not remove the central 1 from Fig. 6.1c. UNGER's technique introduced the idea of smoothing by scrutinizing the detailed pictorial structure within a window [6.37] rather than by thresholding the number of 1's [6.59, 79, 155].

Noise can be removed from a visual pattern by spatial frequency filtering, but in commercial character recognition it is not usual to do filtering in the Fourier domain. Instead we ignore very high spatial frequencies by determining some sort of average grey level over each of a set of small areas (e.g., 130 μm diameter) of a document.

6.5 Segmentation and Normalization

6.5.1 Normalization by Measurement

When character recognition is achieved by correlation with reference characters, or by the use of logic circuits as in the IBM 1418 [6.60], it is necessary to bring a character into a predetermined alignment position before a recognition decision can safely be made. It may be possible to determine the vertical position of a character, or of a line of characters, by examining the vertical profile. For this purpose the vertical profile is most simply a column of bits in 1:1 correspondence with the rows of bits of a binarized pattern. In this column a bit is 1 if at least one bit in the corresponding row is 1, and 0 otherwise. Within the vertical profile the vertical positions of the top and bottom of a character (or row of characters) are indicated by the positions of the topmost and bottommost 1 in a connected set that comprises a sufficient number of 1's [6.6, 36, 14, 32, 54], cf. [6.78]. If a line of characters is not quite horizontal, it is better to locate the tops and bottoms of a few successive characters rather than an entire line at a time (e.g. [6.36, 93]).

If characters have been specially written or printed for character recognition purposes there are usually white gaps between neighboring characters, so that the horizontal position of a character can be approximately determined by measuring the leftmost or rightmost black column in the character (e.g. [6.32, 54, 84, 116]). This horizontal position information can be used to shift the character into the horizontal alignment that is required for recognition. When characters are not separated by white gaps, segmentation of text into individual characters may be non-trivial, and we will return to this later.

In many systems that recognize characters which have variable size, it is desirable to standardize the size of a character before recognition. When the topmost and bottommost rows and leftmost and rightmost columns of a character have been determined, the character can be rescanned with the scanner adjusted so that these outermost rows and columns become coincident with the sides of a standard rectangle [6.6, 107]. In the image dissector of GENCHI and YONEYAMA [6.54], size normalization is achieved by adjusting the magnification of the electron image of a character, and this is quicker than bit-by-bit re-scanning.

ANDREWS and KIMMEL [6.5] measured the size [6.92] and position of a binarized character that is held in a two-dimensional binary matrix store. To avoid re-scanning, ANDREWS and KIMMEL transferred the data from this store bit-by-bit into a second two-dimensional binary matrix store. The location to which any given bit is transferred depends on its original position and also on the size and position of the original character, and is such that the second store contains a size and position normalized character after all bits have been processed. A read-only store provides the address to which a given bit is transferred. This store is itself addressed by data derived from the original position of the given bit and the size and position of the original character.

Size normalization may, for instance, render an upper-case C indistinguishable from a lower-case c. To avoid this we can measure the height of a character that is known to be upper-case, and use this to determine a scale factor that is held constant for all subsequent characters on the document, whether they are in upper- or lower-case [6.144].

Hitachi patents [6.166, 171] disclose devices that digitally rotate patterns, possibly for purposes of orientation normalization.

6.5.2 Normalization and Segmentation by Recognition

Size normalization is always based on measurement of a character. To achieve position normalization or alignment, a further method is available. In this method an unknown character is correlated with reference characters, or some other recognition procedure is applied, without the unknown character being previously aligned. The unknown character is then slightly shifted and recognition is again attempted. This is done repeatedly until the recognition circuitry produces a confident decision. At that instant the unknown character is likely to be in the correct position. A convenient method of shifting the input character is illustrated by the following example.

In the IBM 1275 the document moves horizontally while video data are obtained from a single vertical column of 72 photodiodes [6.151]. The tallest character to be recognized is 24 bits high, and the photodiode column has been made three times as tall in order to allow for vertical misregistration. The photodiode outputs are binarized in parallel (cf. [6.64, 63]) and the 72 resulting bits are stored in a 72-bit register that we will denote by $\{x_1, \dots x_i, \dots x_{72}\}$. At time instants $t, t+\tau, t+2\tau, t+3\tau, \dots$

the contents of this register are inclusively ORed into a 72-bit shift register $\{y_1, \dots y_i, \dots y_{72}\}$. Thus, for each $i=1, \dots 72$, $y_i = x_i$ or y_i, and this is done in parallel. During the time τ between successive instants when ORing takes place, the shift register $\{y_1, \dots y_i, \dots y_{72}\}$ shifts down 36 bits, and O's are entered into $\{y_1, \dots y_{36}\}$. Meanwhile the bits from $\{y_{37}, \dots y_{72}\}$ are fed into a further shift register (at 1900000 bits per s.). Indeed $\{y_{37}, \dots y_{72}\}$ can be conveniently regarded as the first 36 bits of a shift register that has altogether 20×36 stages. These stages can be thought of as constituting a matrix of 36 rows and 20 columns, the bottom of each column being connected to the top of the next. The time interval τ is chosen so that the widest character is completely scanned within 20τ units of time. The 36×20 stage shift register contains a binary image of the input character that inevitably comes at some time into any predetermined position of alignment within the 36×20 bit array. This technique tolerates positioning of a character at any vertical level spanned by the 72 photodiodes, yet the columns of the shift register are only 36 bits high [6.61]. The IBM 1275 does not work simply by correlation, and we will return to this later.

Suppose now that recognition is in fact done by correlation and that the inputs to the correlation system are taken from a subset of the elements of a two-dimensional shift register. A character coursing through the shift register eventually comes into the correct alignment for recognition. This is true even if the character is preceded and followed by touching characters. When the character comes into proper alignment the correlation score may exceed a threshold, and a recognition decision can then be made, despite touching characters. A weakness of this idea is that if, for instance, the two characters rn touch, then they may be misrecognized as the single character m. To guard against mistakes such as this, HOLT and HILL [6.73] have developed the following idea. If the pitch, that is, the distance between center lines of successive characters, is known and roughly constant, then the time interval T between instants of proper alignment of successive characters is also known and roughly constant. A time instant t is taken to be the correct instant for recognition if the sum of the maximal correlation scores at $t-T$, t, and $t+T$ exceeds a threshold. HOLT and HILL [6.73] made allowance for slight variations in pitch along a line of characters.

6.5.3 Segmentation by Measurement

In an important class of character recognition systems (e.g. [6.4, 126]) logical tests are applied to a character while it passes through a shift register. These tests may yield positive results at somewhat different times. When the entire character has passed into the shift register a recognition decision is made in accordance with the results of the logical tests. More specifically, a recognition decision is made only when a signal is received that indicates that an entire character has just passed into the shift register. This signal, which is commonly known as a *segmentation* signal, is a prerequisite for recognition. If there are white gaps between successive characters then a segmentation signal can be given when the white gap that follows a character is detected in the shift register, e.g. [6.126]. When there is no clear white column between successive characters, then a more elaborate segmentation technique is required, e.g. [6.12, 38]. By way of example we will briefly consider one such technique.

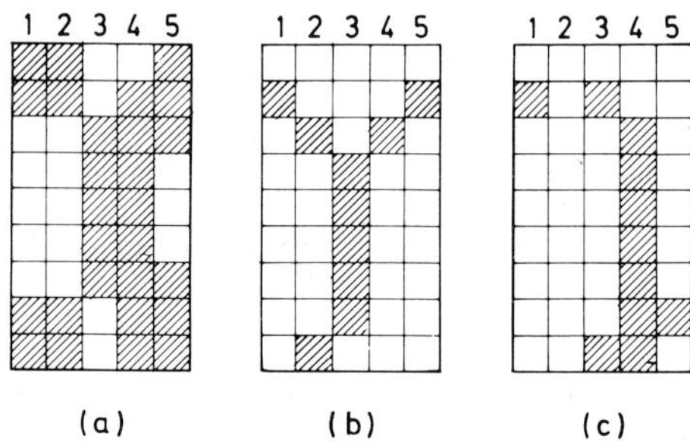

Fig. 6.2a–c. Examples of character portions in a shift register

If a character works through the shift register from left to right we may, for instance, ideally wish to produce a segmentation signal when the leftmost black bit of the character lies in the third-from-left column of the shift register. This column is headed "3" in Fig. 6.2. In Fig. 6.2a the five columns contain the rightmost part of a character c and the leftmost part of an o in the touching pair co. One segmentation procedure is in effect to shift a 3 by 3 window vertically so that its central bit is successively aligned with all bits in turn in column 2. If there is *no* vertical position where either

a) the four bits indicated in row 1 column A of Fig. 6.3 are all black, or
b) the three bits indicated in row 1 column B of Fig. 6.3 are all black, or
c) the four bits indicated in row 1 column C of Fig. 6.3 are all black

then a segmentation signal is produced [6.13]. According to this procedure a segmentation signal is in fact produced in the Fig. 6.2a example. A segmentation signal is also produced in the case of Fig. 6.2b, which is undesirable since Fig. 6.2b shows part of a thin-limbed character y. A segmentation signal is also produced in the case of Fig. 6.2c, which shows the central hump of an n that has a white gap due to very weak printing. To avoid erroneous segmentation, as in Figs. 6.2b and c, we can use row 5 of Fig. 6.3 instead of row 1. In this case a segmentation signal

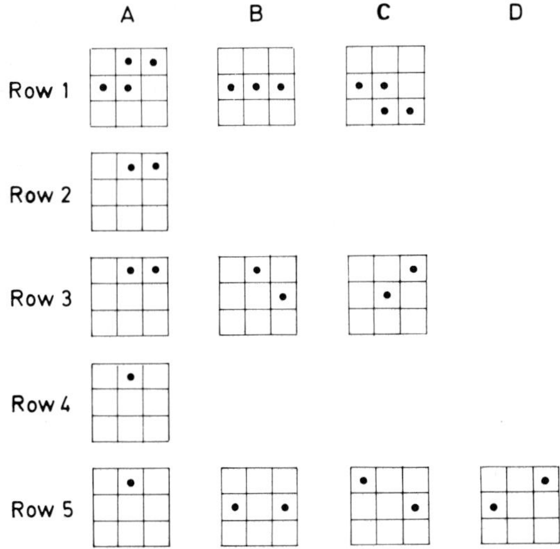

Fig. 6.3A–D. Sets of black bits used in segmentation. In this particular figure blank squares signify "don't care"

is produced if there is *no* vertical position of the 3 by 3 bit window where either

a) the single bit indicated in row 5, column A of Fig. 6.3 is black, or

b) the two bits indicated in row 5, column B of Fig. 6.3 are both black, or

c) the two bits indicated in row 5, column C of Fig. 6.3 are both black, or

d) the two bits indicated in row 5, column D of Fig. 6.3 are both black.

If row 5 of Fig. 6.3 is used in this way instead of row 1, then it is unfortunate that no segmentation signal is produced in the case of Fig. 6.2a. To avoid difficulties such as this, BAUMGARTNER et al. [6.13] use row 1 of Fig. 6.3 for thick-limbed characters and row 5 for thin-limbed characters. Instead of directly measuring limb thickness, BAUMGARTNER et al. used a measure of the average blackness of black, and select the row of Fig. 6.3 according to this measure: the blacker the measure, the lower the row number. Instead of shifting the 3 by 3 bit window into all vertical positions, the window is in fact held fixed while the character passes through all vertical positions in the shift register.

If successive characters touch at many points, it may be necessary to resort to forced segmentation. This is a technique in which the position of the side of a character is computed from measurements of pitch (e.g. [6.114, 12]).

In Fig. 6.2a, column 2 contains black bits of the characters c and o, and this illustrates vertical overlap between neighboring characters. When a segmentation signal is produced it may be desirable to destroy vertical overlap, for instance, by artificially introducing a purely white column into the shift register to separate the characters [6.3].

6.6 Recognition by Correlation

6.6.1 Introduction

Let $X=\{x_1, \dots x_i, \dots x_N\}$ be a rectangular array of grey-scale values obtained by scanning an entire character that is to be recognized, and let $X'=\{x'_1, \dots x'_i, \dots x'_N\}$ be a rectangular array obtained by scanning a reference character. Let us suppose at first that X and X' are not binarized. A simple idea is to recognize X as belonging to the class of the reference character X' such that

$$\sum_{i=1}^{i=N} x_i \cdot x'_i \tag{6.1}$$

is greater than for any other reference character, when X and X' are properly aligned in position. Unfortunately, if X is the character F then the value of (6.1) may be the same whether X' is an E or an F. One way of avoiding this difficulty is in effect to assign X to the class of the reference character X' such that the Euclidean distance between X and X' is less than that between X and any other reference character.

In practice we may find that all patterns $kX=\{kx_1, \dots kx_i, \dots kx_N\}$ belong to the same recognition class, for all positive values of the factor k. For instance, different values of k may correspond to different levels of illumination. In this case it may not be appropriate to use Euclidean distance, because the Euclidean distance between kX and X' depends strongly on k. To gain independence of k we can in effect normalize kX and X' so that they become vectors of unit length [6.69, 89]. However, in practical character recognition the blackness of black may to some extent be independent of the whiteness of white, and in general the value of the factor k may be different at different points within a character. This is not remedied by normalization of vectors to unit length. Instead the usual practice is to binarize characters, thereby gaining at least limited invariance to distortion of the grey scale. A binarized character X may then be assigned to the class of the binarized reference character X' such that the Hamming distance between X and X' is less than the

Hamming distance between X and any other binarized reference character. In this case there may be no confusion between pairs of characters such that one is included in the other, for instance F, E and P, B.

Another common technique is to determine the Hamming distance between corresponding subsets of elements of X and X'. To clarify this, let $x_i \forall x_i'$ be the exclusive OR of x_i and x_i'. Thus

$$\sum_{i=1}^{i=N} (x_i \forall x_i')$$

is the Hamming distance between X and X'. Let $C_r = \{c_{r1}, \ldots c_{ri}, \ldots c_{rN}\}$ be a vector whose components correspond 1:1 to the elements of X. The pattern X can be assigned to the class of X' such that

$$\sum_{i=1}^{i=N} c_{ri}(x_i \forall x_i') \tag{6.2}$$

is minimal. The set of elements $\{x_i'\}$ such that $c_{ri}=1$ is known as a *peephole mask* [6.9, 142]. EVEY [6.47], WADA et al. [6.153], and STOFFEL [6.143] have contributed methods for automatically determining peephole masks.

One of the underlying ideas is that a character recognition system is only required to recognize characters in a fixed set of, say, 2^m recognition classes, and a correct recognition is not required for any character that does not belong to any of these 2^m recognition classes. For instance, a machine for recognizing numerals is not expected to recognize letters correctly. To distinguish between 2^m recognition classes, only m bits should ideally be necessary. To capitalize on this, the character shapes are chosen in [6.80, 95, 152, 154] so that the same vector C_r, which consists of m 1's and $N-m$ 0's, can be used for all recognition classes, because the resulting m-bit peephole masks for all classes are different. Economy is gained by ignoring the remaining $N-m$ bits, but this simple technique presupposes that the m bits that are retained are not subject to noisy variation.

In binarized printed characters, there is generally more noise at the black/white edges of limbs than at other points. Instead of simply using Hamming distance, we may be able to reduce recognition error rates by ignoring pattern elements near black/white edges, or in other words, setting $c_{ri}=0$ for all such elements [6.153]. The use of a peephole mask in this way is a cruder technique than the use of a more general linear discriminant function [6.46, 146], but it has the advantage of simplicity.

BENE and NELSON [6.16] used more than one peephole mask per class. PATTERSON [6.105] employed piecewise linear discriminants in which there are many possible values for each weight.

6.6.2 Implementation

Neither holographic nor non-holographic optical correlation techniques [6.147] appear to be much used in commercial character recognition. Possibly one of the most important reasons for this is that optical correlation techniques do not provide invariance to distortion of the grey scale. It is well known that spatial frequency filtering can be done in holographic correlation [6.71, 91] but it is not clear that this is really useful in practical character recognition. If we attenuate low spatial frequencies, we tend thereby to enhance the importance of noise. Noise can be removed by filtering out high spatial frequencies, and we have already commented on this at the end of Section 6.4.

In character recognition "correlation" often means either the determination of Hamming distance or the evaluation of (6.2), and this is usually accomplished electronically, although the process of opto-electronic transduction may itself be a major source of noise. For each reference pattern X' an analogue signal proportional to

$$N-(\text{Hamming distance between } X \text{ and } X')$$

can be obtained by a resistor mask technique [6.146], and circuits are available for determining which such signal is greater than any of the others by at least a predetermined margin [6.74, 97, 134]. If there is any such signal, then the input character can be assigned to the class of the reference character associated with this signal. If there is no analogue signal greater than all others by at least a predetermined amount, the input character can be rejected. The magnitude of this predetermined amount controls the trade-off between reject and substitution error rates. In practice very careful engineering is required in order to obtain reliable sensitivity to small differences in Hamming distance, and to avoid this difficulty we can turn to purely digital correlation techniques.

Digital correlation techniques were disclosed in [6.15, 79, 141, 166, 167], and by way of example we will briefly consider just one such technique. IRVIN and RIDER [6.84] found the horizontal and vertical profiles of a character and used this information to shift the character into a predetermined position in a 20 row by 10 column register. A digitally stored reference pattern X' consists of 18 rows and 10 columns, and corresponding to each reference pattern a separate and generally different 18 by 10 bit care/don't care vector C_r is also digitally stored. The right-hand column of the unknown character X is transferred in parallel into a shift register, and the right-hand columns of X' and C_r are transferred into further shift registers. These three shift registers are shifted synchronously while three counters respectively determine

$$\sum c_{ri}(x_i \forall x_i') \quad \text{and} \quad \sum c_{ri}(x_{i-1} \forall x_i') \quad \text{and} \quad \sum c_{ri}(x_{i+1} \forall x_i')$$

for the entire column. This process is repeated for all columns in turn so that the three counters finally contain

$$\sum c_{ri}(x_i \forall x_i') \quad \text{and} \quad \sum c_{ri}(x_{i-1} \forall x_i') \quad \text{and} \quad \sum c_{ri}(x_{i+1} \forall x_i')$$

for the whole of the input character. By taking the three counts, IRVIN and RIDER allowed for the character to be displaced one row above or below its nominal position. To allow for displacement by one column to the left or right, the matching process is repeated with the unknown character shifted one column to the left and then to the right with respect to X' and C_r. The entire process is repeated for each different reference pattern [6.84].

One way to speed up the recognition process is by reducing the number of reference patterns. For instance, in the recognition of a word, after some of the characters have been recognized, a further character is likely to belong to one of only a few recognition classes, and therefore it can be compared first with reference patterns in these few classes [6.79]. If a character is known to be upper-case, it need not be compared with lower-case reference characters; and if format dictates that a particular character is a numeral, then this character need not be matched with reference letters [6.16].

6.7 Boolean Recognition Techniques

6.7.1 Introduction

The small squares in Fig. 6.4a represent the elements of a two-dimensional shift register that contains a binarized character. The following is a very simple example of a logic statement for recognizing 4 printed in a single font. A character is recognized as 4 if there is at least one step of the shift register when at least five of the elements labelled A *and* at least five of the elements labelled B *and* at least five of the elements labelled C in Fig. 6.4a are black *and* at least four of the elements labelled D *and* all of the elements labelled E in Fig. 6.4a are white.

Parallel hardware (e.g. [6.61, 126]) can rapidly determine the truth value of a logic statement such as this, and recognition can be accomplished far more quickly than in a bit-by-bit digital correlation technique. However, speed and ease of implementation are not the only reasons for using logic statements in recognition.

To introduce a further reason, let us consider a general linear method for recognizing characters. In this method there is at least one weight

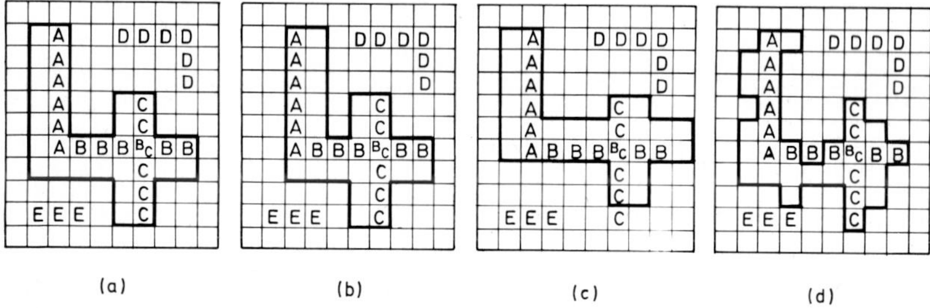

Fig. 6.4a–d. Zones used in logic statement for 4

vector $W_r = \{w_{r0}, w_{r1}, \ldots, w_{ri}, \ldots, w_{rN}\}$ for each recognition class, and an unknown character $X = \{x_1, \ldots, x_i, \ldots, x_N\}$ is assigned to the class associated with maximal

$$w_{r0} + \sum_{i=0}^{i=N} x_i \cdot w_{ri} \,. \tag{6.3}$$

BRYAN and TEACHER [6.24] found that if the weight vector for the character R has high weight for the slant leg, to facilitate the distinction between R and P, then the rest of the weight vector becomes less significant, so that it may be difficult to distinguish reliably between R and K. To remedy this, BRYAN and TEACHER in effect defined a variable $v_r = 1$ or 0 according to whether the value of (6.3) did or did not exceed a prescribed threshold; and a character was assigned to a particular class only if an AND function of a predetermined set of variables $\{v_r\}$ was true. A similar idea is disclosed in [6.82] where, for instance, a character is recognized as a U if it has smaller Hamming distance to the U in Fig. 6.5 than to any other reference pattern *and* more of the arrowed bits are white than black. This AND condition is intended to facilitate the distinction between O and U, cf. [6.112].

The usefulness of AND, or of more complicated Boolean functions, is not surprising from a theoretical point of view. For instance, BRYAN and TEACHER [6.24] started by requiring ideally that $v_r = 1$ for all patterns in one particular class, and $v_r = 0$ for all patterns in other classes, but this requirement for linear separability is unlikely to be satisfied if the recognition classes contain many different patterns [6.35]. Recognition by logic statements has no such prerequisite.

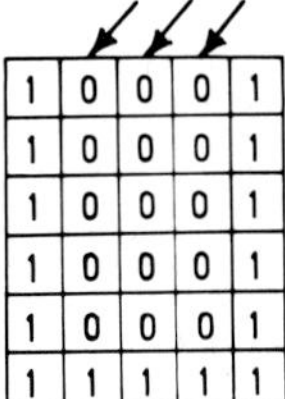

Fig. 6.5. A binary reference U

6.7.2 Alignment Counting Techniques

GREANIAS [6.60] has given examples of logic statements used in the IBM 1418 reading machine, and these are very much more complicated than in our Fig. 6.4 example. In this very simple example the logic statement has been chosen so that the 4's outlined in Fig. 6.4b, c, and d would correctly be recognized as 4. Due to imperfections of printing or scanning, the vertical limbs of the 4 may turn out to be closer together or further apart, as in Figs. 6.4b and c, and there may be similar distortion in the vertical direction. It is important that the logic statement is tolerant of this limited distortion and also of the raggedness that is illustrated in Fig. 6.4d.

A corollary of this tolerance is that the logic statement is still true when the 4 outlined in Fig. 6.4a is shifted up, down, left or right by one row or column. Van Heddegem's system [6.150] does not recognize a character as a 2 unless the logic statement for 2 is true for at least a predetermined number of alignments of the input character with the recognition logic, and similar conditions are used for all other recognition classes. To recognize a character as a 2, the IBM 1275 [6.151] requires that the logic statement for 2 be

a) true in at least a predetermined number of alignments, and

b) true in more alignments than the statement for any other class by at least a predetermined threshold amount. (By changing this threshold the reject/substitution trade-off can be controlled.)

It may not matter if, for instance, the logic statement for 3 is true in one or two alignments when the input character is in fact a 2. Therefore it is somewhat easier to design logic statements in the IBM 1275 than in the IBM 1418. The IBM 1418 [6.60] did not require a logic statement to be true in more than one alignment, and instead it required that, for instance, a 2 should not in any alignment satisfy the logic statement for 3. This is a stricter requirement than in the IBM 1275, and it is therefore more difficult to satisfy.

VAN STEENIS [6.151] mentions that in the IBM 1275 the logic statement for a particular character is generally the OR of five further statements. One of these five statements is designed to match characters that have fat limbs, and the other four are designed, respectively, to match characters that have the top, bottom, left, or right side missing due to defective printing.

6.7.3 Logic Statements for Feature Detection

In the IBM 1275 [6.151] there is a separate overall logic statement for each character, but, for instance, part of the logic statement for 3 may also be part of the logic statement for 7, since the top parts of 3 and 7 are similar. The IBM 1275 recognizes only the numerals and four special symbols in the fonts ISO-A or ISO-B. One approach to the design of machines that recognize many complete fonts, including upper and lower case letters, numerals, and punctuation marks, is to concentrate first on the recognition of features. By way of example we will now consider a much simplified version of the feature detector that recognizes the top left corner of B, D, E, F, P, R, 5 in a Scan-Data system [6.126].

The small squares in Fig. 6.6 represent some of the elements of a shift register through which an input character passes. It is convenient to define AND functions $a_1, a'_1, b_1, b'_1, \ldots$ as follows. Function $a_1 = 1$ if all elements labelled A_1 are 1, and $a_1 = 0$ otherwise. Functions $a_2, a_3, b_1, b_2, b_3, \ldots$ are defined similarly. Function $a'_1 = 1$ if all elements labelled A_1 are 0, and $a'_1 = 0$ otherwise. Functions $a'_2, a'_3, b'_1, b'_2, \ldots$ are defined similarly. We shall, for instance, write $\bar{a}_1$ to mean "*not* a_1". In this simplified example the feature is deemed present if and only if $(a'_1 + a'_2 + a'_3 \geqq 2)$ *and* $(a'_4 + a'_5 + a'_6 \geqq 2)$ *and* $(b'_4 + b'_5 + b'_6 \geqq 2)$ *and* $(c'_1 + c'_2 + c'_3 \geqq 2)$ *and* $(d'_1 + d'_2 + d'_3 \geqq 2)$ *and* $(\bar{e}_1$ *and* $\bar{e}_2)$ *and* $(\bar{g}'_1 + \bar{g}'_2 + \bar{g}'_3 + \bar{g}'_4 \geqq 3)$ *and* $(\bar{h}'_1 + \bar{h}'_2 + \bar{h}'_3 + \bar{h}'_4 \geqq 3)$ *and* this condition is satisfied at a time when the top left-hand part of the character is aligned with the elements shown in Fig. 6.6. The timing condition is such that the feature can be detected anywhere within a top left zone in the region that contains the unknown character. Note that $\bar{g}'_1 = 1$ if at least one of the elements labelled G_1 is 1: we can always implement an OR of black elements by negating the AND of the corresponding white elements. Note also that a term such as $[(a'_1 + a'_2 + a'_3 \geqq 2)$ and $(a'_4 + a'_5 + a'_6 \geqq 2)]$ can be used not only in the recognition of

	A_1	A_1	A_1	A_4	A_4	A_4	B_1	B_1	B_4	B_4	B_4	
	A_2	A_2	A_2	A_5	A_5	A_5	B_2	B_2	B_5	B_5	B_5	
	A_3	A_3	A_3	A_6	A_6	A_6	B_3	B_3	B_6	B_6	B_6	
	C_1	C_2	C_3									
	C_1	C_2	C_3		G_1	GH_{11}	H_2	H_3	H_4			
	C_1	C_2	C_3		G_2	GH_{21}	H_2	H_3	H_4			
	D_1	D_2	D_3		G_3	G_3						
	D_1	D_2	D_3		G_4	G_4		E_1	E_1			
	D_1	D_2	D_3					E_2	E_2			

Fig. 6.6. Labelling of elements for corner feature logic statement

the Fig. 6.6 feature but also in the recognition of further features. This is why the elements labelled $B_1, B_2, \ldots$ are not instead labelled $A_7, A_8, \ldots$.

In [6.126], when a feature is detected by logic circuits a bistable element is set to state 1. There is one such bistable element corresponding to each feature, and when the white gap that follows a character is detected in the shift register, this set of bistables contains a binary pattern that we can denote by $X=\{x_1, \ldots, x_i, \ldots, x_N\}$. In this, $x_i=1$ or 0 according to whether the i-th feature has or has not been detected in the input character. A typical condition for recognition in this Scan-Data system [6.126] is: recognize the character as B if $\bar{x}_{101}+\bar{x}_{60}+\bar{x}_{29}+x_{264}+x_{228}+x_{198}+x_{166}+x_{153}+x_{127}+(x_{115}$ or $x_{117})+x_{45}+(x_7$ or $x_8)\geqq 11$. Apart from the ORs this is essentially a peephole mask recognition condition applied to X.

The IBM 1975 is a further example of a multifont print recognition system where feature detection yields a binary pattern $X=\{x_1, \ldots, x_i, \ldots, x_N\}$ in which $x_i=1$ or 0 according to whether the i-th feature has or has not been detected [6.4]. A character is recognized by matching X with peephole masks, and there is more than one peephole mask per recognition class [6.16]. Each peephole mask generally has a different care/don't care pattern C_r, and (6.2) is evaluated digitally [6.15].

6.7.4 Microprogramming Implementation

Recognition by peephole masks is essentially linear or piecewise linear, and we have previously mentioned that there are severe limitations on the possible accomplishments of linear decision making. To avoid such limitations, ANDREWS and ATRUBIN [6.2], like VAN STEENIS [6.151] and ROHLAND et al. [6.121], use logic statements for the final stage of recognition. In the system of ANDREWS and ATRUBIN [6.2] recognition is a multi-stage logical process that is not confined simply to two stages comprising firstly feature detection and secondly decision making. ANDREWS and ATRUBIN claimed that the total dedicated hardware cost is lower, and detailed logic modification is easier, when microprogramming is used instead of a hardwired implementation of the complex logical structure. Microprogramming [6.77] provides flexibility like that of ordinary computer simulation, and speed comparable to that of hardwired circuitry. In a microprogramming implementation, the amount of dedicated hardware that must be checked for serviceability (cf. [6.90]) is less than in the hardwired case. The IBM 3886, which is the most recent IBM reading machine, is microprogrammed [6.81].

In the system of ANDREWS and ATRUBIN [6.2], a binarized character passes through a shift register. Two hundred and forty AND gates connected to various elements of the shift register detect subfeatures such as a black line or a black curve or a white area. Examples are not given, so we do not know whether these AND gates typically have more inputs than $a_1, a'_1, b_1, b'_2, \ldots$ in the Fig. 6.6 example. Anyway, for each AND gate there is a separate bistable that is set when the corresponding AND gate is activated. After a character has passed into the shift register, the states $y_1, \ldots, y_i, \ldots, y_{240}$ of the 240 bistables indicate which of the 240 AND gates have at some time been activated. A further 16 bistables $y_{241}, \ldots, y_{256}$ can be set by the microprogrammed process. This process takes as its data the states of the bistables $y_1, \ldots, y_{256}$, and it takes as its program a set of blocks of bytes stored in read-only memory.

Each such block is used for the evaluation of a specific logic function. This function is restricted to being an AND of OR functions of threshold functions of some of $y_1, \ldots, y_{256}$. Here a *threshold function* is a binary function that has the value 1 if and only if at least a threshold number of variables in a prescribed subset of $y_1, \ldots, y_{256}$ have prescribed states. For simplicity in the following example we will consider only threshold functions that are in fact AND functions because the threshold number is equal to the total number of elements in the subset of $y_1, \ldots, y_{256}$.

As an example we will consider the evaluation of

$$(y_7 \cdot y_8 \vee y_{21} \vee \bar{y}_{22}) \cdot (y_{15} \cdot \bar{y}_{30} \vee y_{16} \cdot \bar{y}_{42} \cdot \bar{y}_{43}) \tag{6.4}$$

where "." denotes AND and "$\vee$" denotes inclusive OR. To avoid parentheses, ANDREWS and ATRUBIN [6.2] used different symbols for AND at different logical levels. In particular, they used the symbol "b" for AND at the lowest level and ")(" for AND at the next higher level. They used "$++$" instead of "$\vee$", and ")b" served as a statement-terminator symbol. Thus

$$(y_7 b y_8 + + y_{21} + + \bar{y}_{22}) \cdot (y_{15} b \bar{y}_{30} + + y_{16} b \bar{y}_{42} b \bar{y}_{43}) b \tag{6.5}$$

Table 6.1. A microprogram for evaluating a logic statement

y address	1 or 0	Connective
7	1	b
8	1	$++$
21	1	$++$
22	0	$)($
15	1	b
30	0	$++$
16	1	b
42	0	b
43	0	$)b$

is equivalent to (6.4). The microprogram block that evaluates (6.5) starts with a control word with which we shall be concerned later. The control word is followed by a succession of bytes, and each of these contains the information indicated in one of the rows of Table 6.1. Successive bytes, and thus successive rows in Table 6.1, correspond to successive operand-operator pairs in (6.5), the subscripts of y serving as addresses of bits in $\{y_1, \ldots, y_{256}\}$. The column "1 or 0" in Table 6.1 indicates whether the addressed bit is required to be 1 or 0.

When any given byte is processed, the addressed one of the bits $\{y_1, \ldots, y_{256}\}$ is compared with the bit in the "1 or 0" column of Table 6.1. If these two bits are identical we define $e=1$, and $e=0$ otherwise. The decoded connectives and a signal that represents e provide inputs to the logic circuit that is shown in Fig. 6.7. Figure 6.7 is not the same as Fig. 5 in [6.2]; for instance, we have, for simplicity, omitted the means for computing threshold functions. At the start of processing a logic statement, or in other words, at the start of processing a block of bytes, the bistables L_1 and L_3 in Fig. 6.7 are initially set to state 1, and L_2 is initially in state 0. During the processing of a block of bytes, AND gate G_4 is activated by the occurrence of a sequence such as $e++$ or $ebe++$ or $ebe)$ or $ebebebe)($ or $ebebebe++$ in which every $e=1$. For instance, for the first byte in our example, if $y_7=0$ then gate G_3 is activated via inverter G_1, so that at the next byte G_4 is not activated even if $y_8=1$. On the other hand if both $y_7=1$ and $y_8=1$ then bistable L_2 is set to state 1. AND gate G_6 is activated by the occurrence of a sequence such as $\underline{e}++\underline{e}$ or $\underline{e}++\underline{e}++\underline{e})($ or $\underline{ebe}++\underline{e}++\underline{ebe})b$ or $\underline{ebebe}++\underline{ebe}++\underline{ebebebe})b$ in which there is at least one $e=0$ in every underlined term. For instance, if $y_7=0$ but $y_{22}=0$ then G_6 is not activated at the fourth byte in our example. On the other hand, if $y_7=y_{21}=0$ and $y_{22}=1$, then G_6 is

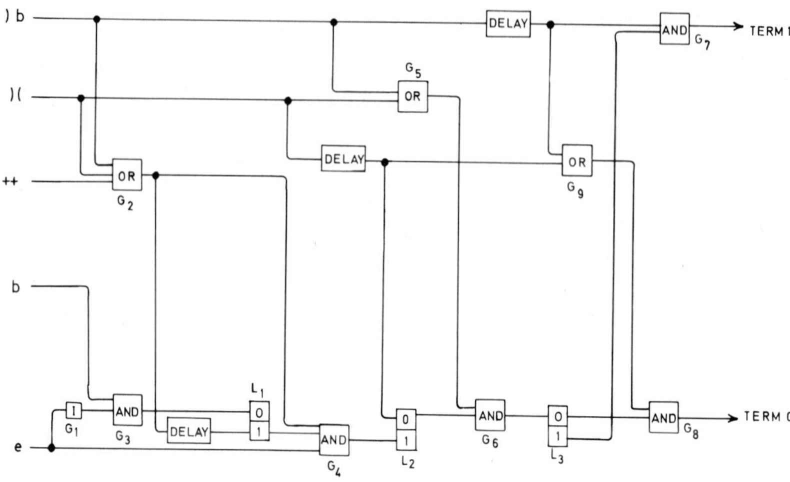

Fig. 6.7. Logic unit similar to but not the same as that of ANDREWS and ATRUBIN [6.2]

activated, bistable L_3 is reset to state 0, and the TERM 0 output signal is produced, indicating that the logic statement is *false*. AND gate G_7 is activated, and the TERM 1 output signal is produced, only if the logic statement is true. The delays in the circuit are sufficient to allow for settling of transients.

The control word that heads a block of bytes is processed as soon as a TERM 1 or TERM 0 signal is produced. This control word contains eight bits that can be read out of the machine to indicate the identity of an input character. It also contains four bits that can be used to address one of $y_{241}, \ldots, y_{256}$. Further it contains two bits that tell the system whether to enter a 1 into the addressed one of $y_{241}, \ldots, y_{256}$ or whether to read out the 8-bit identifying code. Finally it contains the 11-bit starting addresses of two further blocks; one of these addresses is selected if the TERM 0 signal is produced, and the other is selected if the TERM 1 signal is produced. Via bits $y_{241}, \ldots, y_{256}$ the truth value of a logic statement can serve as an input to a further logic statement; and it is generally necessary to process many logic statements before a character is recognized [6.2]. TERM 1 may select an address that leads to termination of the process of recognizing one character.

References [6.79] and [6.99] describe very different examples of digital pattern recognition devices in which logic elements are controlled by instructions that are read out of memory. Reference [6.166] describes a general purpose digital pattern processor.

6.7.5 Associative Structures

When a binary pattern is stored in a shift register or indeed in any binary register, we can denote the elements of the register by $X=\{x_1, \ldots, x_N\}$. A binary pattern is then regarded as a *state* of X, or in other words, a binary vector value of X. Let $V_1, \ldots, V_j, \ldots, V_u$ be u arbitrarily chosen subsets of X, and let A_{rj} be the set of all states of V_j that have occurred in at least one pattern in a training set of patterns in the r-th recognition class. Let a_{rj} be a binary variable defined by $a_{rj}=1$ or 0 according to whether a pattern that is to be recognized does or does not contain a state of V_j that belongs to A_{rj}. BLEDSOE and BROWNING's method of recognition [6.19] is to assign an unknown pattern to the r-th recognition class if

$$\sum_{j=1}^{j=u} a_{rj}$$

is greater than the corresponding sum for any other class.

To determine whether $a_{rj}=1$, we can use the state of V_j to address an associative memory in which A_{rj} is stored [6.145]. If the state of V_j matches any state stored in this memory, then $a_{rj}=1$. Suppose, for instance, that V_j consists of six elements of X, and that A_{rj} consists of the fourteen 6-bit states that are listed in Table 6.2a. In this case Table 6.2a contains all the *true* rows of the truth table for the Boolean function a_{rj}. To gain implementational economy we can apply the Quine-McCluskey procedure [6.96] to Table 6.2a and thereby obtain a set of prime implicants, as shown in Table 6.2b, where "–" signifies "don't care". It is easy to check that $a_{rj}=1$ if and only if the state of V_j matches one of the three rows of Table 6.2b. Instead of storing Table 6.2a in an ordinary associative memory, we can gain economy by storing Table 6.2b in an associative memory in which each element has three possible states 1, 0, and – [6.118]. For each r, the u different tables corresponding to $j=1, \ldots, u$ can be stored in a single associative memory, each table having a distinct key [6.118]. If X is in fact a shift register, we can use the same key for several different shift positions of a binarized character, and determine whether $a_{rj}=1$ for at least one of these positions, to allow for distortion.

In Fig. 6.6, the set of elements labelled $A_1, \ldots, A_6$ constitute a subfeature that may be common to several different features. Insofar as features are composed of subfeatures, feature detection is a two-level process in the Fig. 6.6 example. Feature detection also has several logical levels in, for instance, [6.104, 105, 76]. BARTHOLOMEW et al. [6.8] have patented a multi-level feature detection system in which subfeatures are detected by an associative memory technique. The output of the subfeature detection stage serves as the input to a further associative memory

Table 6.2a and b

a) Example of a set of states A_{rj}	b) Prime implicants for a_{rj} in Table 6.2 example
1 1 0 1 0 1	
1 0 0 1 0 1	
1 1 0 1 1 1	
1 0 0 1 1 1	
0 0 0 0 0 1	
0 0 0 0 1 1	1 – 0 1 – 1
0 0 0 1 0 1	0 0 – – – 1
0 0 0 1 1 1	– 1 1 0 1 0
0 0 1 0 0 1	
0 0 1 0 1 1	
0 0 1 1 0 1	
0 0 1 1 1 1	
1 1 1 0 1 0	
0 1 1 0 1 0	

that detects features, and so on through several further stages. This recognition system has a hierarchical logical structure in which the circuitry at all levels is similar, cf. [6.48].

Like microprogramming techniques, associative memory techniques offer the advantage that by reloading the memory we can change the set of recognizable classes without altering the hardware. Serviceability can be checked by means of a memory test rather than a logic circuit test.

6.7.6 Logical Dichotomization Schemes

In the simplest linear recognition systems, there is one reference pattern or weight vector per recognition class. This exemplifies one/many dichotomization in that a single weight vector is employed to separate patterns in one class from patterns in many other classes [Ref. 6.146, Section 3.9]. It is more likely that we will be able to find a weight vector that reliably dichotomizes between only two classes, and this is why "pairwise" or one/one dichotomization has been used, for instance, in [6.83, 124, 156, 182]. The use of a single weight vector to separate many classes from many others is unusual in character recognition, but see, e.g. [6.34].

In the IBM 1418 [6.60] for each recognition class there is one Boolean function that is used to dichotomize between patterns that do or do not belong to the corresponding class, and this exemplifies one/many logical dichotomization. For instance, the Boolean function for the r-th recognition class is intended to dichotomize between patterns in the r-th recognition class and patterns in many (indeed all) other classes. In the IBM 1975 [6.4] and in a Scan-Data system [6.126] a feature is deemed present if and only if a predetermined Boolean function is true. This Boolean function is intended to perform many/many dichotomization in that it is ideally true for all patterns in one set of recognition classes and false for all patterns in another set of recognition classes.

STEARNS [6.139] described a primitive many/many logical dichotomization scheme in which m Boolean functions were used in the recognition of 2^m recognition classes. In designing the logic, STEARNS started by arbitrarily providing a different m-bit word for each recognition class, and we will call these m-bit words *signature* words. Next, STEARNS chose m Boolean functions of the binary input pattern. The i-th of these functions was true for all classes that had 1 as the i-th bit of their signature word, and false for all others. This was the case for each $i=1, \ldots, m$.

To allow for design failures whereby a Boolean function is true when it should be false, or vice versa, HOWARD [6.75] has extended STEARNS' idea by providing each of the 2^m classes with a signature word that has more than m bits. As in [6.109, 34], HOWARD chose the 2^m signature words *so that* there is at least a predetermined Hamming distance between any two of them. Following STEARNS [6.139], HOWARD [6.75] chose one Boolean function corresponding to each bit in a signature word. The i-th of these Boolean functions is ideally true for all patterns in classes for which the i-th bit in the signature word is 1. A disadvantage of this approach is that it does not necessarily attempt to implement the easiest dichotomies. For instance, a relatively easy dichotomy is between characters that do or do not possess a feature such as the Fig. 6.6 feature. It is presumably more difficult to dichotomize between two groups of classes when the classes within at least one group have no common feature. For instance, it might be unnatural and difficult to dichotomize between O, M, X, B, T and Q, H, Y, P, I, but Howard's technique may demand such dichotomies.

In recognition, the signature word of an unknown pattern is determined by evaluating one Boolean function per bit, and HOWARD finds the class whose reference signature word has the smallest Hamming distance to the signature word of the unknown pattern. If the smallest Hamming distance is less than a prescribed value, then the unknown pattern is assigned to this nearest class. It is important that Howard's system [6.75] is constructed so that the nearest class can be determined without determining the Hamming distances between the unknown signature and reference signature words. Using a general-purpose computer Howard's procedure for finding the nearest class is faster than a conventional procedure that computes the 2^m Hamming distances that correspond to the 2^m recognition classes.

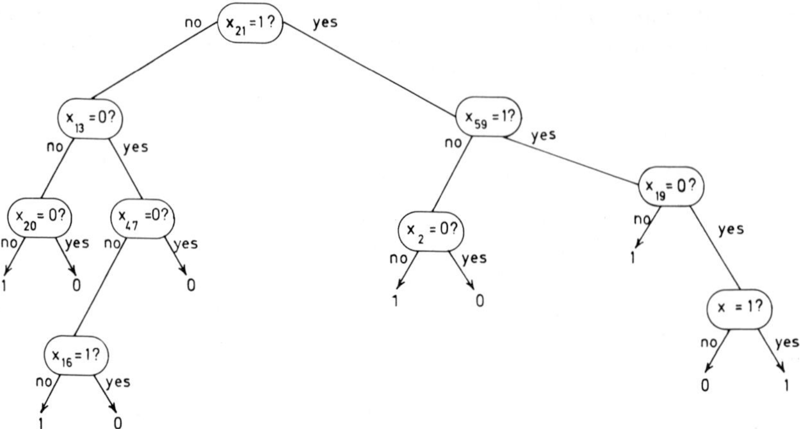

Fig. 6.8. Simple example of a binary tree

In Howard's system the Boolean functions that evaluate the signature bits are implemented by means of trees like the one shown in Fig. 6.8. At each non-terminal node the branching depends on whether a prescribed bit in the input pattern is 1 or 0. When a terminal node is reached the Boolean function is found to be true or false, and the terminal nodes are accordingly labelled 1 or 0 in Fig. 6.8. NAYLOR [6.101] has described a man-machine interactive system that can efficiently design these trees. These trees, and indeed the whole of Howard's system, can conveniently be implemented on a general purpose digital computer.

6.8 Recognition of Distorted Characters

6.8.1 The Problem of Distortion

The shapes of handprinted characters have been systematically studied by WRIGHT [6.159]. It is common experience that if, for example, the character 5 is freely handprinted by two different people, then the two 5's generally have somewhat different shapes. By applying a very complicated distortion to one of these 5's, we may be able to render it identical to the other 5. In practice a handprint recognition system is required to be invariant to many such distortions, but not all possible distortions, for instance, not to a distortion that would straighten out a 3 into a 1.

Some of the techniques mentioned above in Section 6.7 are used commercially for handprint recognition, and further handprint recognition techniques will be briefly considered in Section 6.9. The present section differs from Sections 6.7 and 6.9 in that it sets out to deal with distortion *per se*.

6.8.2 Character Topology

For any given character we can construct a symmetric matrix $[\alpha_{ij}]$ in which α_{ij} is the number of limbs that interconnect the i-th and j-th ends or junctions of limbs. The matrix $[\alpha_{ij}]$ has one row corresponding to each end or junction of three or more limbs, but the sequence of the rows has no significance. For instance, for the character shown in Fig. 6.9a the matrix $[\alpha_{ij}]$ is

$$\begin{matrix} 1 & 1 & 1 \\ 1 & 0 & 0 \\ 1 & 0 & 0. \end{matrix}$$

If we disregard the thickness of limbs, then $[\alpha_{ij}]$ is an expression of the topology of the character. The topology of a character, and therefore its matrix $[\alpha_{ij}]$, is unchanged when the character is continuously distorted. A *distortion* is a continuous function that has an inverse which is also a continuous function [6.46].

To achieve invariance to distortion, SHERMAN [6.133] considered assigning to the same recognition class all characters whose matrices $[\alpha_{ij}]$ could be made identical by permutation of corresponding rows and columns. The permutation was necessitated by the arbitrariness of the order of the rows of $[\alpha_{ij}]$. Sherman's technique would assign to the same recognition class all characters whose topology was the same, but unfortunately all of the characters shown in Fig. 6.9b have the same topology. To distinguish between these characters we can use additional information such as slopes of limbs, e.g. [6.133, 43], or relative positions of limb-ends and junctions, e.g. [6.18, 76].

All of the characters in Fig. 6.9b have two limb-ends and one four-limb junction. This fact alone is sufficient to distinguish the Fig. 6.9b numerals from numerals such as 0 that have different topology. To distinguish between numerals that have different topology we need not construct any matrix $[\alpha_{ij}]$. Instead, for any character we can merely list the numbers of limb-ends, junctions of three limbs, and junctions of four or more limbs. Because of their simplicity, characters that have the same list generally have the same topology. Figure 6.9c shows a counter-example, but any difficulty that this might cause can be resolved by additional information, such as slopes of limbs and relative positions of limb-ends and junctions, which we will need to use anyway. In practice, such additional information, together with the numbers of limb-ends and junctions, should be sufficient for distinguishing between Arabic numerals.

Nevertheless, the literature describes several recognition techniques in which topology is fully expressed. For instance, in [6.43, 57, 138] the topology is expressed essentially graphically. In [6.42, 123, 129] a character is squashed and somewhat deformed until it just fits into a small rectangle. This squashing process is designed to preserve the significant topology of a character but to remove differences between characters that belong to the same recognition class, so that a squashed character can easily be recognized, for instance by correlation.

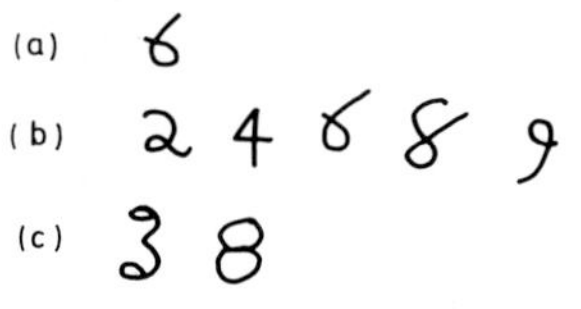

Fig. 6.9a–c. Examples to illustrate discussion of character topology

6.8.3 Recognition Techniques that Use Junctions and Ends of Limbs

In a recognition system described by BEUN [6.18] an input character is smoothed and skeletonized [6.125]. If a skeletonized numeral is found to contain one limb-end and one 3-limb junction, then it may be a 6 or a 9; and if the end is situated above the junction, then the numeral may be a 6. As a further simple example, if a numeral contains two ends and no junctions then it may be a 1, 2, 3, 5 or 7. If the total number of black bits in the skeletonized character is equal to the total number of rows that each contains at least one black bit, then the character is 1. If the upper end touches the right side of a rectangle that circumscribes the character, then the character is 5, and so on. Skeletonization facilitates the detection of junctions and ends.

There is a considerable literature on skeletonization or thinning. A usual idea is to remove from the black/white edges of a character all black elements whose removal will not change the topology of the character, for instance, by introducing disconnecting white breaks or white holes into black limbs. This is done repeatedly until all that remains is a skeleton in which the limb-width is not more than one bit (e.g. [6.79, 133, 140, 162, 177]). BASKIN [6.11] has patented a skeletonizing technique in which the grey scale is quantized into four levels 0, 1, 2, and 3. Level 0 is white and levels 1, 2, 3 are black, level 3 being the blackest. BASKIN first removes all 1's that can be removed without changing the topology, then removes all 2's that can be removed without changing the topology, then removes all 3's that can be removed without changing the topology. Further patents (e.g. [6.94, 108, 117, 128]) disclose scanning techniques that are intended to extract centerlines of limbs directly from characters on documents, not from electronically stored digitized characters.

The system described by BEUN [6.18] is by no means the only one in which skeletonization has been employed to facilitate the detection of

α	α	α γ	α γ	γ
δ	δ	β	β	β
δ	δ	β	β	β
δ	δ	β	β	β

Fig. 6.10. Labelling of zones for logic statement

ends and junctions of limbs; [6.43, 76, 79, 133] are further examples. HUNT [6.76] divided the rectangle that circumscribes a skeletonized character into 5 by 5 approximately equal zones. To provide a simplified formulation of Hunt's technique we define:

$$y_{i1} = \begin{cases} 1 & \text{if a limb-end regardless of orientation is detected in the } i\text{-th of the 25 zones} \\ 0 & \text{otherwise} \end{cases}$$

$$y_{i2} = \begin{cases} 1 & \text{if a right-angled or acute bend regardless of orientation is detected in the } i\text{-th zone} \\ 0 & \text{otherwise} \end{cases}$$

$$y_{i3} = \begin{cases} 1 & \text{if a 3-limb junction regardless of orientation is detected in the } i\text{-th zone} \\ 0 & \text{otherwise} \end{cases}$$

$$y_{i4} = \begin{cases} 1 & \text{if a 4-limb junction regardless of orientation is detected in the } i\text{-th zone} \\ 0 & \text{otherwise} \end{cases}$$

$$y_{i5} = \begin{cases} 1 & \text{if a roughly horizontal limb is detected in the } i\text{-th zone} \\ 0 & \text{otherwise} \end{cases}$$

$$y_{i6} = \begin{cases} 1 & \text{if a roughly vertical limb is detected in the } i\text{-th zone} \\ 0 & \text{otherwise.} \end{cases}$$

Characters are recognized by means of logic statements whose inputs are $y_{11}, \ldots, y_{ij}, \ldots, y_{25,6}$. As an example we can formulate the logic statement for 4 by defining $Y_{\alpha j}=1$ if $y_{ij}=1$ for at least one zone labelled α in Fig. 6.10, and $Y_{\alpha j}=0$ otherwise; $Y_{\beta j}=1$ if $y_{ij}=1$ for at least one zone labelled β; and so on. A handprinted character is recognized as 4 if

$$Y_{\alpha 1} \cdot (Y_{\beta 3} \vee Y_{\beta 4}) \cdot (\overline{Y_{\gamma 5} \vee Y_{\gamma 2}}) \cdot (Y_{\delta 2} \vee Y_{\delta 3}) \cdot Y_{\beta 1}$$

is true. Note that HUNT, like BOMBA [6.20], has found it worthwhile to use lines of several orientations and also to use bends, as well as limb-ends and junctions.

6.8.4 Recognition Techniques that Use Loops and Ends of Limbs

BEUN [6.18] says that for any given skeleton character,

$$F-E=2(L-S)$$

where S is the number of mutually disconnected black parts of the character, L is the number of closed loops, E is the number of ends, and F is the number of 3-limb junctions. BEUN counts a 4-limb junction as two 3-limb junctions. Beun's equation means that in practice we can determine E, S, and either F or L. It is not necessary to determine both F and L. Whereas BEUN determines F, a Cognitronics system [6.31] determines L.

This Cognitronics system [6.31] uses a stream following technique to chain code the black/white contours of a binarized character. A chain code is a record of directions of movement during the tracing of a contour [6.52]. For instance, if the eight 45° directions are indexed as in Fig. 6.11a, then the chain code for the rectangular array of 1's and 0's in Fig. 6.11b is 11244756 if we start at the leftmost black element. A closed external chain is obtained from a closed external contour, and a closed internal chain is obtained from a closed internal contour. If a character has one external and two internal closed chains then it has two loops and is likely to be 8. If it has one external chain and one internal chain then it has one loop and is likely to be 6, 9 or 0. To distinguish, for instance, between 6, 9 and 0, the Cognitronics system makes use of the positions and orientations of limb-ends.

Limb-ends can be detected by finding a total change of direction of 180 degrees within, say, $\lambda+3$ successive elements of the chain code, where λ is an integer that is an approximate measure of the limb-width of the character [6.31]. This rough measure can be obtained [6.31, 58, 79, 122], for instance, from the ratio

$$\frac{2 \times \text{total number of black bits}}{\text{total number of chain code elements}}.$$

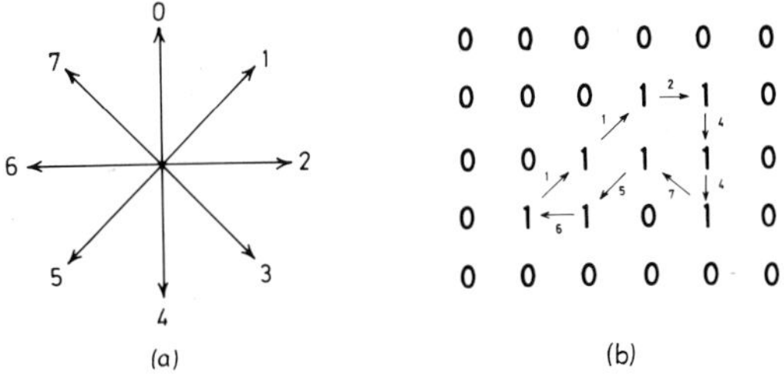

Fig. 6.11a and b. Example of chain coding

A loop or a smudge in a thin-limbed character may be as fat as a *bona fide* end in a fat-limbed character. By using λ in the detection of limb-ends, the Cognitronics system ensures that, for instance, a tight loop in a thin-limbed character is not recognized as an end, but a *bona fide* end in a fat-limbed character is nevertheless detected. To check that an end is not due to an accidental break in a limb, the Cognitronics system [6.31] checks for two ends pointing at each other and separated by only a small distance.

By looking for loops rather than junctions, and by using sophisticated end-detection, the Cognitronics system avoids skeletonization; and the entire recognition process, starting from binarization, appears to involve less computation than in systems that skeletonize and then detect junctions. The Cognitronics system can be implemented commercially in a general purpose computer.

Another system that detects loops is the stream-following system of CASKEY and COATES [6.29], which makes more use of cavities than of ends. For instance, 5 has a right-facing cavity above a left-facing cavity (cf. [6.119, 148, 100, 136, 170, 174]).

Let us now briefly digress into the topic of loop detection. Loops can be detected by stream following or by contour following, or by a technique such as that of MIYAMOTO and HIBI which employs propagation of deactivation in a microcellular array [6.98]. MIYAMOTO and HIBI store a binary pattern in an array of bistable elements. They also store in a further similar array an inverted version of the original pattern in which each 1 has been changed to 0 and each 0 has been changed to 1. The original and inverted arrays are separately scanned bit by bit, row by row. When a 1 is encountered by the scan, all 1's chain connected to it are changed to 0 by grass-fire propagation. Propagation is so fast that the scan only hits one 1 in any connected set of 1's. The number of 1's hit by the scan is counted. When the whole of the positive array has been scanned, this number of hits is equal to the number of mutually disconnected black regions in the pattern. For the inverted array the number of hits minus one is equal to the number of closed loops in the original pattern. By repeated inversion MIYAMOTO and HIBI have extended their technique to count, for instance, the number of loops that contain disconnected black regions; cf. [6.25, 41].

To introduce a further technique, suppose that a binary pattern is shifted successively into all possible alignments with a 2 by 2 bit window. Let N_e be the number of alignments in which the pattern in the window contains exactly one 1, and let N_i be the number of alignments in which the pattern in the window contains exactly one 0. N_e and N_i are, respectively, the total numbers of convex and concave corners of black. For

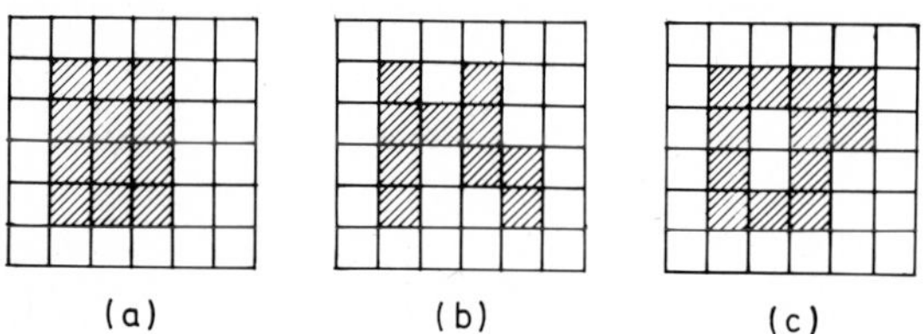

Fig. 6.12a–c. Examples of black figures on a white background

instance, we find that $N_e - N_i = 4$ for Fig. 6.12a. We also find that $N_e - N_i = 4$ for Fig. 6.12b, and indeed for any pattern that contains only one connected black region that has no holes. For Fig. 6.12c, $N_e - N_i = 0$, and in general [6.79, 122]

$$N_e - N_i = 4(S - L)$$

where, as before, S is the number of separate black regions that are not connected to each other by black, and L is the number of loops or holes. The value of $N_e - N_i$ is sensitive to isolated 1's and 0's that result from noise, and this value is used in [6.79] for deciding whether a denoising process should be applied to an input pattern. For instance, denoising may not be required if

$$-2 \leqq \frac{N_e - N_i}{4} \leqq +3 .$$

6.8.5 A Recognition Technique that Uses Edge Curvature

In Subsection 6.8.2 we suggested that in practice the full use of character topology is neither necessary nor sufficient for recognition. It is not sufficient because it does not distinguish between, for instance, the characters in Fig. 6.9b. It is not necessary because for practical purposes a list of ends and junctions or loops of limbs is just as useful as the complete topology that is expressed by a matrix $[\alpha_{ij}]$. In character recognition a basic requirement is to obtain enough information to distinguish between characters in different classes, and to discard unnecessary information. To meet this requirement it may not be necessary to detect either junctions or loops.

PATTERSON [6.105] has patented a handprint recognition system that detects neither junctions nor loops, but instead roughly measures the curvature of black/white contours. PATTERSON shifts a binarized character past a 7 by 7 bit window. In each alignment, if the central bit in the window is black and lies on a black/white edge of a character, then the pattern within the window is assigned by logic circuits to one of 200 classes. Within any one of these 200 classes the shapes of the edge have approximately the same curvature and orientation.

The rectangular array that contains a character is divided into 3 by 3 slightly overlapping zones. To formulate Patterson's method abstractly, let $\{z_1, \ldots, z_h, \ldots, z_{200}\}$ be a vector in which each component corresponds to a different one of the 200 edge-shape classes. For each $h=1, \ldots, 200$, let us say that the component z_h is the number of alignments, within the i-th of the nine zones, at which the h-th edge-shape has been detected, and that this holds for each $h=1, \ldots, 200$. For feature detection PATTERSON in effect employs eighteen 200-component weight vectors; let us denote the j-th of these by $\{w_1, \ldots, w_h, \ldots, w_{200}\}$. The component weights are integers that may be positive, negative or zero. Let us define $y_{ij}=1$ if $\sum_{h=1}^{h=200} z_h w_h$ exceeds a prescribed threshold, and $y_{ij}=0$ otherwise. When $y_{ij}=1$. this means intuitively that the j-th feature has been detected in the i-th of the nine zones. A character is recognized by recognizing its 162-bit vector $\{y_{11}, \ldots, y_{ij}, \ldots, y_{9,18}\}$ piecewise linearly, using altogether approximately 200 162-component weight vectors for the ten numerals.

In Patterson's system four of the eighteen features are horizontal, vertical or diagonal lines, and eight are convex or concave curves in the four quadrants. A further four are sharp bends in the four quadrants. An excessively sharp bend of white around black is classified as an end, which is a separate feature. The final feature is an excessively sharp bend of black around white. If, for instance, y_{ij} is a sharp-bend feature, then w_h is positive and large if the h-th edge shape is sharply bent in the appropriate direction. If instead the h-th edge shape is fairly straight, then w_h may be negative. In the i-th zone, a few very sharply bent edge-shapes can produce the same result as a larger number of less sharply bent edge-shapes.

By working with shapes of edges, PATTERSON [6.105] achieved invariance to the fatness of character limbs. Although $\sum_{h=1}^{h=200} z_h w_h$ is computed digitally, the system could in principle have an unusually highly parallel implementation.

6.8.6 Contour Following

Let us now discuss PATTERSON's technique [6.105] a little further, although it does not involve contour following. To recognize, for instance, 3's that differ from the 3 in Fig. 6.13a by distortion, as in Figs. 6.13c and d, PATTERSON detects features such as bends and ends within 3 by 3 zones. Roughly speaking, Patterson's system is invariant to distortions that move features within zones but not across zone boundaries. To achieve this invariance it would be simpler to deem a zone black if it contained more than a small predetermined amount of black, and white otherwise, so that a character would yield nine black or white bits, one bit per zone. This nine-bit pattern would be invariant to many distortions of the character from which it was derived, but Figs. 6.13a and b would yield the same nine-bit pattern, and such ambiguities would be disastrous. Feature detection is, hopefully, a means of resolving such ambiguities while retaining worthwhile invariance to distortion. However, feature detection is not the only means of doing this, and we can introduce an alternative method in terms of Fig. 6.13.

In Fig. 6.13a no black point in Zone A is an immediate neighbor of any black point in Zone D, but this is not true in Fig. 6.13b. Furthermore, in Fig. 6.13a black in Zone C touches black in Zone F, but this is not true in Fig. 6.13b. We can use these facts to distinguish between Figs. 6.13a and b without using features and without losing invariance to distortions that move black within zones. We can do this by tracing around the external contours, for instance starting clockwise in Zone A. Around Fig. 6.13a the tracing point passes successively through zones ABCFEDEFIHGHIFEDCBA, whereas around Fig. 6.13b the zone sequence is ABCBADEFIHGHFEDA. We can distinguish between Figs. 6.13a and b by recognizing their zone sequences, without looking for detailed features within any zone.

Any character recognition technique that relies on topology or connectivity relies fundamentally on the supposition that characters are subject predominantly to continuous distortion. Figures 6.13c and d show continuously distorted versions of Fig. 6.13a, and because continuous distortions preserve connectivity, the zone sequences for Figs. 6.13c and d are the same as for Fig. 6.13a. Discontinuous distortions would not preserve connectivity and instead might introduce breaks such as those

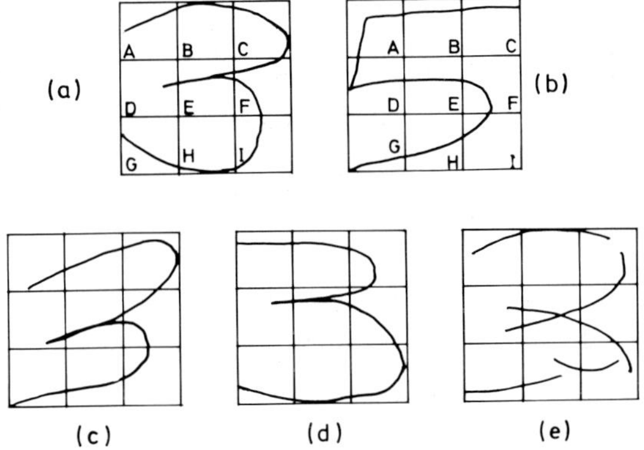

Fig. 6.13a–e. Examples of character shapes

in Fig. 6.13e, which would spoil the zone sequence. The zone sequence would also be spoiled if the cusp of 3 did not protrude leftward into Zone D, yet this would not make the 3 intuitively unrecognizable. To allow for acceptable variations such as this, we must allow several alternative zone sequences for each recognition class. The supposition that characters are subject predominantly to continuous distortion leads to the supposition that we will need to allow only a manageably small number of alternative zone sequences per class.

Instead of tracing round the black/white contours of a character, we could obtain a zone sequence by tracing the centerlines of limbs. Another idea is to record the sequence of zones through which the pen passes while a character is actually being written, e.g. [6.17, 22]. This is known as a *real-time* or *on-line* recognition technique because data processing takes place at the time of writing. The real-time technique of POWERS [6.110] measures approximately the instantaneous angular direction of the pen and uses a direction sequence instead of a zone sequence.

A real-time recognition process can make use of information that is not available in non-real-time character recognition. For instance, the direction and order of formation of character strokes, and the velocity of the pen, have been used experimentally in the recognition of cursive script [6.65]. Burroughs [6.26] has patented an automatic signature validation system that measures the pressure of the pen on the document and makes use of the pressure versus time waveform.

Zone and direction sequences have also been used in non-real-time character recognition. For instance, the IBM 1287 employs a non-real-time contour following technique for recognizing handprinted characters [6.121]. A servo-controlled flying spot traces around the external black/white contour of a character, and the angular direction of the spot is continuously determined. To detect features such as "top of 2" or "cusp of 3", the system employs one sequential circuit per feature. A feature is detected provided that a sequence of conditions is satisfied in a prescribed order. Each condition requires that the scanning spot move for a predetermined time within a predetermined region and within a predetermined range of angular directions. This system has been discussed in [6.146] and will therefore not be further discussed here.

Contour tracing can be carried out either directly on a document, as in the IBM 1287 [6.121], or on a character that has previously been scanned, digitized, and stored, e.g. [6.156]. Recognition can easily be made invariant to the precise position of a character, and a further advantage is that a contour can be discarded if it is not long enough to be a contour of a *bona fide* character. An obvious disadvantage of contour tracing is its sensitivity to accidental breaks and joins in limbs. To some extent this disadvantage is shared also by all systems that look for ends and loops or junctions of limbs.

6.9 Strip Classification Techniques

The order in which we are considering various character recognition techniques does not reflect any judgement as to their relative excellence. Strip classification techniques generally have the advantage of implementational convenience, but, as with other recognition techniques, error rates depend strongly on the detailed practical realization. Strip classification techniques classify successive horizontal or vertical strips of a character, and character recognition is based on the strip classifications.

6.9.1 One Row or Column per Strip

In the simplest case, a strip is simply a row or a column of a character, e.g. [6.40, 161]. In a system of Data Card Corporation [6.39], a stylized character moves horizontally past a vertical column of five sensors. The outputs of the five sensors are binarized, so that at any instant the sensors yield a five-bit pattern. A stylized character is classified according to the sequence of five-bit patterns that it yields. The print recognition system of Zworykin and Flory [6.160] also moves characters past a column of five sensors. The data from each sensor are processed separately to yield four bits after the character has moved completely past the sensors. For the i-th of the five sensors, the first three of the four bits are 1 or 0 according to whether the i-th sensor has crossed black one, two, or three times, respectively. The fourth bit is 1 if the i-th sensor has passed through at least one long stretch of black, and 0 otherwise. The 5 by 4 bit pattern obtained from a character is recognized by matching with reference patterns. An advantage of these techniques is that the exact speed of a character past the sensors does not matter.

DOYLE [6.45] devised a handprint recognition system that performed various tests on a character. In one of the simplest tests a vertical line was moved horizontally across a character, and in each successive position the intersections of the line with black were counted. For instance, moving a vertical line across a hand-printed 2 from left to right, successive numbers of intersections might be 022333210. From this sequence DOYLE would remove successive repetitions to obtain 023210. One way to discriminate between different characters such as hand-printed 2, 5, 9 that may yield this same sequence is to determine whether, for instance, at a position of the vertical line where it passes three times through black, further conditions are satisfied. For example, one condition might be that between the upper two intersections there must be a white element such that a line extending leftward horizontally from it meets no black [6.45, 119, 100, 136]. This condition is likely to be satisfied for a hand-printed 2 but not for a 5, 6, or 9.

When we find, for instance, that a column of a digitized character contains three intersections with black, we may assign this column to a class that comprises all three-intersection columns. This illustrates why crossing-counting techniques [6.146, Section 7.6] can be regarded as belonging to the family of strip classification techniques.

6.9.2 Two Rows or Columns per Strip

In a strip classification scheme patented by HILL and HOLT [6.70], each strip consists of two columns of a binarized character. By comparing two successive columns we can determine whether they contain the beginning of a character limb that was not present in previous columns. We can detect a "fork" by determining whether a column contains two black sections that are adjacent to a common black section in the previous column. Joins and ends of character limbs can be detected similarly. If a pair of columns contain a beginning of a limb that is above a further chunk of black, HILL and HOLT [6.70] classify this pair of columns as "high begin". If a pair of columns contains a join that is below a further chunk of black, HILL and HOLT classify this as "low join". "High join", "high end", "low fork", "low end", etc., are similarly defined. A character is recognized according to the sequence of features such as "high join", "low fork" that are detected in successive pairs of columns.

If a black/white transition in one column is sufficiently close to a black/white transition in the next column, we can assume that the black elements belong to the same character limb. Differences in the vertical positions of such transitions in successive columns can tell us the slope [6.72] (cf. [6.124, 137]) and eventually the curvature [6.111] of a character limb. If the slope of a black/white contour is quantized into, say, eight angular directions, then the sequence of successive slopes of a single contour constitutes a chain code for that contour. This is how chain codes are obtained in the Cognitronics system [6.31] that was mentioned previously in Subsection 6.8.4, and this technique for obtaining chain codes is an example of a stream-following technique.

In a stream-following or "watchbird" technique, while a pair of columns moves horizontally relative to a character, we can obtain a separate stream of data relating to each black/white contour that crosses the two columns. Since at any time the columns are usually crossed by more than one contour, we usually obtain more than one stream of data in parallel, whereas contour following such as in the IBM 1287 [6.121] yields only a single stream of data. Stream following is a general technique for obtaining data from a character or other figure, and it is not a strip classification technique, although in practice it may be combined with a strip classification technique, e.g. [6.106, 149]. Stream following can be used purely for exploring the connectivity of the black parts of a character or other figure [6.25].

KIJI and HOSHINO [6.87] have developed a strip-classification technique in which a strip is a pair of adjacent rows of a binarized character. The classification of a strip is based on the numbers of separate black sections in each of the two rows, and also on the difference in position between the leftmost white/black transitions in the two rows, the difference in position between the leftmost black/white transitions in the two rows, the difference in position between the rightmost white/black transitions in the two rows, and the difference in position between the rightmost black/white transitions in the two rows. These data are used for determining whether a strip contains a feature such as a "cap" where two vertical limbs are joined above by a single black section, a "*T*-shape" where a relatively narrow limb is joined above by horizontal black protruding on both sides of the relatively narrow limb, a bend where a limb is joined above by horizontal black protruding only to one side of the limb, a decrementally diagonal limb, a start of a new limb while a limb that has previously started continues to the left of the start, and so on. KIJI and HOSHINO [6.87] mention altogether fourteen such features, or in other words, fourteen possible classifications of a strip. A handprinted character is classified in accordance with the vertical sequence of strip features that it contains. A system of this type has been developed by the Nippon Electric Company for use in postal sorting in Japan [6.103].

6.9.3 Three or Four Rows or Columns per Strip

GENCHI et al. [6.53, 86, 99] have described a handprint recognition system developed by Toshiba for Japanese postal sorting. In this strip-classification system, a strip consists of three adjacent rows of a binarized and roughly skeletonized character. A 3 by 3 bit window is used in the first stage of recognition. Logic circuits are used to classify the pattern in the 3 by 3 bit window into one of seven classes. These classes are "vertical line", "incrementally sloping line", "decrementally sloping line", "horizontal line high in the window", "horizontal line low in the window", "blank", "black with no clear direction". As the 3 by 3 window is shifted horizontally across a character, the classification of the pattern in the window generally changes. The sequence of classifications of the 3 by 3 bit pattern is used for classifying the 3-row strip that the window sweeps along. For instance, a strip is classified as CAP if it contains two limbs that are joined above by a black section, as at the top of 8, 9, or 0. A strip is classified as PAL if it contains two roughly parallel black segments, and so on. There is provision for refining the strip classification by determining which black sections in one strip are connected to black in the next strip [6.86]. A handprinted character is recognized according to its vertical sequence of strip classifications.

The sequential classification technique of GENCHI et al. [6.53] differs from those of, for instance, KLOOMOK et al. [6.88], ROHLAND et al. [6.121], and HILL and HOLT [6.70]. To formulate this difference, let us arbitrarily denote the possible classifications of any given strip by letters A, B, C, D, ... In HILL and HOLT [6.70] there is a prescribed sequence of strip classifications for each character to be recognized, and let us suppose, for example, that the prescribed sequence for 2 is BEAC. A sequence such as ABFEBADCG or GBFCEADC would be recognized as 2 because it includes BEAC, but BFEGDAG would not be recognized as 2 because it does not include the sequence BEAC. In this example, symbols other than BEAC are ignored, but in the technique of GENCHI et al. [6.53] no symbols are ignored. GENCHI et al. do not merely check that a sequence includes a prescribed sequence.

ROHLAND [6.120] patented a strip-classification technique that worked by checking for the presence of prescribed sequences within longer sequences. A strip consisted of four adjacent columns. In order to classify a strip, a window four bits wide and one bit high was in effect swept up the strip. At any given time the 4-bit pattern in the window was

Table 6.3. Rohland's classification of 4-bit patterns in [6.120]

Pattern	Class
0 0 0 0	0
1 0 – –	1
1 1 0 –	3
1 1 1 0	5
1 1 1 1	7
0 1 – –	2
0 0 1 –	4
0 0 0 1	6

assigned to one of eight classes denoted respectively by numerals 0–7. Table 6.3 gives the definitions of the eight classes, and as usual "–" means "don't care". As the 4-bit window swept up the strip, a sequence of numerals such as 00015772466006277757... was obtained, there being one numeral for each vertical position of the window. For each of 47 possible classifications of a strip, ROHLAND provided a sequential condition. A sequence of numerals was assigned to a particular one of the 47 classes only if the corresponding condition was satisfied. For example, part of a sequential condition might be that a sequence must contain a 1, a 3 and a 5, in that order, regardless of intervening numerals, but the first occurrence of 1 and the first occurrence of 5 must be separated by not more than eight and not less than five numerals. A character was assigned to a particular recognition class if its horizontal sequence of strip-classifications included a prescribed sequence of strip classifications.

The idea of recognizing a pattern by recognizing a sequence of strip classifications has not been confined to character recognition. For instance, in the speech recognition system of GILLI and MEO [6.55] the output of a filter bank constitutes a one-dimensional pattern that we can regard as a strip, and a spoken word is recognized according to its sequence of strip-classifications; cf. [Ref. 6.146, Subsection 9.6.3].

6.10 Combinations of Classification Techniques

MUNSON [Ref. 6.175, p. 126] suggested that two or more different classification methods should be applied to an unknown character. For instance, optical and magnetic techniques might both be applied to characters printed in magnetic ink [6.179, 181]. SPANJERSBERG [6.180] has considered various different ways of combining the outputs of different classification systems. Instead of applying different classification techniques in parallel, ERCM [6.165] starts by applying a contour-tracing technique to an unknown character. If this technique does not yield an unequivocal decision, then a further technique is applied; and if this further technique does not yield an unequivocal decision, then a final technique is applied, cf. [6.176].

6.11 Concluding Remarks

In the fiercely competitive commercial market, the cost of optical character recognition equipment tends to decrease. This decrease is partly due to increased efficiency of design, decreased cost of electronic components, and increased volume of sales. It is also due to the use of modular design, so that a customer need only purchase modules to do his particular job, without paying for facilities that he does not require. According to FIEDELMAN and BERNSTEIN [6.49] a primary reason for increased sales will be the introduction of multimedia data-entry systems in which character recognition equipment works cooperatively with manually operated keyboards (e.g. [6.127, 163]).

Commercial character recognition equipment is commonly designed so that the detailed recognition logic can readily be modified while the equipment is actually in service with a customer. Modifications may be necessitated by changes in a customer's requirements. Modifications may also be desirable for implementing detailed improvements in design that will help to keep the equipment competitive in its performance. The need for modifiability is remarkably important in practice. It means that hardwired recognition logic should be physically designed to be conveniently modifiable, and this to some extent precludes the use of state-of-the-art hardware technology in hardwired logic. Microprogramming, and possibly associative memory techniques, are likely to become increasingly important because they can provide recognition logic flexibility using state-of-the-art hardware.

Available equipment achieves a very high percentage of correct recognition of handprinted characters only if they have been written in conformity with rules of constraint. If the rules of constraint are severe the recognition equipment may be simple (e.g. [6.56, 62, 85, 154, 170, 174]). Particularly for the recognition of a complete alphabet of handprinted characters, no set of rules has yet won practical acceptance by all manufacturers. For instance, it is still questionable whether handprinted characters should be written in preprinted boxes, or whether instead it is better to employ preprinted arrays of dots or curvilinear marks.

At the present state of the art it is probably true to say that humans can recognize a far wider diversity of character shapes and qualities than any known machine can recognize. After many years of effort, this gap between human and machine recognition performance still appears to be so wide that we can reasonably doubt whether it will ever be closed purely by the natural evolutionary process of commercial development without the help of basic research. In other words, it appears that there are still significant basic research problems in the area of character recognition. It would be utterly wrong to say that character recognition has been "done" and is no longer of research interest.

Character recognition has a rich literature of patents, but only a few of them are included in the following list of references. This list and the references in [6.146, 147] emphatically do *not* provide a complete bibliography of character recognition. Plenty of good work has been quite arbitrarily left out.

References

6.1 P.L. Anderson: Datamation **17**, 22 (1971)
6.2 D.R. Andrews, A.J. Atrubin (IBM): US Pat. 3573730 (1971) and UK Pat. 1293831 (1972)
6.3 D.R. Andrews, A.J. Atrubin, R.J. Baumgartner, M.F. Bond, K.-C. Hu (IBM): US Pat. 3517387 (1970) and UK Pat. 1179916 (1970)
6.4 D.R. Andrews, A.J. Atrubin, K.-C. Hu: IBM J. Res. Dev. **12**, 364 (1968)
6.5 D.R. Andrews, M.J. Kimmel (IBM): US Pat. 3710323 (1973)
6.6 R. Bakis (IBM): US Pat. 3350505 (1967) and UK Pat. 1030835 (1966)
6.7 P.F. Bargh (Farrington): US Pat. 3522437 (1970)
6.8 G.E. Bartholomew, D.J. Kostuch, T.E. Robinson, W.S. Rohland (IBM): US Pat. 3697951 (1972). *See also* UK Pat. 1271620 (1972)
6.9 G.J. Balm: Pattern Recognition **2**, 151 (1970)
6.10 M.R. Bartz: IBM J. Res. Dev. **12**, 364 (1968). *See also* UK Pat. 1219334 (1971)
6.11 H.B. Baskin (IBM): US Pat. 3196398 (1965)
6.12 R.J. Baumgartner, M.F. Bond (IBM): US Pat. 3526876 (1970), UK Pat. 1144319 (1969). *See also* UK Pat. 1304429 (1973)
6.13 R.J. Baumgartner, J.L. Lovgren, J.W. McCullough (IBM): UK Pat. 1337159 (1973)
6.14 A.H. Beiser, L.J. Nunley, I. Sheinberg (REI): US Pat. 3509533 (1970). *See also* UK Pats. 1115909 and 1115910 (1968)
6.15 J.F. Bene, G.A. Garry (IBM): US Pat. 3492646 (1970)
6.16 J.F. Bene, P.E. Nelson (IBM): US Pat. 3384875 (1968), UK Pat. 1102359 (1968)
6.17 M.I. Bernstein: In *Pattern Recognition*, ed. by L.N. Kanal (Thompson Book Co., Washington, D.C. 1968) pp. 109–114
6.18 M. Beun: Philips Tech. Rev. **33**, 89 and 130 (1973)
6.19 W.W. Bledsoe, I. Browning: *Proc. Eastern Joint Computer Conf.* (1959), p. 225
6.20 J.S. Bomba: *Proc. Eastern Joint Computer Conf.* (1959), p. 218

6.21 British Computer Society: *Character Recognition 1971* (British Computer Society, London 1971)
6.22 R. M. Brown: IEEE Trans. EC-**13**, 750 (1964)
6.23 G. Brust (International Standard Electric Corp.): US Pat. 3234513 (1966)
6.24 J. S. Bryan, C. F. Teacher (Philco): US Pat. 3167745 (1965)
6.25 O. P. Buneman: In *Machine Intelligence 5*, ed. by B. Meltzer, D. Michie (Edinburgh University Press 1969) pp. 383–393
6.26 Burroughs Corporation: UK Pat. 1227643 (1971)
6.27 Burroughs Corporation: US Pat. 3675201 (1972), UK Pat. 1279375 (1972). *See also* UK Pat. 1290595 (1972)
6.28 Business Press: *Optical Character Recognition and the Years Ahead* (Business Press, Elmhurst, Ill. 1969)
6.29 D. L. Caskey, C. L. Coates: *First Int. Joint Conf. on Pattern Recognition* (Washington D.C., Oct. 1973, IEEE Catalogue No. 73 CHO 821-9C) pp. 41–49
6.30 J. B. Chatten, C. F. Teacher: In *Optical Character Recognition*, ed. by G. L. Fischer et al. (Spartan Books, New York 1962) pp. 51–59
6.31 Cognitronics Corporation: UK Pat. 1287604 (1972)
6.32 Compagnie Internationale Pour L'Informatique: UK Pat. 1352048 (1974)
6.33 Control Data Corporation: *Control Data 921* OCR *Document Reader* (Control Data Corporation, Rockville, Maryland)
6.34 A. W. M. Coombs: In *Machine Intelligence 4*, ed. by B. Meltzer, D. Michie (Edinburgh University Press 1969) pp. 385–401
6.35 T. M. Cover: IEEE Trans. EC-**14**, 326 (1965)
6.36 A. Cutaia (IBM): US Pat. 3587047 (1971)
6.37 A. Cutaia (IBM): US Pat. 3737855 (1973), UK Pat. 1381970 (1975)
6.38 A. Cutaia, J. W. McCullough, D. W. Piller (IBM): UK Pat. 1304880 (1973)
6.39 Data Card Corporation: UK Pat. 1354847 (1974), US Pat. 3714631 (1973)
6.40 Data Recognition Corporation: UK Pat. 1352611 (1974)
6.41 D. W. Davies: In *Machine Perception of Patterns and Pictures* (Institute of Physics, Conf. Series No. 13, London 1972) pp. 303–310
6.42 F. M. Demer (IBM): US Pat. 3629833 (1971), UK Pat. 1294710 (1972)
6.43 E. S. Deutsch: US Pat. 3609685 (1971), UK Pat. 1171627 (1969)
6.44 G. P. Dinneen: *Proc. Western Joint Computer Conf.* (1955) pp. 94–100
6.45 W. Doyle: *Proc. Western Joint Computer Conf.* (1960) pp. 133–142
6.46 R. O. Duda, P. E. Hart: *Pattern Classification and Scene Analysis* (Wiley Interscience, New York, London 1973)
6.47 R. J. Evey: *Proc. Eastern Joint Computer Conf.* (1959) pp. 205–211
6.48 V. S. Fain: Radio Engng. **15**, No. 3, 16 (1960)
6.49 L. Fiedelman, G. B. Bernstein: Datamation **19**, 44 (1973)
6.50 A. I. Frank, J. A. Angeloni, J. J. McIntyre, R. L. Baracka (Scan-Data): US Pat. 3593284 (1971)
6.51 M. D. Freedman (Bendix Corp): US Pat. 3694806 (1972)
6.52 H. Freeman: Computing Surveys **6**, 57 (1974)
6.53 H. Genchi, K.-I. Mori, S. Watanabe, S. Katsuragi: Proc. IEEE **56**, 1292 (1968). *See also* UK Pat. 1180290 (1970), US Pat. 3541511 (1970)
6.54 H. Genchi, T. Yoneyama (Toshiba): UK Pat. 1354993 (1974)
6.55 L. Gilli, A. R. Meo: Acustica **19**, 38 (1967/68). *See also* Intern. J. Man Machine Studies **2**, 317 (1970)
6.56 H. E. Goldberg: US Pat. 1117184 (1914)
6.57 M. Gössel: Elektronische Informationsverarbeitung **6**, 145 (1970). English Translation: ASTIA Document AD 745238 (1970)
6.58 S. B. Gray: IEEE Trans. C-**20**, 551 (1971)

6.59 E.C. GREANIAS (IBM): US Pat. 2928073 (1960). *See also* US Pat. 2959769 (1960)
6.60 E.C. GREANIAS: In *Optical Character Recognition*, ed. by G.L. FISCHER et al. (Spartan Books, New York 1962) pp. 129–146
6.61 E.C. GREANIAS, A. HAMBURGEN (IBM): US Pat. 3105956 (1963)
6.62 M.L. GREENOUGH, M.L. KUDER: US Pat. 3699518 (1972)
6.63 R.E. HALL, L.P. SEGAR (IBM): UK Pat. 1280915 (1972)
6.64 C.C. HANSON (IBM): UK Pat. 1251117 (1971)
6.65 L.D. HARMON: Proc. IEEE **60**, 1165 (1972)
6.66 J.D. HARR (IBM): UK Pat. 1329109 (1973), US Pat. 3599151 (1971)
6.67 R.D. HAXBY, G.O. NORRIE (Farrington): US Pat. 3496541 (1970)
6.68 P.G. HENDRICKSON: In *Automatic Pattern Recognition* (National Security Industrial Association, Washington, D.C. 1969) pp. 9–17
6.69 W.H. HIGHLEYMAN: IRE Trans. EC-**10**, 501 (1961). *See also* US Pat. 2978675 (1961)
6.70 J.D. HILL, A.W. HOLT (CDC): US Pat. 3178688 (1965)
6.71 M.M. HOLEMAN: In *Pattern Recognition*, ed. by L.N. KANAL (Thompson Book Co., Washington, D.C. 1968) pp. 63–78. *See also* UK Pat. 1151190 (1969)
6.72 A.W. HOLT (CDC): US Pat. 3142818 (1964)
6.73 A.W. HOLT, J.D. HILL (CDC): US Pat. 3519990 (1970)
6.74 L.P. HORWITZ, O. MOND (IBM): US Pat. 3293452 (1966)
6.75 P.H. HOWARD (IBM): UK Pat. 1338287 (1973)
6.76 D.J. HUNT: In *Machine Perception of Patterns and Pictures* (Institute of Physics, Conf. Series No. 13, London 1972) pp. 28–33. *See also* UK Pat. 1345032 (1974)
6.77 S.S. HUSSON: *Microprogramming: Principles and Practices* (Prentice-Hall, Englewood Cliffs, N.J. 1970)
6.78 T. IIJIMA, I. YAMAZAKI, S. MORI, H. GENCHI, S. KATSURAGI (Kogyo Gijutsuin and Toshiba): UK Pat. 1319158 (1973)
6.79 Information International Incorporated: UK Pat. 1353912 (1974)
6.80 Information Technology Incorporated: UK Pat. 1289984 (1972)
6.81 International Business Machines: *The* IBM *3886 Optical Character Reader* (IBM, New York). *See also The* IBM *3886 Optical Character Reader Model 1 Component Reference Manual*, GA21-9147-1 (IBM)
6.82 International Standard Electric Corporation: UK Pat. 1156266 (1969)
6.83 International Standard Electric Corporation: UK Pat. 1201399 (1970)
6.84 D.L. IRVIN, A.J. RIDER (REI): UK Pat. 1349303 (1974)
6.85 R.B. JOHNSON (IBM): US Pat. 2741312 (1956)
6.86 S. KATSURAGI, H. GENCHI, K. MORI, S. WATANABE: In *Proc. Intern. Joint Conf. on Artificial Intelligence*, ed. by D.E. WALKER, L.M. NORTON (The Mitre Corp., Bedford, Mass. 1969) pp. 161–170
6.87 K. KIJI, Y. HOSHINO (NEC): US Pat. 3585592 (1971), UK Pat. 1253302 (1971)
6.88 M. KLOOMOK, E.C. GREANIAS, C.H. HOPPEL, J.S. OSBORNE: Automation Progress **2**, 159 (1957)
6.89 F.H. LANGE: *Correlation Techniques* (Iliffe, London, and Van Nostrand, Princeton, N.J. 1967)
6.90 J.D. LINEROOTH, E.W. ROSEN (IBM): US Pat. 3713097 (1973), UK Pat. 1363593 (1974)
6.91 A.W. LOHMANN, W.H. WERLICH: Appl. Opt. **10**, 670 (1971)
6.92 D.L. MALABY (IBM): UK Pat. 1060920 (1967)
6.93 D.L. MALABY (IBM): US Pat. 3506807 (1970), UK Pat. 1156229 (1969)
6.94 Matsushita Electric Industrial Company: UK Pat. 1120194 (1968)
6.95 M. MAUL (IBM): US Pat. 2000403 (1935). *See also* US Pat. 2294679 (1942)
6.96 E.J. MCCLUSKEY: *Introduction to the Theory of Switching Circuits* (McGraw-Hill, New York and London 1965)

6.97 R.E. MILFORD (GEC): US Pat. 3092732 (1963)
6.98 Y. MIYAMOTO, M. HIBI (Hitachi): US Pat. 3593238 (1971)
6.99 K.-I. MORI, H. GENCHI, S. WATANABE, S. KATSURAGI: Pattern Recognition **2**, 175 (1970)
6.100 R.A. NASHLJUNAS, R.A. ZHLABIS, K.G.K. BUCHJUNAS, P.P. SHVAGZHDIS: UK Pat. 1234066 (1971)
6.101 W.C. NAYLOR (IBM): UK Pat. 1327325 (1973)
6.102 Nippon Electric Company: UK Pat. 1287205 (1972)
6.103 Y. OHTA, Y. NAKAGAWA, S. YOSHIMURA, T. YOSHIDA, K. KIJI: NEC Res. Dev. **20**, 100 (1971)
6.104 J.R. PARKS, D.A. BELL, R.S. WATSON, G. COWIN, S.E. OLDING: *Second Intern. Joint Conf. on Pattern Recognition* (Copenhagen, August 1974, IEEE Catalogue No. 74 CH0885-4C) pp. 416–420
6.105 J.V. PATTERSON (REI): UK Pat. 1320243 (1973)
6.106 P.G. PEROTTO: IEEE Trans. EC-**12**, 521 (1963). *See also* US Pat. 3178687 (1965)
6.107 Philco Corporation: UK Pats. 1033531 and 1033532 (1966)
6.108 Philips Electronic and Associated Industries Ltd: UK Pat. 1310293 (1973)
6.109 R.I. POLONNIKOV, V.V. ALEKSANDROV: Eng. Cybernetics **1**, 92 (1967)
6.110 V.M. POWERS: Pattern Recognition **5**, 291 (1973)
6.111 D.J. QUARMBY, J. RASTALL: IEEE Trans. SMC-**1**, 331 (1971). *See also* UK Pat. 1333916 (1973), US Pat. 3813646 (1974)
6.112 J. RABINOW (CDC): US Pat. 3167744 (1965)
6.113 J. RABINOW (CDC): US Pat. 3201751 (1965)
6.114 J. RABINOW (CDC): US Pat. 3219974 (1965)
6.115 J. RABINOW: In *Pattern Recognition*, ed. by L.N. KANAL (Thompson Book Co., Washington, D.C. 1968) pp. 3–29
6.116 J. RABINOW, A.W. HOLT, W. FISCHER, L.W. MADER: US Pat. 3104369 (1963)
6.117 T.P. REEDE (Philips): US Pat. 3673566 (1972)
6.118 T.E. ROBINSON (IBM): US Pat. 3737852 (1973), UK Pat. 1242039 (1971)
6.119 N. ROCHESTER, J.R. JOHNSON, G.M. AMDAHL, W.E. MUTTER (IBM): US Pat. 2889535 (1959)
6.120 W.S. ROHLAND (IBM): US Pat. 3274551 (1966)
6.121 W.S. ROHLAND, P.J. TRAGLIA, P.J. HURLEY: *1968 Fall Joint Computer Conf.* (AFIPS Conf. Proceedings **33**, Pt. 2) pp. 1151–1162. *See also* US Pat. 3303465 (1967)
6.122 K. SAKAI, S. KATSURAGI, S. WATANABE (Toshiba): US Pat. 3668637 (1972), UK Pat. 1271039 (1972)
6.123 M. SAMI, R. STEFANELLI: Pattern Recognition **5**, 133 (1973)
6.124 J. SAMMON, J. SANDERS (Pattern Analysis and Recognition Inc.): US Pat. 3755780 (1973)
6.125 P. SARAGA, D.J. WOOLLONS (Mullards): UK Pat. 1255653 (1971). *See also* UK Pat. 1153703 (1969)
6.126 Scan-Data Corporation: UK Pat. 1271705 (1972), US Pat. 3613080 (1971)
6.127 Scan-Data Corporation: *Scan-Data Data Entry Systems* (Scan-Data Corp., Norristown, PA)
6.128 G.G. SCARROTT, A.D. COOK (ICL): UK Pat. 1127741 (1968). *See also* UK Pat. 1262080 (1972)
6.129 H. SCHUMANN: UK Pat. 1223365 (1971)
6.130 H. SHADE (IBM): UK Pat. 1174340 (1969)
6.131 G.L. SHELTON (IBM): UK Pat. 1093079 (1967)
6.132 D.H. SHEPARD: US Pat. 2978590 (1961)
6.133 H. SHERMAN: In *Information Processing*, Proc. UNESCO Conf., Paris 1959 (Butterworths, London 1960) pp. 232–238

6.134 D. H. SKRENES (IBM): UK Pat. 1354040 (1974)
6.135 Société D'Electronique et D'Automatisme: UK Pat. 1143585 (1969)
6.136 A. A. SPANJERSBERG: UK Pat. 1345686 (1974), US Pat. 3858180 (1974)
6.137 W. SPRICK, K. GANZHORN: In *Information Processing*, Proc. UNESCO Conf., Paris 1959 (Butterworths, London 1960) pp. 238–244
6.138 W. STALLINGS: Computer Graphics and Image Processing **1**, 47 (1972)
6.139 S. D. STEARNS: IRE Trans. EC-**9**, 48 (1960)
6.140 R. STEFANELLI, A. ROSENFELD: J.A.C.M. **18**, 255 (1971)
6.141 N. STEINBERGER (Compuscan): US Pat. 3576534 (1971)
6.142 M. E. STEVENS: "Automatic Character Recognition, A State of the Art Report" (Tech. Note No. 112, National Bureau of Standards, Washington, D.C. 1961)
6.143 J. C. STOFFEL: IEEE Trans. C-**23**, 428 (1974)
6.144 A. I. TERSOFF: IRE National Convention Record **6**, Pt. 4, 318 (1958)
6.145 J. R. ULLMANN: Pattern Recognition **3**, 297 (1971)
6.146 J. R. ULLMANN: *Pattern Recognition Techniques* (Butterworths, London, and Crane Russak, New York 1973)
6.147 J. R. ULLMANN: Opto-Electronics **6**, 319 (1974)
6.148 S. H. UNGER: Proc. IRE **46**, 1744 (1958); Proc. IRE **47**, 1737 (1959)
6.149 G. U. UYEHARA: 1963 IEEE Intern. Convention Record, Pt. 4, pp. 64–74 (1963)
6.150 L. H. E. VAN HEDDEGEM (International Standard Electric Corporation): UK Pat. 1337368 (1973)
6.151 H. VAN STEENIS: In *Pattern Recognition in Biological and Technical Systems*, ed. by O.-J. GRÜSSER, R. KLINKE (Springer Berlin, Heidelberg, New York 1971) pp. 253–261. *See also* US Pat. 3618016 (1971) *and* IBM *system/360 Component Description. IBM 1275 Optical Reader Sorter*, File No. S 360-03, Form A19-0034-0, IBM (1969)
6.152 E. VROOM (Bell Telephone Labs.): US Pat. 2723308 (1955)
6.153 H. WADA, S. TAKAHASHI, T. IIJIMA, Y. OKUMURA, K. IMOTO: In *Information Processing*, Proc. UNESCO Conf., Paris 1959 (Butterworths, London 1960) pp. 227–232
6.154 A. WEAVER: US Pat. 1815996 (1931)
6.155 J. A. WEAVER, D. J. WOOLLONS (Mullards): UK Pats. 1106974 and 1106975 (1968)
6.156 J. A. WEAVER: *Reading Machines* (Mills and Boon, London 1972). *See also* UK Pat. 1157991 (1969)
6.157 J. S. WESZKA, R. N. NAGEL, A. ROSENFELD: University of Maryland Computer Science Center Tech. Rep. TR 243, May 1973
6.158 M. G. WILSON (IBM): US Pat. 3626092 (1971), UK Pat. 1255759 (1971)
6.159 G. G. N. WRIGHT: *The Writing of Arabic Numerals* (University of London Press 1952)
6.160 V. K. ZWORYKIN, L. E. FLORY (RCA): US Pat. 2616983 (1952)
6.161 W. R. BEALL, A. F. HERMANN, G. J. MURPHY (Input Business Machines Inc.): US Pat. 3840856 (1974)
6.162 M. BEUN, P. REIJNIERSE (Philips): US Pat. 3735349 (1973). *See also* US Pat. 3753229 (1973), UK Pat. 1375992 (1974)
6.163 W. F. BROK, A. A. SPANJERSBERG, J. VAN STAVEREN: US Pat. 3832682 (1974), UK Pat. 1366009 (1974)
6.164 W. H. COCHRAN, G. M. HEILING (IBM): US Pat. 3800078 (1974)
6.165 ECRM, Inc.: UK Pats. 1365992, 1365991 (1974)
6.166 H. ENDOU, J. KAWASAKI, Y. FUJIMOTO, K. NAKANE, Y. KITAZUME (Hitachi): US Pat. 3849760 (1974)
6.167 Y. FUJIMOTO, T. HANANOI, M. YASUDA, H. MAKIHARA, M. KURAMIZU, M. OTA, H. KOKIDO, S. KADOTA (Hitachi): US Pat. 3849762 (1974). *See also* US Pat. 3829831 (1974)
6.168 G. A. HART, M. G. WILSON (IBM): UK Pat. 1366616 (1974)
6.169 G. HAUPT, W. KOCHERT: US Pat. 3833833 (1974)

6.170 A. W. HOLT: Computer Design **13**, 85–89 (1974)
6.171 F. INOSE, Y. KITA (Hitachi): US Pat. 3811110 (1974). *See also* US Pat. 3760357 (1973)
6.172 D. H. MCMURTRY (IBM): US Pat. 3836958 (1974)
6.173 W. D. MCNEILL, W. S. ROHLAND (IBM): US Pat. 3800079 (1974)
6.174 N. MORIMOTO (Fujitsu): US Pat. 3852715 (1974), UK Pat. 1370375 (1974)
6.175 J. H. MUNSON: In *Pattern Recognition*, ed. by L. N. KANAL (Thompson Book Co., Washington, D.C. 1968) pp. 115–139
6.176 M. NADLER, C. MASSON (Société Honeywell Bull): US Pat. 3832683 (1974)
6.177 M. OKA, M. YASUDA (Hitachi): US Pat. 3846754 (1974)
6.178 Philips Electronic and Associated Industries: UK Pat. 1357652 (1974)
6.179 Philips Electronic and Associated Industries: UK Pat. 1351214 (1974)
6.180 A. A. SPANJERSBERG: *2nd Intern. Joint Conf. on Pattern Recognition* (Copenhagen, August 1974, IEEE Catalogue No. 74 CH0885-4C) pp. 208–209
6.181 R. M. TYBURSKI, D. W. RUSSELL, B. D. MAYBERRY, J. R. KENNEY (Optical Recognition Systems Inc.): US Pat. 3764978 (1973)
6.182 M. YASUDA, H. MAKIHARA (Hitachi): US Pat. 3810093 (1974)

Subject Index